Partial Differential Equations with Applications: Analytical Methods

Partial Differential Equations with Applications: Analytical Methods

Editors

Almudena del Pilar Márquez Lozano
Vladimir Iosifovich Semenov

MDPI Basel • Beijing • Wuhan • Barcelona • Belgrade • Novi Sad • Cluj • Manchester

Editors
Almudena del Pilar Márquez Lozano
University of Cadiz
Cadiz
Spain

Vladimir Iosifovich Semenov
Immanuel Kant Baltic Federal University
Kaliningrad
Russia

Editorial Office
MDPI AG
Grosspeteranlage 5
4052 Basel, Switzerland

This is a reprint of articles from the Special Issue published online in the open access journal *Mathematics* (ISSN 2227-7390) (available at: https://www.mdpi.com/si/mathematics/PDE_analy_methods).

For citation purposes, cite each article independently as indicated on the article page online and as indicated below:

Lastname, A.A.; Lastname, B.B. Article Title. *Journal Name* **Year**, *Volume Number*, Page Range.

ISBN 978-3-7258-2683-4 (Hbk)
ISBN 978-3-7258-2684-1 (PDF)
doi.org/10.3390/books978-3-7258-2684-1

Contents

mathematics

Article

Asymptotic Stability for the 2D Navier–Stokes Equations with Multidelays on Lipschitz Domain

Ling-Rui Zhang [1], Xin-Guang Yang [1,*] and Ke-Qin Su [2]

[1] Department of Mathematics and Information Science, Henan Normal University, Xinxiang 453007, China
[2] College of Information and Management Science, Henan Agricultural University, Zhengzhou 450046, China
* Correspondence: yangxinguang@hotmail.com or 011137@htu.edu.cn

Abstract: This paper is concerned with the asymptotic stability derived for the two-dimensional incompressible Navier–Stokes equations with multidelays on Lipschitz domain, which models the control theory of 2D fluid flow. By a new retarded Gronwall inequality and estimates of stream function for Stokes equations, the complete trajectories inside pullback attractors are asymptotically stable via the restriction on the generalized Grashof number of fluid flow. The results in this presented paper are some extension of the literature by Yang, Wang, Yan and Miranville in 2021, as well as also the preprint by Su, Yang, Miranville and Yang in 2022

Keywords: Navier–Stokes equations; multidelays; Lipschitz domain

MSC: 35B40; 35B41; 35Q30; 76D03; 76D05

Citation: Zhang, L.-R.; Yang, X.-G.; Su, K.-Q. Asymptotic Stability for the 2D Navier–Stokes Equations with Multidelays on Lipschitz Domain. *Mathematics* **2022**, *12*, 4561. https://doi.org/10.3390/math10234561

Academic Editors: Almudena del Pilar Marquez Lozano and Vladimir Iosifovich Semenov

Received: 17 October 2022
Accepted: 29 November 2022
Published: 1 December 2022

1. Introduction

The 2D incompressible Navier–Stokes equations govern the conservation law of fluid flow for momentum and mass on a bounded domain with smooth boundary, which can be described by

$$\begin{cases} \frac{\partial}{\partial t}u - \nu\Delta u + (u\cdot\nabla)u + \nabla p = F(t,x), \\ \mathrm{div}u = 0, \end{cases} \tag{1}$$

where u and p are the velocity field and pressure for incompressible fluid flow such as water, $\nu > 0$ denotes the viscosity of fluid and $F(t,x)$ is the external force.

A bounded domain $\Omega \subset \mathbb{R}^d$ is said to be Lipschitz if $\partial\Omega$ can be covered by finite many balls $B_i = B(Q_i, r_0)$ with $Q_i \in \partial\Omega$, such that for any ball B_i there is a rectangular coordinate system and a Lipschitz function $\Psi_i : \mathbb{R}^{d-1} \to \mathbb{R}$ with

$$B(Q_i, 3r_0) \cap \Omega = \{(x_1, x_2, \cdots, x_d) | x_d > \Psi_i(x_1, x_2, \cdots, x_{d-1})\} \cap \Omega,$$

which can be seen in [1]. The 2D incompressible Navier–Stokes equations defined on the Lipschitz domain have been studied in Brown, Perry and Shen [1], which presented the well-posedness and finite fractal dimensional global attractor for an autonomous system, which has been extended to a non-autonomous case in [2] and some related literature.

The delay on differential equations originates from the controller on boundary in engineering, which can be described by evolutionary partial differential equations with delayed term, and were first investigated for ordinary differential equations, such as in [3]. The Navier–Stokes equations with delay have also become interesting topics in the recent two decades, which are important dominant physical models for fluid mechanics, such as the wind tunnel model. The research on the well-posedness and dynamics of Navier–Stokes equations with delay can be seen in [4–12] and the literature therein. For the Navier–Stokes system with time-varying delay, the tempered pullback dynamics are obtained by energy equation approach to achieve compactness, such as in Caraballo and Real [5–7], García-Luengo and Marín-Rubio [9] and Yang, Wang, Yan and Miranville [12]. Recently, Su, Yang, Miranville and Yang [11] considered (2) and derived the well-posedness, regularity, pullback dynamics and robustness. Since stability, observability and controllability are crucial in the control theory and applications in engineering, the asymptotic stability of complete trajectories is an important basis for the research on controllability and dynamic systems. To the best of our knowledge, there are fewer results on the asymptotic stability and reduction in trajectories inside pullback attractors of 2D incompressible Navier–Stokes equations defined on Lipschitz domain which are non-smooth, this is our motivation for this presented research.

This paper investigates the asymptotic stability of trajectories inside pullback attractors for the two-dimensional incompressible Navier–Stokes equations with multidelays on Lipschitz domain $\Omega \subset \mathbb{R}^2$ with inhomogeneous boundary, which reads as

$$\begin{cases} \frac{\partial}{\partial t}u - \nu\Delta u + (u(t-\rho(t))\cdot\nabla)u + \nabla p = f(t,u(t-\rho(t))) + g(t,x), \ (t,x)\in\Omega_\tau, \\ divu = 0, \ (t,x)\in\Omega_\tau, \\ u(t,x) = \varphi, \ \varphi\cdot\mathbf{n} = 0, \ (t,x)\in\partial\Omega_\tau, \\ u(\tau,x) = u(\tau), \ x\in\Omega, \\ u(\tau+\theta,x) = \phi(\theta,x), \ (\theta,x)\in\Omega_h, \end{cases} \tag{2}$$

where $\Omega_\tau = (\tau,+\infty)\times\Omega$, $\partial\Omega_\tau = (\tau,+\infty)\times\partial\Omega$, $\Omega_h = [-h,0]\times\Omega$, $\tau\in\mathbb{R}$ is the initial time and $h > 0$ is a positive constant. ν is the kinematic viscosity of the fluid, $u = (u_1(t,x), u_2(t,x))$ is the unknown velocity field of the fluid, p denotes the unknown pressure and ν is the kinematic viscosity of the fluid. The non-autonomous external forces contain $g(t,x)$ and continuous delay $f(t,u(t-\rho(t)))$, where $\rho(t)$ is the delay in $[0,h]$. The function ϕ denotes the initial state in $[-h,0]$ with $u(\tau) = \phi(0)$. The forcing boundary condition $\varphi\in L^\infty(\partial\Omega)$, where $\mathbf{n}$ is the outward unit normal to the boundary $\partial\Omega$.

Originated from [13–16], based on the results in [11], the asymptotic stability of trajectories inside pullback attractors for (2) are investigated in this presented paper with features and difficulties as follows.

(I) The problem (2) contains an inhomogeneous boundary on a Lipschitz-like domain; using the stream function ψ for the corresponding Stokes equations subject to the same boundary condition, the inhomogeneous problem (2) can be transformed into an equivalent homogeneous system (10). For the model (2), the delays on external force $f(\cdot,\cdot)$ and convective term $(u(\cdot)\nabla)u(\cdot)$ can be different as $\rho_1(t)$ and $\rho_2(t)$, which have the same difficulty under some appropriate hypotheses in Section 2.2. For simplicity, we assume they are the same as the case $\rho_1(t) = \rho_2(t) = \rho(t)$.
Based on the global well-posedness of weak solutions and pullback attractors in [11], the asymptotic stability of complete trajectories inside pullback attractor $\mathcal{A}^{M_H}$ of (18) has been achieved by using a new retard Gronwall inequality and some estimates on stream function for Stokes equations. Since there are two delays contained in (2), the energy estimates cannot be obtained by using the technique as in [15,17] to achieve the desired estimate for using differential Gronwall inequalities, which is the main difficulty here. By introducing a new retard Gronwall inequality in [13], and using the iteration technique, one sufficient condition (12) on generalized Grashof number guarantees our asymptotic stability; see Theorem 5 in Section 2.4.

(II) The results in this presented paper are a further research of [15], which is a special case of (2). The asymptotic stability of (2) is an extension of the recent work [11]. Our work also implies the exponentially attracting property for the existence of invariant manifold although the inertial manifolds for 2D Navier–Stokes equation is still open.

The outline is organized as follows. The main results are stated in Section 2 and proved in the third part, which is based on the preliminary in Section 3.

2. Main Results

2.1. Preliminary

Let $E := \{u | u \in (\mathfrak{D}(\Omega))^2, \mathrm{div} u = 0\}$, H and V are the closure of E in $(L^2(\Omega))^2$ and $(H^1(\Omega))^2$ topology, respectively; the norm and inner product of H is defined as

$$\|u\|_H^2 = (u,u), \ (u,v) = \sum_{j=1}^{2} \int_\Omega u_j(x) v_j(x) dx$$

for $u, v \in H$, and for V as $\|u\|^2 = ((u,u))$ and

$$((u,v)) = \sum_{i,j=1}^{2} \int_\Omega \frac{\partial u_j}{\partial x_i} \frac{\partial v_j}{\partial x_i} dx$$

for $u, v \in V$. It is easy to check that H and V are Hilbert spaces, $V \hookrightarrow H \equiv H' \hookrightarrow V'$, and the injections are dense and continuous. $\|\cdot\|_*$ and $\langle \cdot, \cdot \rangle$ denote the norm in V' and the dual product between V and V', respectively, and also H to itself.

Let P_L be the Helmholz–Leray orthogonal projection in $(L^2(\Omega))^2$ onto H, and $A := -P_L \Delta$ the Stokes operator. The bilinear and trilinear operators are defined as $B(u,v) := P_L((u \cdot \nabla)v)$, $b(u,v,w) = \langle B(u,v), w \rangle$, which satisfies $b(u,v,v) = 0$, $b(u,v,w) = -b(u,w,v)$, and hence

$$|b(u,v,w)| \le C \|u\|_H^{\frac{1}{2}} \|u\|^{\frac{1}{2}} \|v\| \|w\|_H^{\frac{1}{2}} \|w\|^{\frac{1}{2}}, \ \forall\, u,v,w \in V.$$

For any $t \in (\tau, T)$, we define $u : (\tau - h, T) \to (L^2(\Omega))^2$, and the delayed functional space as follows

$$C_X = C([-h,0]; X), \ \|u\|_{C_X} = \sup_{\theta \in [-h,0]} \|u(t+\theta)\|_X, \ X = H, \ V,$$

which are Banach spaces. Moreover, the p-power delayed integrable space can be defined as $L_X^p = L^p(-h,0;X)$, $1 < p \le +\infty$, and the norm is similar as the general Lebesgue space in delayed interval $[-h,0]$. Moreover, the product space is defined well as $M_H = H \times (C_H \cap L_V^2)$ with norm

$$\|(u(t), u_t)\|_{M_H}^2 = \|u(t)\|_H^2 + \|u_t\|_{C_H}^2 + \|u_t\|_{L_V^2}^2.$$

2.2. Hypotheses

For the well-posedness and pullback dynamics of (2), we force assumptions on $\rho(t)$ and $f(\cdot,\cdot)$ as follows.

(H-a) There exists $m > 0$ such that the external force $g(\cdot,\cdot) \in L^2_{loc}(\mathbb{R}, V')$ satisfies

$$\int_{-\infty}^{t} e^{ms} \|g(s,\cdot)\|_{V'}^2 ds < \infty, \ \ \forall t \in \mathbb{R}. \tag{3}$$

(H-b) The function $f(\cdot, u) : [\tau, +\infty) \to H$ is measurable for all $u \in H$, and $f(t,\cdot) : C_H \to H$ is continuous for all $t \ge \tau$. The delay $\rho \in C^1([0,+\infty); [0,h])$, and there exists a positive constant $\rho^* < 1$ such that $|\frac{d\rho}{dt}| \le \rho^*$.

(H-c) There exist functions $\alpha, \ \beta : [\tau, +\infty) \to [0, +\infty)$, where $\alpha(\cdot) \in L^\infty(\tau, T)$ and $\beta(\cdot) \in L^1(\tau, T)$ with $\limsup\limits_{\tau \to -\infty} \dfrac{1}{t-\tau} \displaystyle\int_\tau^t \beta(s) ds = \tilde{\beta}_0 \in (0, +\infty)$, such that $|f(t,u)|^2 \le \alpha(t)|u|^2 + \beta(t)$, $\forall\, t \ge \tau$.

In addition, there exists a constant $L(r) > 0$ such that $|f(t,w_1) - f(t,w_2)| \le L(r)\gamma^{1/2}(t)$ $|w_1 - w_2|$ for $\|w_1\|_H \le r$, $\|w_2\|_H \le r$ with $\tilde{\gamma}(t) \in L^\infty(\tau, T)$.

(H-d) $\nu - \frac{\|\alpha(t)\|_{L^\infty(\tau,T)}}{\nu \lambda_1 (1-\rho^*)} > 0$.

(H-e) Denoting

$$\limsup_{\tau \to -\infty} \frac{1}{t-\tau} \int_{\tau}^{t} \alpha(r)dr = \alpha_0 \in [0, +\infty), \tag{4}$$

for arbitrary $t \in \mathbb{R}$, then there exists some $\delta > 0$ such that

$$\frac{e^{\nu\lambda_1 h}\alpha_0}{\nu\lambda_1(1-\rho^*)} + \frac{Ce^{\frac{\nu\lambda_1 h}{2}}\|\varphi\|_{L^\infty(\partial\Omega)}}{\nu(1-\rho^*)} + \delta < \nu\lambda_1. \tag{5}$$

(H-f) Assume that

$$\kappa_\delta(t,s) = \Big(\nu\lambda_1 - \frac{Ce^{\frac{\nu\lambda_1 h}{2}}\|\varphi\|_{L^\infty(\partial\Omega)}}{\nu(1-\rho^*)} - \delta\Big)(t-s) - \frac{e^{\nu\lambda_1 h}}{\nu\lambda_1(1-\rho^*)}\int_s^t \alpha(r)dr, \tag{6}$$

where

$$\kappa_\delta(0,t) - \kappa_\delta(0,s) = -\kappa_\delta(t,s) \tag{7}$$

and

$$\kappa_\delta(0,r) \le \kappa_\delta(0,t) + \Big(\nu\lambda_1 - \frac{Ce^{\frac{\nu\lambda_1 h}{2}}\|\varphi\|_{L^\infty(\partial\Omega)}}{\nu(1-\rho^*)} - \delta\Big)h \tag{8}$$

if $\nu\lambda_1 - \frac{Ce^{\frac{\nu\lambda_1 h}{2}}\|\varphi\|_{L^\infty(\partial\Omega)}}{\nu(1-\rho^*)} - \delta > 0$ for $r \in [t-h, t]$.

The function $\beta(\cdot)$ satisfies the pullback tempered condition

$$\int_{-\infty}^{t} e^{-\kappa_\delta(t,s)}\beta(s)ds < +\infty. \tag{9}$$

2.3. Well-Posedness and Pullback Dynamics

The problem (2) can be transformed into the following equivalent homogeneous system in abstract form

$$\begin{cases} \frac{\partial v}{\partial t} + \nu Av + B(v(t-\rho(t)), v) + B(v(t-\rho(t)), \psi) + B(\psi, v) & \\ \qquad = P_L(g(t,x) + f(t, u(t-\rho(t))) + \nu F) - B(\psi), & (t,x) \in \Omega_\tau, \\ \mathrm{div} v = 0, & (t,x) \in \Omega_\tau, \\ v = 0, & (t,x) \in \partial\Omega_\tau, \\ v(\tau, x) = v(\tau), & x \in \Omega, \\ v(\theta) = \eta(\theta, x) = \eta(\theta), & (\theta, x) \in \Omega_h. \end{cases} \tag{10}$$

Theorem 1. *(Global weak solution) Let $(v(\tau), \eta) \in M_H$, and the hypotheses (H-a)-(H-d) hold. Then, there exists at least one global weak solution $v(t,x)$ to system (10) on $[\tau - h, T]$.*

Proof. See, e.g., the details in Su, Yang, Miranville and Yang [11]. □

Theorem 2. *(Uniqueness) Assume the hypotheses in Theorem 1 hold. Moreover, we assume that for any $r > 0$, there exists a constant $L(r) > 0$ such that*

$$|f(t, w_1) - f(t, w_2)| \le L(r)\gamma^{1/2}(t)|w_1 - w_2|, \ \forall\, t \ge \tau, \ \|w_1\|_H \le r, \ \|w_2\|_H \le r, \tag{11}$$

where $\gamma \in L^\infty(\tau, T) : [\tau, T) \to [0, +\infty)$. Then, the global weak solution in Theorem 1 is unique, which generates a continuous process $\{S(t,\tau)\}$ in M_H.

Proof. See, e.g., the details in Su, Yang, Miranville and Yang [11]. □

Remark 1. *Originated from the idea to deal with uniform attractors in Chepyzhov and Vishik [18], based on global well-posedness in the phase space M_H, the global solution generates a process $S(t,\tau): M_H \to M_H$, which has the similar property of skew product flow as in [18].*

The existence of a minimal family of pullback attractors for problem (18) can be stated as follows.

Theorem 3. *(Tempered pullback dynamics) Suppose that $f: \mathbb{R} \times C_H \to H$ satisfies the hypotheses (H-a)–(H-d); let the functions $\alpha(\cdot)$ and $\beta(\cdot)$ satisfy (H-e)–(H-f). Then, for any $(v(\tau),\eta) \in M_H$, the process $(S(t,\tau); M_H)$ generated by the global weak solutions of problem* (10) *possesses a minimal family of tempered pullback attractors $\mathcal{A}_{\kappa_\delta}$ in $H \times C_H$, for all $\kappa_\delta \in (0, \kappa_\delta(t,\tau)]$. Moreover, if we choose fixed κ_δ^F for fixed universe to achieve pullback attractors as $\mathcal{A}^F_{\kappa_\delta^F}$, then we have the relation $\mathcal{A}^F_{\kappa_\delta^F} \subset \mathcal{A}_{\kappa_\delta} \subset \mathcal{A}_{\kappa_\delta(t,\tau)} \subset \mathcal{D}^{H\times C_H}_{\kappa_\delta(t,\tau)}$.*

Proof. See, e.g., the details in Su, Yang, Miranville and Yang [11]. □

Theorem 4. *Assume $(v(\tau),\eta) \in M_H$ and $\frac{\nu}{4} > \frac{Ce^{\frac{\nu\lambda_1}{2}h}\|\varphi\|_{L^\infty(\partial\Omega)}}{\nu(1-\rho^*)} + \frac{4e^{\frac{\nu\lambda_1}{2}h}\|\alpha(t)\|_{L^\infty(\tau,T)}}{\nu(1-\rho^*)}$, the process $S(t,\tau): M_H \to M_H$ generated by the system* (10) *possesses a minimal family of $\mathcal{D}$-pullback attractors $\mathcal{A} = \{A(t)\}_{t\in\mathbb{R}}$ in M_H.*

Proof. See, e.g., the details in Su, Yang, Miranville and Yang [11]. □

2.4. Asymptotic stability

Definition 1. *The pullback attractors are asymptotically stable if the trajectories inside the attractors reduce to a single orbit as $\tau \to -\infty$, which also demonstrates the exponentially tracking property.*

Based on the global well-posedness and the existence of tempered and $\mathcal{D}$-pullback attractors for problems (2) and (18) in [11], we present our main result as the following theorem.

Theorem 5. *Assume the external force $g \in L^2_{loc}(\mathbb{R}; V')$ and the hypothesis (H-a)-(H-d) hold, the initial data $(u(\tau),\phi) \in M_H$. Then, the trajectories' pullback attractor $\mathcal{A} = \{A(t)\}|_{t\geq\tau}$ is asymptotically stable if $G(t) + K_0 \leq \frac{2\nu}{7(1+\gamma)}$, where $G^2(t) = \frac{\langle \|g\|^2_{V'}\rangle|_{\leq t}}{\nu^2\lambda_1}$ is a generalized Grashof number for the fluid flow, and*

$$\begin{aligned}
\frac{2}{7(1+\gamma)}K_0 = & \Big[\frac{C\|\varphi\|_{L^\infty(\partial\Omega)}}{\nu} + \frac{C}{\nu\lambda_1}\|\alpha(t)\|_{L^\infty}\Big]\frac{C|\Omega|}{\nu\lambda_1}\|\beta\|_{L^1(\tau,T)} \\
& + \frac{C|\Omega|}{\nu\lambda_1}\tilde{\beta}_0 + \frac{C|\Omega|\|\varphi\|^2_{L^\infty(\partial\Omega)}}{2\nu\lambda_1}\|\alpha(t)\|_{L^\infty(\tau,T)} \\
& + \Big[\frac{C\|\varphi\|_{L^\infty(\partial\Omega)}}{\nu^2\lambda_1} + \frac{C}{\nu^2\lambda_1^2}\|\alpha(t)\|_{L^\infty} + 1\Big]\Big[\frac{C\nu|\partial\Omega|}{\varepsilon}\|\varphi\|^2_{L^\infty(\partial\Omega)} + \frac{C\varepsilon\|\varphi\|^4_{L^\infty(\partial\Omega)}|\partial\Omega|}{\nu}\Big], \qquad (12)
\end{aligned}$$

where $\gamma > 0$ is defined by the retard Gronwall inequality determined by the parameters in our problem.

3. The Proof of Theorem 5

3.1. A Retarded Gronwall Inequality

Lemma 1. *(See [13]) Considering the following retarded integral inequalities for*

$$y(t) \leq E(t,\tau)\|y_\tau\|_X + \int_\tau^t K_1(t,s)\|y_s\|_X ds + \int_t^\infty K_2(t,s)\|y_s\|_X ds + \rho, \ \forall\, t \geq \tau \geq 0, \qquad (13)$$

where E, K_1 and K_2 are non-negative measurable functions on $\mathbb{R}^2$, and $\rho \geq 0$ denotes a constant. Let X be a Banach space with a spatial variable, then we use $\|\cdot\|$ to denote the norm of space $C([-h,0];X)$ for some $h \geq 0$, $y(t) \geq 0$ is a continuous function defined on $C([-h,T];X)$, $y_t(s) = y(t+s)$ for $s \in [-h,0]$. Let

$$\mathcal{L}(E,K_1,K_2,\rho) = \{y \in C([-h,T];X) | y \geq 0 \text{ and satisfies the inequality (13)}\},$$

and

$$\kappa(K_1,K_2) = \sup_{t\geq\tau}\Big(\int_\tau^t K_1(t,s)ds + \int_t^\infty K_2(t,s)ds\Big)$$

with $\kappa(K_1,K_2) < +\infty$. Assume that $\lim_{t\to+\infty} E(t+s,s) = 0$ uniformly with respect to $s \in \mathbb{R}^+$, and denote $\vartheta = \sup_{t\geq s\geq\tau} E(t,s)$ and $\kappa = \kappa(K_1,K_2)$, then we have the following estimates:

(1) If $\kappa < 1$, then for any R, $\varepsilon > 0$, there exists $\tilde{T} > 0$ such that

$$\|y_t\|_X < \mu\rho + \varepsilon, \tag{14}$$

for $t > \tilde{T}$ and all bounded functions $y \in \mathcal{L}(E,K_1,K_2,\rho)$ with $\|y_0\| \leq R$, where $\mu = \frac{1}{1-\kappa}$.

(2) If $\kappa < \frac{1}{1+\vartheta}$, then there exist parameters $M > 0$ and $\lambda > 0$, which are independent on ρ such that

$$\|y_t\|_X \leq M\|y_0\|_X e^{-\lambda t} + \gamma\rho, \quad t \geq \tau \tag{15}$$

for all bounded functions $y \in \mathcal{L}(E,K_1,K_2,\rho)$, where $\gamma = \frac{\mu+1}{1-\kappa c}$ and $c = \max\{\frac{\vartheta}{1-\kappa}, 1\}$.

3.2. The Stokes Problem on Lipschitz Domains

From [1], the stream function ψ solves the following Stokes system on the Lipschitz domain

$$\begin{cases} -\Delta u + \nabla q = 0, \ x \in \Omega, \\ \text{div} u = 0, \ x \in \Omega, \\ u = \varphi \text{ a.e. } x \in \partial\Omega \text{ in the sense of non-tangential convergence.} \end{cases} \tag{16}$$

Assume that $u = (u_1, u_2)$ is the solution to (16) with $\varphi \in L^\infty(\partial\Omega)$ and $\varphi \cdot n = 0$, then we define the stream function ψ satisfying (16) and

$$\begin{cases} \|\psi\|_{L^\infty(\Omega)} \leq C\|\varphi\|_{L^\infty(\partial\Omega)}, \\ \sup_{x\in\Omega}|\psi(x)| + \sup_{x\in\Omega}|\nabla\psi(x)|\text{dist}(x,\partial\Omega) \leq C\|\varphi\|_{L^\infty(\partial\Omega)}, \\ \||\nabla\psi|\text{dist}(\cdot,\partial\Omega)^{1-\frac{1}{p}}\|_{L^p(\Omega)} \leq C\|\varphi\|_{L^p(\partial\Omega)}, \ 2 \leq p \leq \infty. \end{cases}$$

In addition, the stream function ψ can be written as the following form $\Delta\psi = \nabla(q\eta_\varepsilon) + F$, where $\text{supp}F \subset \{x \in \Omega;\ C_1'\varepsilon \leq \text{dist}(x,\partial\Omega) \leq C_2'\varepsilon\}$ and $|F| \leq \frac{C}{\varepsilon^{3/2}}\|\varphi\|_{L^2(\partial\Omega)}$. The above estimate is based on the singular operator and Hardy's inequality as

$$\int_\Omega \frac{|u(x)|^2}{[\text{dist}(x,\partial\Omega)]^2}dx \leq C\int_\Omega |\nabla u(x)|^2 dx, \ \forall u \in V. \tag{17}$$

3.3. Proof of Main Results

Proof. By an equivalent system as (18) and stationary equation as (16), the trajectories in pullback attractors of systems (2) and (18) are synchronous, which implies we only need to consider the asymptotic stability of trajectories inside the pullback attractor for (18). The proofs are divided into the following steps.

Step 1: Some estimates of differencing equations

Setting $v = u - \psi$ and $(v(\tau), \eta) \in H \times (C_H \cap L^2_V)$, then (2) can be transformed into the following equivalent abstract functional evolutionary differential equations with homogeneous boundary condition

$$\begin{cases} \frac{\partial v}{\partial t} + \nu A v + B(v(t-\rho(t)), v) + B(v(t-\rho(t)), \psi) + B(\psi, v) \\ \qquad\qquad = P_L(f(t, u(t-\rho(t))) + g(t,x) + \nu F) - B(\psi), \\ \text{div} v = 0, \\ v|_{\partial\Omega} = 0, \\ v(\tau, x) = v(\tau), \\ v(\theta) = \eta(\theta), \ \theta \in [\tau - h, \tau]. \end{cases} \tag{18}$$

Let $v(t)$ and $\tilde{v}(t)$ be two global solution orbits for problem (18) inside the $\mathcal{D}$-pullback attractor with initial data

$$v(\tau+\theta)|_{\theta\in[-h,0]} = \eta(\theta), \quad v|_{t=\tau} = v(\tau),$$
$$\tilde{v}(\tau+\theta)|_{\theta\in[-h,0]} = \tilde{\eta}(\theta), \quad \tilde{v}|_{t=\tau} = \tilde{v}(\tau),$$

respectively.

By the procedure in achieving the $\mathcal{D}$-pullback attractors in [11], the global weak solution for (18) generates a continuous process $S(t,\tau)$ in $M_H = H \times (C_H \cap L^2_V)$ as

$$(v, v_t) = S(t,\tau)(v(\tau), \eta) \ \text{and} \ (\tilde{v}, \tilde{v}_t) = S(t,\tau)(\tilde{v}(\tau), \tilde{\eta}), \tag{19}$$

which are also two trajectories inside the pullback attractors $\mathcal{A} = \{A(t)\}_{t\in\mathbb{R}}$ in M_H, here, $v_t = v(t+s)$ for $s \in [-h, 0]$.

Denoting $w = v(t) - \tilde{v}(t)$ and $w_t = v_t - \tilde{v}_t$ by some simple computation, it is easy to check that w satisfies the following initial and boundary value problem for functional evolutionary partial differential equations as

$$\begin{cases} \frac{\partial w}{\partial t} + \nu A w + B(w(t-\rho(t)), v) + B(\tilde{v}(t-\rho(t)), w) + B(w(t-\rho(t)), \psi) + B(\psi, w) \\ \qquad\qquad = P_L(f(v(t-\rho(t))) + \psi) - f(\tilde{v}(t-\rho(t)) + \psi)), \\ \text{div} w = 0, \\ w|_{\partial\Omega} = 0, \\ w(t=\tau) = v(\tau) - \tilde{v}(\tau), \\ w(\tau+\theta) = \eta(\theta) - \tilde{\eta}(\theta), \ \theta \in [-h, 0]. \end{cases} \tag{20}$$

Multiplying (20) by w at both sides, using Poincaré's inequality, noting the property of the trilinear operator $(B(\tilde{v}(t-\rho(t)), w), w) = 0$ and $(B(\psi, w), w) = 0$, we derive that

$$\begin{aligned} \frac{1}{2}\frac{d}{dt}\|w\|^2_H + \nu\|w\|^2 \ &\leq \ \Big|(B(w(t-\rho(t)), v) + B(w(t-\rho(t)), \psi), w)\Big| \\ &\quad + \Big|(P_L(f(v(t-\rho(t)) + \psi) - f(\tilde{v}(t-\rho(t)) + \psi)), w)\Big|. \end{aligned} \tag{21}$$

Using the Hardy and Hölder inequalities, we have

$$\begin{aligned} \Big|(B(w(t-\rho(t)), v, w)\Big| \ &\leq \ \int_\Omega |(w(t-\rho(t))||\nabla w||v| dx \\ &\leq \ \frac{1}{2}\|v\|^2\|w(t)\|^2 + \frac{C}{2}\|w(t-\rho(t))\|^2_H \end{aligned} \tag{22}$$

and

$$\begin{aligned} \Big|(B(w(t-\rho(t), \psi), w)\Big| \ &\leq \ C\|\varphi\|_{L^\infty(\partial\Omega)} \int_{\text{dist}(x,\partial\Omega)\leq C_2'\varepsilon} \frac{|w(t)||w(t-\rho(t))|}{\text{dist}(x,\partial\Omega)} dx \\ &\leq \ \frac{\nu}{4}\|w\|^2 + \frac{C\|\varphi\|_{L^\infty(\partial\Omega)}}{\nu}\|w(t-\rho(t))\|^2_H \end{aligned} \tag{23}$$

and

$$\begin{aligned} &|(f(t,v(t-\rho(t))+\psi)-f(t,\tilde{v}(t-\rho(t))+\psi),w)| \\ \leq\ &\frac{\nu}{4}\|w(t)\|^2+\frac{1}{\nu\lambda_1}L^2(r)\tilde{\gamma}(t)\|w(t-\rho(t))\|_H^2. \end{aligned} \tag{24}$$

We can use the Poincaré and Gronwall inequalities to achieve the hypothesis in Lemma 1 for the asymptotic stability of trajectories inside $\mathcal{D}$-pullback attractors $\mathcal{A}$ in [11], provided that

$$\nu\lambda_1-\|v\|_V^2>0, \tag{25}$$

then, we can obtain

$$\begin{aligned} \|w\|_H^2 \leq\ & e^{\int_\tau^t -(\nu\lambda_1-\|v\|_V^2)ds}\|v(\tau)-\tilde{v}(\tau)\|_H^2+ \\ &+\Big[\frac{C}{2}+\frac{C\|\varphi\|_{L^\infty(\partial\Omega)}}{\nu}+\frac{1}{\nu\lambda_1}L^2(r)\|\tilde{\gamma}\|_{L^\infty(\tau,T)}\Big]\int_\tau^t e^{-\int_s^t(\nu\lambda_1-\|v\|_V^2)d\sigma}\|w_t\|_H^2 ds. \end{aligned} \tag{26}$$

Denoting

$$\begin{aligned} &E(t,\tau)=e^{-\int_\tau^t(\nu\lambda_1-\|v\|_V^2)ds}, \\ &K_1(t,s)=\Big[\frac{C}{2}+\frac{C\|\varphi\|_{L^\infty(\partial\Omega)}}{\nu}+\frac{1}{\nu\lambda_1}L^2(r)\|\tilde{\gamma}(t)\|_{L^\infty}\Big]e^{-\int_s^t(\nu\lambda_1-\|v\|_V^2)d\sigma}, \\ &\Theta=\sup_{t\geq s\geq\tau}E(t,s),\quad \kappa(K_1,0)=\sup_{t\geq\tau}\int_\tau^t K_1(t,s)ds, \end{aligned}$$

noting the assumption and inequality in Lemma 1, choosing $\kappa(K_1,0)<\frac{1}{1+\Theta}$. In fact, since $v\in L^\infty(\tau,T;H)\cap L^2(\tau,T;V)$, we have

$$\sup_{t\geq\tau}\int_\tau^t K_1(t,s)ds\leq\sup_{t\geq\tau}\frac{M\Big(\frac{C}{2}+\frac{C\|\varphi\|_{L^\infty(\partial\Omega)}}{\nu}+\frac{1}{\nu\lambda_1}L^2(r)\|\tilde{\gamma}(t)\|_{L^\infty}\Big)}{\nu\lambda_1}[1-e^{-\nu\lambda_1(t-\tau)}] \tag{27}$$

and there exists a pullback time $\bar{\tau}<<\tau$ such that $\kappa(K_1,0)<\frac{1}{2}$, which implies the assumption in Lemma 1 holds.

Hence, from Lemma 1, there exist $\bar{M}>0$ and $\mu>0$, such that we can obtain the following estimate

$$\|w(t-\rho(t))\|_H^2\leq M\Big[\|v(\tau)-\tilde{v}(\tau)\|_H^2+\|\eta(\theta)-\tilde{\eta}(\theta)\|_{L_V^2}^2\Big]e^{-\mu(t-\tau)}. \tag{28}$$

Substituting (28) into (21), we can conclude the following estimate

$$\begin{aligned} \|w\|_H^2 \leq\ & e^{\int_\tau^t -(\nu\lambda_1-\|v\|_V^2)ds}\|v(\tau)-\tilde{v}(\tau)\|_H^2+ \\ &+M\Big[\|v(\tau)-\tilde{v}(\tau)\|_H^2+\|\eta(\theta)-\tilde{\eta}(\theta)\|_{L_V^2}^2\Big]e^{-\mu(t-\tau)} \\ &\times\Big[\frac{C}{2}+\frac{C\|\varphi\|_{L^\infty(\partial\Omega)}}{\nu}+\frac{1}{\nu\lambda_1}L^2(r)\|\tilde{\gamma}(t)\|_{L^\infty}\Big]\int_\tau^t e^{-\int_s^t(\nu\lambda_1-\|v\|_V^2)d\sigma}ds. \end{aligned} \tag{29}$$

Step 2: The sufficient condition on asymptotic stability via generalized Grashof number

Combining (28) with (29), considering the trajectories represented by (19) for fixing initial data, and letting $\tau\to-\infty$, we can then conclude that the trajectories inside pullback attractors reduce to a single one, which implies the asymptotic stability provided that $\nu\lambda_1 > \langle \|v\|_V^2 \rangle_{\le t}$, where $\langle h \rangle_{\le t}$ is defined as

$$\langle h \rangle_{\le t} = \limsup_{\tau \to -\infty} \frac{1}{t-\tau} \int_\tau^t h(r) dr. \tag{30}$$

Since v and $\tilde{v}$ are two global weak solutions for (18), we use Lemma 1 for a iteration procedure and some delicate estimates to present a sufficient condition for asymptotic stability of trajectories inside the pullback attractors by virtue of the uniform boundedness of stream function.

Taking the inner product of (18) with u in H, this yields

$$\begin{aligned} & \frac{1}{2}\frac{d}{dt}\|v\|_H^2 + \nu\|v\|^2 \\ \le\; & |(P_L(f(t, v(t-\rho(t)) + \psi) + \nu F), v)| + |(B(\psi,\psi), v)| \\ & + |(B(v(t-\rho(t)), \psi), v)| + |\langle g, v\rangle|. \end{aligned} \tag{31}$$

Using the Hardy and Hölder inequalities, by virtue of estimates for stream function in Section 3.2 and $\|\varphi\|_{L^2(\partial\Omega)} \le C|\partial\Omega|^{1/2}\|\varphi\|_{L^\infty(\partial\Omega)}$ from [1], we obtain

$$\begin{aligned} |b(v(t-\rho(t)), \psi, v)| &\le C\|\varphi\|_{L^\infty(\partial\Omega)} \int_{\mathrm{dist}(x,\partial\Omega)\le C_2'\varepsilon} \frac{|v(t)||v(t-\rho(t))|}{[\mathrm{dist}(x,\partial\Omega)]} dx \\ &\le \frac{\nu}{14}\|v\|^2 + \frac{C\|\varphi\|_{L^\infty(\partial\Omega)}}{\nu}\|v(t-\rho(t))\|_H^2, \end{aligned} \tag{32}$$

$$\begin{aligned} |(B(\psi,\psi), v)| &\le C\|\varphi\|_{L^\infty(\partial\Omega)} \int_\Omega \frac{|v|}{\mathrm{dist}(x,\partial\Omega)}|\psi| dx \\ &\le C\varepsilon^{1/2}\|\varphi\|_{L^\infty(\partial\Omega)}^2 |\partial\Omega|^{1/2}\|v\| \\ &\le \frac{\nu}{14}\|v\|^2 + \frac{C\varepsilon\|\varphi\|_{L^\infty(\partial\Omega)}^4|\partial\Omega|}{\nu} \end{aligned} \tag{33}$$

and

$$\nu|\langle F, v\rangle| \le \frac{C\nu}{\sqrt{\varepsilon}}\|\varphi\|_{L^2(\partial\Omega)}\|v\| \le \frac{\nu}{14}\|v\|^2 + \frac{C\nu|\partial\Omega|}{\varepsilon}\|\varphi\|_{L^\infty(\partial\Omega)}^2, \tag{34}$$

$$|\langle g, v\rangle| \le \frac{\nu}{14}\|v\|^2 + \frac{7/2}{\nu}\|g(t)\|_{V'}^2. \tag{35}$$

By hypotheses (H-a)-(H-d), the estimates of stream function and the Minkowski inequality, we can derive that

$$\begin{aligned} & (f(t, v(t-\rho(t)) + \psi), v(t)) \\ \le\; & \alpha^{\frac{1}{2}}(t)\|v(t-\rho(t))\|_H\|v(t)\|_H + \alpha^{\frac{1}{2}}(t)|\psi|\|v(t)\|_H + \beta^{\frac{1}{2}}(t)\|v(t)\|_H \\ \le\; & \frac{C}{\nu\lambda_1}\alpha(t)\|v(t-\rho(t))\|_H^2 + \frac{\nu}{14}\|v(t)\|^2 + \frac{C|\Omega|\|\varphi\|_{L^\infty(\partial\Omega)}^2}{2\nu\lambda_1}\alpha(t) + \frac{C|\Omega|}{\nu\lambda_1}\beta(t). \end{aligned} \tag{36}$$

Combining (31)–(36), we obtain

$$\begin{aligned} & \frac{d}{dt}\|v\|^2 + \nu\|v\|^2 \\ \le\; & \Big[\frac{C\|\varphi\|_{L^\infty(\partial\Omega)}}{\nu} + \frac{C}{\nu\lambda_1}\|\alpha(t)\|_{L^\infty}\Big]\|v(t-\rho(t))\|_H^2 + \frac{C|\Omega|\|\varphi\|_{L^\infty(\partial\Omega)}^2}{2\nu\lambda_1}\alpha(t) + \frac{C|\Omega|}{\nu\lambda_1}\beta(t) \\ & + \frac{C\nu|\partial\Omega|}{\varepsilon}\|\varphi\|_{L^\infty(\partial\Omega)}^2 + \frac{C\varepsilon\|\varphi\|_{L^\infty(\partial\Omega)}^4|\partial\Omega|}{\nu} + \frac{7/2}{\nu}\|g(t)\|_{V'}^2. \end{aligned} \tag{37}$$

By using the Poincaré inequality and Lemma 1, we can conclude that

$$
\begin{aligned}
\|v\|_H^2 \leq\ & e^{-\nu\lambda_1(t-\tau)}\|v(\tau)\|_H^2 + \frac{C|\Omega|\|\varphi\|^2_{L^\infty(\partial\Omega)}}{\nu\lambda_1}\int_\tau^t e^{-\nu\lambda_1(t-s)}\alpha(s)ds \\
& + \frac{C|\Omega|}{\nu\lambda_1}\int_\tau^t e^{-\nu\lambda_1(t-s)}\beta(s)ds \\
& + \Big[\frac{C\|\varphi\|_{L^\infty(\partial\Omega)}}{\nu} + \frac{C}{\nu\lambda_1}\|\alpha(t)\|_{L^\infty}\Big]\int_\tau^t e^{-\nu\lambda_1(t-s)}\|v(s-\rho(s))\|_H^2 ds \\
& + \frac{1}{\nu\lambda_1}\Big[\frac{C\nu|\partial\Omega|}{\varepsilon}\|\varphi\|^2_{L^\infty(\partial\Omega)} + \frac{C\varepsilon\|\varphi\|^4_{L^\infty(\partial\Omega)}|\partial\Omega|}{\nu}\Big](1-e^{-\nu\lambda_1(t-\tau)}) \\
& + \frac{7/2}{\nu}\int_\tau^t e^{-\nu\lambda_1(t-s)}\|g\|^2_{V'}ds.
\end{aligned} \tag{38}
$$

Denoting

$$
\begin{aligned}
& E(t,\tau) = e^{-\nu\lambda_1(t-\tau)}, \\
& K_1(t,s) = \Big[\frac{C\|\varphi\|_{L^\infty(\partial\Omega)}}{\nu} + \frac{C}{\nu\lambda_1}\|\alpha(t)\|_{L^\infty}\Big]e^{-\nu\lambda_1(t-s)}, \\
& \rho = \frac{C|\Omega|\|\varphi\|^2_{L^\infty(\partial\Omega)}}{\nu\lambda_1}\int_\tau^t e^{-\nu\lambda_1(t-s)}\alpha(s)ds + \frac{C|\Omega|}{\nu\lambda_1}\int_\tau^t e^{-\nu\lambda_1(t-s)}\beta(s)ds \\
& \quad + \frac{1}{\nu\lambda_1}\Big[\frac{C\nu|\partial\Omega|}{\varepsilon}\|\varphi\|^2_{L^\infty(\partial\Omega)} + \frac{C\varepsilon\|\varphi\|^4_{L^\infty(\partial\Omega)}|\partial\Omega|}{\nu}\Big](1-e^{-\nu\lambda_1(t-\tau)}) \\
& \quad + \frac{7/2}{\nu}\int_\tau^t e^{-\nu\lambda_1(t-s)}\|g\|^2_{V'}ds, \\
& \Theta = \sup_{t\geq s\geq\tau} E(t,s), \quad \kappa(K_1,0) = \sup_{t\geq\tau}\int_\tau^t K_1(t,s)ds,
\end{aligned}
$$

choosing a small enough $\tilde{\tau} << \tau$ such that $\kappa(K_1,0) < \frac{1}{1+\Theta}$, then by using Lemma 1, there exist parameters $\hat{M} > 0$, $\gamma > 0$ and $\hat{\mu} > 0$, such that we can obtain the estimate

$$
\begin{aligned}
\|v(t-\rho(t))\|_H^2 \leq\ & \hat{M}\Big[\|v(\tau)\|_H^2 + \|\eta\|^2_{L^2_V}\Big]e^{-\hat{\mu}(t-\tau)} + \gamma\Big[\frac{7/2}{\nu}\int_\tau^t e^{-\nu\lambda_1(t-s)}\|g\|^2_{V'}s \\
& + \frac{C|\Omega|\|\varphi\|^2_{L^\infty(\partial\Omega)}}{\nu\lambda_1}\int_\tau^t e^{-\nu\lambda_1(t-s)}\alpha(s)ds + \frac{C|\Omega|}{\nu\lambda_1}\int_\tau^t e^{-\nu\lambda_1(t-s)}\beta(s)ds \\
& + \frac{1}{\nu\lambda_1}\Big(\frac{C\nu|\partial\Omega|}{\varepsilon}\|\varphi\|^2_{L^\infty(\partial\Omega)} + \frac{C\varepsilon\|\varphi\|^4_{L^\infty(\partial\Omega)}|\partial\Omega|}{\nu}\Big)(1-e^{-\nu\lambda_1(t-\tau)})\Big]
\end{aligned} \tag{39}
$$

Substituting (39) into (38), integrating (37) over $[\tau, t]$, we can obtain

$$
\begin{aligned}
\|v\|_H^2 \leq\ & e^{-\nu\lambda_1(t-\tau)}\|v(\tau)\|_H^2 + C_1\Big[\|v(\tau)\|_H^2 + \|\eta\|^2_{L^2_V}\Big]e^{-\hat{\mu}(t-\tau)} \\
& + C_2\Big[\|g(t)\|^2_{L^2(\tau,T;V')} + \|\alpha(t)\|_{L^\infty(\tau,T)} + \|\beta(t)\|_{L^1(\tau.T)} + 1\Big]
\end{aligned} \tag{40}
$$

and

$$
\begin{aligned}
\frac{\nu}{t-\tau}\int_\tau^t \|v(r)\|_V^2 dr \leq\ & \Big[\frac{C\|\varphi\|_{L^\infty(\partial\Omega)}}{\nu}+\frac{C}{\nu\lambda_1}\|\alpha(t)\|_{L^\infty}\Big]\Big\{\hat{M}\Big[\|v(\tau)\|_H^2+\|\eta\|_{L_V^2}^2\Big]e^{-\tilde{\mu}(t-\tau)}\\
&+\gamma\Big[\frac{C|\Omega|\|\varphi\|^2_{L^\infty(\partial\Omega)}}{\nu\lambda_1}\int_\tau^t e^{-\nu\lambda_1(t-s)}\alpha(s)ds+\frac{C|\Omega|}{\nu\lambda_1}\int_\tau^t e^{-\nu\lambda_1(t-s)}\beta(s)ds\\
&+\frac{1}{\nu\lambda_1}\Big(\frac{C\nu|\partial\Omega|}{\varepsilon}\|\varphi\|^2_{L^\infty(\partial\Omega)}+\frac{C\varepsilon\|\varphi\|^4_{L^\infty(\partial\Omega)}|\partial\Omega|}{\nu}\Big)\Big]\Big\}\\
&+\frac{C|\Omega|\|\varphi\|^2_{L^\infty(\partial\Omega)}}{2\nu\lambda_1}\|\alpha(t)\|_{L^\infty(\tau,T)}+\frac{C|\Omega|}{\nu\lambda_1}\tilde{\beta}_0\\
&+\frac{C\nu|\partial\Omega|}{\varepsilon}\|\varphi\|^2_{L^\infty(\partial\Omega)}+\frac{C\varepsilon\|\varphi\|^4_{L^\infty(\partial\Omega)}|\partial\Omega|}{\nu}+\frac{7(1+\gamma)}{2\nu}\|g(t)\|^2_{V'}.
\end{aligned}
\tag{41}
$$

Combining (37)–(41), we conclude the asymptotic stability holds, provided that

$$
\begin{aligned}
\langle\|v\|_V^2\rangle|_{\leq t} \leq\ & \Big[\frac{C\|\varphi\|_{L^\infty(\partial\Omega)}}{\nu}+\frac{C}{\nu\lambda_1}\|\alpha(t)\|_{L^\infty}\Big]\frac{C|\Omega|}{\nu^2\lambda_1}\|\beta\|_{L^1(\tau,T)}+\frac{C|\Omega|}{\nu^2\lambda_1}\tilde{\beta}_0\\
&+\frac{C|\Omega|\|\varphi\|^2_{L^\infty(\partial\Omega)}}{\nu^2\lambda_1}\|\alpha(t)\|_{L^\infty(\tau,T)}+\frac{7(1+\gamma)}{2\nu^2}\langle\|g\|^2_{V'}\rangle|_{\leq t}\\
&+\Big[\frac{C\|\varphi\|_{L^\infty(\partial\Omega)}}{\nu^3\lambda_1}+\frac{C}{\nu^3\lambda_1^2}\|\alpha(t)\|_{L^\infty}+1\Big]\Big[\frac{C\nu|\partial\Omega|}{\varepsilon}\|\varphi\|^2_{L^\infty(\partial\Omega)}+\frac{C\varepsilon\|\varphi\|^4_{L^\infty(\partial\Omega)}|\partial\Omega|}{\nu}\Big]\\
\leq\ & \nu\lambda_1.
\end{aligned}
\tag{42}
$$

If the generalized Grashof number is defined as $G(t)=\left(\frac{\langle\|g\|^2_{V'}\rangle|_{\leq t}}{\nu^2\lambda_1}\right)^{1/2}$, then we can deduce a sufficient condition for the asymptotic stability of trajectories inside pullback attractors as

$$
G(t)+K_0\leq\frac{2\nu}{7(1+\gamma)},
\tag{43}
$$

which completes the proof for our work. □

4. Conclusions and Further Research

Based on the well-posedness and pullback dynamics for 2D Navier–Stokes equations with double time-varying delays defined on a Lipschitz-like domain in [11], this presented work investigated the asymptotic stability of complete trajectories inside a pullback attractor of problem (2), which is an extension of [11,12]. However, when the delay is infinite, the dynamics and asymptotic stability are still open, which is our interest in the future.

Author Contributions: Conceptualization, X.-G.Y., L.-R.Z. and K.-Q.S.; formal analysis, X.-G.Y., L.-R.Z. and K.-Q.S.; investigation, X.-G.Y., L.-R.Z. and K.-Q.S.; writing—original draft preparation, X.-G.Y., L.-R.Z. and K.-Q.S.; writing—review and editing, X.-G.Y.; visualization, X.-G.Y., L.-R.Z. and K.-Q.S.; supervision, X.-G.Y.; funding acquisition, X.-G.Y., L.-R.Z. and K.-Q.S. All authors have read and agreed to the published version of the manuscript.

Funding: Xin-Guang Yang was partly supported by the Key project of Henan Education Department (No. 22A110011) and Incubation Fund Project of Henan Normal University (No. 2020PL17), Henan Overseas Expertise Introduction Center for Discipline Innovation (No. CXJD2020003). Lingrui Zhang is partly supported by innovation fund of Ph.D in Henan Normal Univesity.

Data Availability Statement: Not applicable.

Acknowledgments: The authors thank referees by his/her comments, which led to improvements in the presentation of this paper.

Conflicts of Interest: The authors declare no conflict of interest.

References

1. Brown, R.M.; Perry, P.A.; Shen, Z. On the dimension of the attractor of the non-homogeneous Navier-Stokes equations in non-smooth domains. *Indiana Univ. Math. J.* **2000**, *49*, 1–34. [CrossRef]
2. Yang, X.-G.; Qin, Y.; Lu, Y.; Ma, T.F. Dynamics of 2D incompressible non-autonomous Navier-Stokes equations on Lipschitz-like domains. *Appl. Math. Optim.* **2021**, *83*, 2129–2183. [CrossRef]
3. Nave, O.; Gol'dshtein, V.; Dan, E. The Delay Phenomena in Thermal Explosion of Polydisperse Fuel Spray: Asymptotic Analysis. *At. Sprays* **2011**, *21*, 69–85. [CrossRef]
4. Barbu, V.; Sritharan, S.S. Navier-Stokes equations with hereditary viscosity. *Z. Angew. Math. Phys.* **2003**, *54*, 449–461. [CrossRef]
5. Caraballo, T.; Real, J. Navier-Stokes equations with delays. *R. Soc. Lond. Proc. Ser. A Math. Phys. Eng. Sci.* **2001**, *457*, 2441–2453. [CrossRef]
6. Caraballo, T.; Real, J. Asymptotic behavior for two-dimensional Navier-Stokes equations with delays. *R. Soc. Lond. Proc. Ser. A Math. Phys. Eng. Sci.* **2003**, *459*, 3181–3194. [CrossRef]
7. Caraballo, T.; Real, J. Attractors for 2D Navier-Stokes models with delays. *J. Differ. Equ.* **2004**, *205*, 271–297. /j.jde.2004.04.012. [CrossRef]
8. García-Luengo, J.; Marín-Rubio, P.; Real, J. Pullback attractors for 2D Navier-Stokes equations with delays and their regularity. *Adv. Nonlinear Study* **2013**, *13*, 331–357. [CrossRef]
9. García-Luengo, J.; Marín-Rubio, P. Attractors for a doubled time-delayed 2D-Navier-Stokes model. *Discret. Contin. Dyn. Syst.* **2014**, *34*, 4085–4105. [CrossRef]
10. Marín-Rubio, P.; Real, J. Pullback attractors for 2D-Navier-Stokes equations with delays in continuous and sub-linear operators. *Discret. Contin. Dyn. Syst.* **2010**, *26*, 989–1006. [CrossRef]
11. Su, K.; Yang, X.-G.; Miranville, A.; Yang, H. Dynamics and Robustness for the 2D Navier-Stokes Equations with Multi-Delays on Lipschitz domain. 2022, Preprint. Available online: https://www.researchgate.net/publication/358457930 (accessed on 1 September 2022).
12. Yang, X.-G.; Wang, R.; Yan, X.; Miranville, A. Dynamics of the 2D Navier-Stokes equations with sublinear operators in Lipschitz-like domains. *Discret. Contin. Dyn. Syst.* **2021**, *41*, 3343–3366. [CrossRef]
13. Li, D.; Liu, Q.; Ju, X. Uniform decay estimates for solutions of a class of retarded integral inequalities. *J. Differ. Equ.* **2021**, *271*, 1–38. [CrossRef]
14. Miranville, A.; Wang, X. Upper bounded on the dimension of the attractor for nonhomogeneous Navier-Stokes equations. *Discret. Contin. Dyn. Syst.* **1996**, *2*, 95–110. [CrossRef]
15. Yang, X.-G.; Li, L.; Yan, X.; Ding, L. The structure and stability of pullback attractors for 3D Brinkman-Forchheimer equation with delay. *Electron. Res. Arch.* **2020**, *28*, 1395–1418. [CrossRef]
16. Miranville, A.; Wang, X. Attractors for non-autonomous nonhomogenerous Navier-Stokes equations. *Nonlinearity* **1997**, *10*, 1047–1061. [CrossRef]
17. Carvalho, A.N.; Langa, J.A.; Robinson, J.C. *Attractors for Infinite-Dimensional Non-Autonomous Dynamical Systems*; Springer: New York, NY, USA; Heidelberg, Germany; Dordrecht, The Netherlands; London, UK, 2013.
18. Chepyzhov, V.V.; Vishik, M.I. *Attractors for Equations of Mathematical Physics*; American Mathematical Society: Providence Rhode Island, RI, USA, 2002.

mathematics

MDPI

Article

An Integrable Evolution System and Its Analytical Solutions with the Help of Mixed Spectral AKNS Matrix Problem

Sheng Zhang [1,*], **Jiao Gao** [1] **and Bo Xu** [1,2]

1 School of Mathematical Sciences, Bohai University, Jinzhou 121013, China
2 School of Educational Sciences, Bohai University, Jinzhou 121013, China
* Correspondence: szhang@bhu.edu.cn

Abstract: In this work, a novel integrable evolution system in the sense of Lax's scheme associated with a mixed spectral Ablowitz-Kaup-Newell-Segur (AKNS) matrix problem is first derived. Then, the time dependences of scattering data corresponding to the mixed spectral AKNS matrix problem are given in the inverse scattering analysis. Based on the given time dependences of scattering data, the reconstruction of potentials is carried out, and finally analytical solutions with four arbitrary functions of the derived integrable evolution system are formulated. This study shows that some other systems of integrable evolution equations under the resolvable framework of the inverse scattering method with mixed spectral parameters can be constructed by embedding different spectral parameters and time-varying coefficient functions to the known AKNS matrix spectral problem.

Keywords: integrable evolution system; mixed spectral AKNS matrix problem; spectral parameter; scattering data; analytical solution; inverse scattering method

MSC: 35Q58; 35P30; 35G30; 35A20

Citation: Zhang, S.; Gao, J.; Xu, B. An Integrable Evolution System and Its Analytical Solutions with the Help of Mixed Spectral AKNS Matrix Problem. *Mathematics* **2022**, *10*, 3975. https://doi.org/10.3390/math10213975

Academic Editors: Almudena del Pilar Marquez Lozano and Vladimir Iosifovich Semenov

Received: 18 September 2022
Accepted: 24 October 2022
Published: 26 October 2022

1. Introduction

In nonlinear mathematical physics, the derivation, solution and integrability of equations are important topics [1–14]. Generally, an evolution equation is called integrable in the sense of Lax if it can be written as the compatibility condition between the related linear spectral problem and the adjoint time evolution equation [2]. For example [5], the well-known Korteweg-de Vries (KdV) equation $u_t + 6uu_x + u_{xxx} = 0$ has the Lax integrability owing to the compatibility condition [8]:

$$[L, N - \partial_t] \equiv L(N - \partial_t) - (N - \partial_t)L = 0, \tag{1}$$

of a pair of given linear problems:

$$L\phi = \lambda\phi,\ L = \partial_x^2 + u, \tag{2}$$

$$L\phi = \lambda\phi,\ N = -4\partial_x^3 - 6u\partial_x - 3u_x - \partial_t, \tag{3}$$

where the eigenfunction ϕ and the potential function u are dependent on the space variable x and the time variable t, and the spectral parameter λ is a constant.

Since the isospectral AKNS matrix problem [2]:

$$\phi_x = M\phi,\ M = \begin{pmatrix} -\lambda & q \\ r & \lambda \end{pmatrix},\ \phi = \begin{pmatrix} \phi_1(x,t) \\ \phi_2(x,t) \end{pmatrix},\ \lambda = \mathrm{i}k,\ \frac{\mathrm{d}k}{\mathrm{d}t} = 0 \tag{4}$$

and its adjoint time evolution equation:

$$\phi_t = N\phi,\ N = \begin{pmatrix} A & B \\ C & -A \end{pmatrix} \tag{5}$$

were proposed in 1974, a large number of important integrable equations [1–9] have been derived from the compatibility condition of Equations (4) and (5):

$$M_t - N_x + [M, N] = 0, \ [M, N] \equiv MN - NM, \tag{6}$$

such as the KdV equation, the modified KdV (mKdV) equation, the nonlinear Schrödinger (NLS) equation, and the sine-Gordon equation. In Equations (4) and (5), q and r are two smooth potential functions of x and t; A, B and C are three undetermined functions of x, t, q, r and λ; and i is the imaginary unit. The findings of a large number of integrable equations are due to the pioneering work of Lax's scheme [1], including Equations (1) and (6) and their generalizations [5–9]. The generalizations of Equations (4) and (5) can be summarized as follows: (i) extending the isospectrum λ, which is independent of t, to the nonisospectral case depending on t; (ii) embedding some coefficient functions into the evolution equation satisfied by the nonisospectrum λ and/or the function A for the derivation of time-varying nonisospectral equations or isospectral equations with time-varying coefficient functions; (iii) coupling isospectral equations and nonisospectral equations to mixed spectral equations; (iv) modifying local equations to nonlocal equations; (v) extension of the equations with integer-order derivatives to fractional-order equations.

From the view of physics, the variable-coefficient equations and nonisospectral equations can be used to describe solitary waves in nonuniform media, and they have their own advantages [8] in being more suitable for approaching the essence of nonlinear phenomena than the constant-coefficient equations or isospectral equations. This work aims at generalizing Equations (4) and (5) to other different forms by proposing that the spectral parameter $\lambda = ik$ and the undetermined function A satisfy the following time evolution equation:

$$\mathrm{i}\frac{\mathrm{d}k}{\mathrm{d}t} = \frac{1}{2}[\delta(t) + 2\mathrm{i}k\beta(t)], \tag{7}$$

and assumption:

$$A = \partial^{-1}(r, q)\begin{pmatrix} -B \\ C \end{pmatrix} - \frac{1}{2}[\delta(t) + 2\mathrm{i}k\beta(t)]x - \frac{1}{2}\alpha(t)(2\mathrm{i}k)^3 - \frac{1}{2}\gamma(t), \tag{8}$$

respectively. Here $\alpha(t)$, $\beta(t)$, $\gamma(t)$ and $\delta(t)$ are time-varying integrable functions, and B and C are supposed as:

$$\begin{pmatrix} -B \\ C \end{pmatrix} = \alpha(t)L^2\begin{pmatrix} -q \\ r \end{pmatrix} + \beta(t)\begin{pmatrix} -xq \\ xr \end{pmatrix} + 2\mathrm{i}k\alpha(t)L\begin{pmatrix} -q \\ r \end{pmatrix} + \alpha(t)(2\mathrm{i}k)^2\begin{pmatrix} -q \\ r \end{pmatrix}, \tag{9}$$

with

$$L = \sigma\partial + 2\begin{pmatrix} q \\ -r \end{pmatrix}\partial^{-1}(r, q), \ \partial = \frac{\partial}{\partial x}, \ \partial^{-1} = \frac{1}{2}\left(\int_{-\infty}^{x} - \int_{x}^{+\infty}\right)\mathrm{d}x, \ \sigma = \begin{pmatrix} -1 & 0 \\ 0 & 1 \end{pmatrix}. \tag{10}$$

As a results, a novel system of integrable evolution equations:

$$\begin{pmatrix} q \\ r \end{pmatrix}_t = \begin{pmatrix} \alpha(t)q_{xxx} - 6\alpha(t)qrq_x + \beta(t)q + \beta(t)xq_x - \delta(t)xq - \gamma(t)q \\ \alpha(t)r_{xxx} - 6\alpha(t)qrr_x + \beta(t)r + \beta(t)xr_x + \delta(t)xr + \gamma(t)r \end{pmatrix}, \tag{11}$$

is derived in Section 2 for the first time. Equation (11) is a mixed spectral system and this, due to Equation (7), contains two kinds of spectra. One is isospectrum under the case of $\delta(t) + 2\mathrm{i}k\beta(t) = 0$, and the other becomes nonisospectrum when $\delta(t) + 2\mathrm{i}k\beta(t) \neq 0$. Thus, we call such a parameter ik in Equation (7) a mixed spectrum. Meanwhile, Equation (11) is called a mixed spectral system. Several special cases of Equation (11) and their corresponding simplified forms of Equations (7) and (8) can be found in Section 3. In Section 4, the inverse scattering method [2,3,9] combined with the mixed spectral parameter ik satisfying Equation (7) is established to solve Equation (11), and implicit solutions are obtained.

Considering the reflectionless potential, the explicit unified formulae are reduced from the obtained implicit analytical solutions in Section 4. As a conclusion, we summarize the results of this article in Section 5.

2. Derivation of Equation (11) by Lax's Scheme

Substituting the matrices M and N in Equations (4) and (5) into Equation (6), we have:

$$-\mathrm{i}\frac{\mathrm{d}k}{\mathrm{d}t}-A_x+qC-rB=0, \tag{12}$$

$$q_t-B_x-2\mathrm{i}kB-2qA=0, \tag{13}$$

$$r_t-C_x+2\mathrm{i}kC+2rA=0. \tag{14}$$

Then, the substitution of Equations (7) and (8) into Equations (12)–(14) shows that Equation (12) holds automatically, and Equations (13) and (14) are converted as follows:

$$\begin{pmatrix} q \\ r \end{pmatrix}_t = L\begin{pmatrix} -B \\ C \end{pmatrix} - 2\mathrm{i}k\begin{pmatrix} -B \\ C \end{pmatrix} + [\alpha(t)(2\mathrm{i}k)^3+\gamma(t)]\begin{pmatrix} -q \\ r \end{pmatrix} + [\delta(t)+2\mathrm{i}k\beta(t)]\begin{pmatrix} -xq \\ xr \end{pmatrix}. \tag{15}$$

Further, we suppose that:

$$\begin{pmatrix} -B \\ C \end{pmatrix} = \sum_{i=1}^{4}\begin{pmatrix} -b_i \\ c_i \end{pmatrix}(2\mathrm{i}k)^{4-i}, \tag{16}$$

where b_i and c_i are all undetermined functions of x and t. Substituting Equation (16) into Equation (15) and comparing the coefficients of the same powers of $2\mathrm{i}k$ yields:

$$(2\mathrm{i}k)^0\colon\ \begin{pmatrix} q \\ r \end{pmatrix}_t = L\begin{pmatrix} -b_4 \\ c_4 \end{pmatrix} + \delta(t)\begin{pmatrix} -xq \\ xr \end{pmatrix} + \gamma(t)\begin{pmatrix} -q \\ r \end{pmatrix}, \tag{17}$$

$$(2\mathrm{i}k)^1\colon\ \begin{pmatrix} -b_4 \\ c_4 \end{pmatrix} = L\begin{pmatrix} -b_3 \\ c_3 \end{pmatrix} + \beta(t)\begin{pmatrix} -xq \\ xr \end{pmatrix}, \tag{18}$$

$$(2\mathrm{i}k)^2\colon\ \begin{pmatrix} -b_3 \\ c_3 \end{pmatrix} = L\begin{pmatrix} -b_2 \\ c_2 \end{pmatrix}, \tag{19}$$

$$(2\mathrm{i}k)^3\colon\ \begin{pmatrix} -b_2 \\ c_2 \end{pmatrix} = L\begin{pmatrix} -b_1 \\ c_1 \end{pmatrix} + \alpha\begin{pmatrix} -q \\ r \end{pmatrix}, \tag{20}$$

$$(2\mathrm{i}k)^4\colon\ \begin{pmatrix} -b_1 \\ c_1 \end{pmatrix} = 0. \tag{21}$$

Using Equations (18)–(21) we have:

$$\begin{pmatrix} -b_2 \\ c_2 \end{pmatrix} = L\begin{pmatrix} -b_1 \\ c_1 \end{pmatrix} + \alpha(t)\begin{pmatrix} -q \\ r \end{pmatrix} = \alpha(t)\begin{pmatrix} -q \\ r \end{pmatrix}, \tag{22}$$

$$\begin{pmatrix} -b_3 \\ c_3 \end{pmatrix} = L\begin{pmatrix} -b_2 \\ c_2 \end{pmatrix} = \alpha(t)L\begin{pmatrix} -q \\ r \end{pmatrix}, \tag{23}$$

$$\begin{pmatrix} -b_4 \\ c_4 \end{pmatrix} = L\begin{pmatrix} -b_3 \\ c_3 \end{pmatrix} + \beta(t)\begin{pmatrix} -xq \\ xr \end{pmatrix} = \alpha(t)L^2\begin{pmatrix} -q \\ r \end{pmatrix} + \beta(t)\begin{pmatrix} -xq \\ xr \end{pmatrix}, \tag{24}$$

and then Equation (17) gives:

$$\begin{pmatrix} q \\ r \end{pmatrix}_t = \alpha(t)L^3\begin{pmatrix} -q \\ r \end{pmatrix} + \beta(t)L\begin{pmatrix} -xq \\ xr \end{pmatrix} + \delta(t)\begin{pmatrix} -xq \\ xr \end{pmatrix} + \gamma(t)\begin{pmatrix} -q \\ r \end{pmatrix}. \tag{25}$$

Employing Equation (10), we easily find:

$$L\begin{pmatrix} -xq \\ xr \end{pmatrix} = \begin{pmatrix} q + xq_x \\ r + xr_x \end{pmatrix} + 2\begin{pmatrix} q \\ -r \end{pmatrix}\partial^{-1}(-rxq + qxr) = \begin{pmatrix} q + xq_x \\ r + xr_x \end{pmatrix}, \tag{26}$$

$$L^3\begin{pmatrix} -q \\ r \end{pmatrix} = L\begin{pmatrix} -q_{xx} + 2q^2r \\ r_{xx} - 2qr^2 \end{pmatrix} = \begin{pmatrix} q_{xxx} - 6qrq_x \\ r_{xxx} - 6qrr_x \end{pmatrix}, \tag{27}$$

and finally arrive at Equation (11) by means of Equations (25)–(27).

It should be noted that Equation (11) or Equation (25) cannot be included in the known mixed spectral AKNS hierarchy [7]:

$$\begin{pmatrix} q \\ r \end{pmatrix}_t = L^{2n+1}\begin{pmatrix} -xq \\ xr \end{pmatrix} + L^{2n}\begin{pmatrix} -q \\ r \end{pmatrix}, \ (n = 0, 1, 2, \cdots). \tag{28}$$

In fact, Equation (25) contains one sum of two nonisospectral terms:

$$L\begin{pmatrix} -xq \\ xr \end{pmatrix} = \begin{pmatrix} q + xq_x \\ r + xr_x \end{pmatrix}, \ \begin{pmatrix} -xq \\ xr \end{pmatrix}, \tag{29}$$

which cannot occur simultaneously in Equation (28). Similarly, Equation (28) cannot contain the other sum of two isospectral terms:

$$L^3\begin{pmatrix} -q \\ r \end{pmatrix} = \begin{pmatrix} q_{xxx} - 6qrq_x \\ r_{xxx} - 6qrr_x \end{pmatrix}, \ \begin{pmatrix} -q \\ r \end{pmatrix}. \tag{30}$$

In addition, all the four time-varying coefficient functions $\alpha(t)$, $\beta(t)$, $\gamma(t)$ and $\delta(t)$ are absent in Equation (28).

3. Special Cases of Equation (11)

Proper selections of $\alpha(t)$, $\beta(t)$, $\gamma(t)$ and $\delta(t)$ can give some special cases of Equation (11), including the known equations.

Special case 1. Constant-coefficient mixed spectral AKNS equations under the case of $\alpha(t) = \beta(t) = \gamma(t) = \delta(t) = 1$:

$$\begin{pmatrix} q \\ r \end{pmatrix}_t = \begin{pmatrix} q_{xxx} - 6qrq_x + q + xq_x - xq - q \\ r_{xxx} - 6qrr_x + r + xr_x - xr - r \end{pmatrix}, \tag{31}$$

associated with:

$$\mathrm{i}\frac{\mathrm{d}k}{\mathrm{d}t} = \frac{1}{2} + \mathrm{i}k, \tag{32}$$

$$A = \partial^{-1}(r, q)\begin{pmatrix} -B \\ C \end{pmatrix} - \frac{1}{2}(1 + 2\mathrm{i}k)x - \frac{1}{2}(2\mathrm{i}k)^3 - \frac{1}{2}, \tag{33}$$

$$\begin{pmatrix} -B \\ C \end{pmatrix} = L^2\begin{pmatrix} -q \\ r \end{pmatrix} + \begin{pmatrix} -xq \\ xr \end{pmatrix} + 2\mathrm{i}kL\begin{pmatrix} -q \\ r \end{pmatrix} + (2\mathrm{i}k)^2\begin{pmatrix} -q \\ r \end{pmatrix}. \tag{34}$$

Special case 2. Constant-coefficient isospectral AKNS equations [5] under the case of $\alpha(t) = 1$ and $\beta(t) = \gamma(t) = \delta(t) = 0$:

$$\begin{pmatrix} q \\ r \end{pmatrix}_t = \begin{pmatrix} q_{xxx} - 6qrq_x \\ r_{xxx} - 6qrr_x \end{pmatrix}, \tag{35}$$

associated with:

$$\mathrm{i}\frac{\mathrm{d}k}{\mathrm{d}t} = 0, \tag{36}$$

$$A = \partial^{-1}(r, q)\begin{pmatrix} -B \\ C \end{pmatrix} - \frac{1}{2}(2ik)^3, \begin{pmatrix} -B \\ C \end{pmatrix} = L^2\begin{pmatrix} -q \\ r \end{pmatrix} + 2ikL\begin{pmatrix} -q \\ r \end{pmatrix} + (2ik)^2\begin{pmatrix} -q \\ r \end{pmatrix}. \quad (37)$$

Special case 3. Constant-coefficient nonisospectral AKNS equations [5] under the case of $\alpha(t) = \gamma(t) = \delta(t) = 0$ and $\beta(t) = 1$:

$$\begin{pmatrix} q \\ r \end{pmatrix}_t = \begin{pmatrix} q + xq_x \\ r + xr_x \end{pmatrix}, \quad (38)$$

associated with:

$$i\frac{dk}{dt} = ik, \quad (39)$$

$$A = \partial^{-1}(r, q)\begin{pmatrix} -B \\ C \end{pmatrix} - ikx, \begin{pmatrix} -B \\ C \end{pmatrix} = \begin{pmatrix} -xq \\ xr \end{pmatrix}. \quad (40)$$

Special case 4. Variable-coefficient mixed spectral KdV equation under the case of $q = 1$ and $r = -u$:

$$u_t = \alpha(t)u_{xxx} + 6\alpha(t)uu_x + \beta(t)u + \beta(t)xu_x - \delta(t)xu - \gamma(t)u, \quad (41)$$

associated with Equation (7) and:

$$A = \partial^{-1}(-u, 1)\begin{pmatrix} -B \\ C \end{pmatrix} - \frac{1}{2}[\delta(t) + 2ik\beta(t)]x - \frac{1}{2}\alpha(t)(2ik)^3 - \frac{1}{2}\gamma(t), \quad (42)$$

$$\begin{pmatrix} -B \\ C \end{pmatrix} = \alpha(t)L^2\begin{pmatrix} -1 \\ -u \end{pmatrix} + \beta(t)\begin{pmatrix} -x \\ -xu \end{pmatrix} + \alpha(t)2ikL\begin{pmatrix} -1 \\ -u \end{pmatrix} + \alpha(t)(2ik)^2\begin{pmatrix} -1 \\ -u \end{pmatrix}. \quad (43)$$

Special case 5. Constant-coefficient isospectral KdV equation [5] under the case of $q = 1$ and $r = -u$, $\alpha(t) = 1$ and $\beta(t) = \gamma(t) = \delta(t) = 0$:

$$u_t = u_{xxx} + 6uu_x, \quad (44)$$

associated with Equation (36) and:

$$A = \partial^{-1}(-u, 1)\begin{pmatrix} -B \\ C \end{pmatrix} - \frac{1}{2}(2ik)^3, \quad (45)$$

$$\begin{pmatrix} -B \\ C \end{pmatrix} = L^2\begin{pmatrix} -1 \\ -u \end{pmatrix} + 2ikL\begin{pmatrix} -1 \\ -u \end{pmatrix} + (2ik)^2\begin{pmatrix} -1 \\ -u \end{pmatrix}. \quad (46)$$

Special case 6. Constant-coefficient isospectral mKdV equation [5] under the case of $q = v$ and $r = \mp v$, $\alpha(t) = 1$ and $\beta(t) = \gamma(t) = \delta(t) = 0$:

$$v_t = v_{xxx} - 6v^2v_x, \quad (47)$$

associated with Equation (36) and:

$$A = \partial^{-1}(\mp v, v)\begin{pmatrix} -B \\ C \end{pmatrix} - \frac{1}{2}(2ik)^3, \quad (48)$$

$$\begin{pmatrix} -B \\ C \end{pmatrix} = L^2\begin{pmatrix} -v \\ \mp v \end{pmatrix} + 2ikL\begin{pmatrix} -v \\ \mp v \end{pmatrix} + (2ik)^2\begin{pmatrix} -v \\ \mp v \end{pmatrix}. \quad (49)$$

Special case 7. Constant-coefficient isospectral sine-Gordon equation [5] under the case of $q = u_x/2$ and $r = -u_x/2$, $\alpha(t) = 1$ and $\beta(t) = \gamma(t) = \delta(t) = 0$:

$$u_{xt} = \sin u, \tag{50}$$

associated with Equation (36) and:

$$A = \partial^{-1}\left(-\frac{1}{2}u_x, \frac{1}{2}u_x\right)\begin{pmatrix}-B\\ C\end{pmatrix} - \frac{1}{2}(2ik)^3, \tag{51}$$

$$\begin{pmatrix}-B\\ C\end{pmatrix} = L^2\begin{pmatrix}-\frac{1}{2}u_x\\ -\frac{1}{2}u_x\end{pmatrix} + 2ikL\begin{pmatrix}-\frac{1}{2}u_x\\ -\frac{1}{2}u_x\end{pmatrix} + (2ik)^2\begin{pmatrix}-\frac{1}{2}u_x\\ -\frac{1}{2}u_x\end{pmatrix}. \tag{52}$$

Special case 8. Variable-coefficient nonisospectral mKdV equation [5] under the case of $q = v$ and $r = \mp v$, $\beta(t) = 1$ and $\alpha(t) = \gamma(t) = \delta(t) = 0$:

$$v_t = v + xv_x, \tag{53}$$

associated with Equations (39) and (40).

Special case 9. Variable-coefficient nonisospectral mKdV equation [5] under the case of $q = 1$ and $r = -u$, $\beta(t) = \gamma(t) = 1$ and $\alpha(t) = \delta(t) = 0$:

$$u_t = xu_x, \tag{54}$$

associated with Equations (39) and:

$$A = \partial^{-1}(-u, 1)\begin{pmatrix}-B\\ C\end{pmatrix} - ikx - \frac{1}{2}, \begin{pmatrix}-B\\ C\end{pmatrix} = \begin{pmatrix}-x\\ -xu\end{pmatrix}. \tag{55}$$

4. Implicit Solutions of Equation (11)

In what follows, we assume that the potentials q, r and their derivatives of each order with respect to x and t are smooth functions, and when $|x| \to \infty$, they all tend to 0.

Theorem 1. *Let us assume that $q(x,t)$ and $r(x,t)$ evolve according to Equation (11). Then, the time-dependences of scattering data:*

$$\left\{\mathrm{Im}k = 0,\ R(t,k) = \frac{b(t,k)}{a(t,k)},\ \kappa_j(t),\ c_j(t),\ j = 1,2,\cdots,n\right\}, \tag{56}$$

$$\left\{\mathrm{Im}k = 0,\ \overline{R}(t,k) = \frac{\overline{b}(t,k)}{\overline{a}(t,k)},\ \overline{\kappa}_j(t),\ \overline{c}_j(t),\ j = 1,2,\cdots,\overline{n}\right\}, \tag{57}$$

which correspond to the mixed spectral AKNS matrix problem:

$$\phi_x = M\phi,\ M = \begin{pmatrix}-\lambda & q\\ r & \lambda\end{pmatrix},\ \phi = \begin{pmatrix}\phi_1(x,t)\\ \phi_2(x,t)\end{pmatrix},\ \mathrm{i}\frac{\mathrm{d}k}{\mathrm{d}t} = \frac{1}{2}[\delta(t) + 2ik\beta(t)]. \tag{58}$$

are as follows:

$$\kappa_j(t) = \mathrm{e}^{\int_0^t \beta(\tau)\mathrm{d}\tau}\left[\kappa_j(0) - \frac{\mathrm{i}}{2}\int_0^t \mathrm{e}^{\int_0^\tau \beta(w)\mathrm{d}w}\delta(\tau)\mathrm{d}\tau\right], \tag{59}$$

$$c_j(t) = c_j(0)\mathrm{e}^{\frac{1}{2}\int_0^t [\alpha(\tau)(2ik_j(\tau))^3 + \beta(\tau) + \gamma(\tau)]\mathrm{d}\tau}, \tag{60}$$

$$a(t,k) = a(0,k),\ b(t,k) = b(0,k)\mathrm{e}^{\int_0^t [\alpha(\tau)(2i\kappa_j(\tau))^3 + \gamma(\tau)]d\tau}, \tag{61}$$

$$\overline{\kappa}_j(t) = \mathrm{e}^{\int_0^t \beta(\tau)\mathrm{d}\tau}\left[\overline{\kappa}_j(0) - \frac{\mathrm{i}}{2}\int_0^t \mathrm{e}^{\int_0^\tau \beta(w)\mathrm{d}w}\delta(\tau)\mathrm{d}\tau\right], \tag{62}$$

$$\overline{c}_j(t) = \overline{c}_j(0)\mathrm{e}^{-\frac{1}{2}\int_0^t [\alpha(2\mathrm{i}\overline{k}_j(\tau))^3 + \beta(\tau) + \gamma(\tau)]\mathrm{d}\tau}, \tag{63}$$

$$\overline{a}(t,k) = \overline{a}(0,k),\ \overline{b}(t,k) = \overline{b}(0,k)\mathrm{e}^{-\int_0^t [\alpha(\tau)(2\mathrm{i}\overline{\kappa}_j(\tau))^3 + \gamma(\tau)]\mathrm{d}\tau}, \tag{64}$$

where $c_j(0), \overline{c}_j(0)$, $R(0,k) = b(0,k)/a(0,k)$ and $\overline{R}(0,k) = \overline{b}(0,k)/\overline{a}(0,k)$ are the normalization factors and reflection coefficients when $q(x,0)$ and $r(x,0)$ are potentials of the mixed spectral AKNS matrix problem (59).

Proof of Theorem 1. Since $\phi(x,k)$ satisfies Equation (58), $P(x,k) = \phi_t(x,k) - N\phi(x,k)$ also satisfies Equation (58) and then can be expressed by a pair of linearly independent basic solutions $\phi(x,k)$ and $\widetilde{\phi}(x,k)$ [5] of Equation (58):

$$P(x,k) = \phi_t(x,k) - N\phi(x,k) = \zeta(t,k)\phi(x,k) + \tau(t,k)\widetilde{\phi}(x,k), \tag{65}$$

where $\zeta(t,k)$ and $\tau(t,k)$ are two undetermined functions.

Firstly, we start from the discrete spectrum $k = \kappa_j$ $(\mathrm{Im}\kappa_j > 0)$. Because when $x \to +\infty$, $\phi(x,k)$ decreases exponentially while $\widetilde{\phi}(x,k)$ increases exponentially, $\tau(t,k) = 0$. In this case, Equation (65) becomes:

$$\phi_t(x,\kappa_j(t)) - N\phi(x,\kappa_j(t)) = \zeta(t,\kappa_j(t))\phi(x,\kappa_j(t)). \tag{66}$$

Using $(\phi_2(x,\kappa_j(t)), \phi_1(x,\kappa_j(t)))$ to multiply the left-hand side of Equation (66), we have:

$$[\phi_1(x,\kappa_j(t))\phi_2(x,\kappa_j(t))]_t - [C\phi_1^2(x,\kappa_j(t)) + B\phi_2^2(x,\kappa_j(t))] = 2\zeta(t,\kappa_j(t))\phi_1(x,\kappa_j(t))\phi_2(x,\kappa_j(t)). \tag{67}$$

Integrating Equation (67) with respect to x from $-\infty$ to $+\infty$, and considering the assumption [5]:

$$2\int_{-\infty}^{\infty} c_j^2(t)\phi_1(x,\kappa_j(t))\phi_2(x,\kappa_j(t))\mathrm{d}x = 1 \tag{68}$$

between the normalization function $\phi(x,\kappa_j(t))$ and the normalization factor $c_j(t)$, we can find:

$$\zeta(t,\kappa_j(t)) = -c_j^2(t)\int_{-\infty}^{\infty} [C\phi_1^2(x,\kappa_j(t)) + B\phi_2^2(x,\kappa_j(t))]\mathrm{d}x, \tag{69}$$

which has the inner product form:

$$\zeta(t,\kappa_j(t)) = -c_j^2(t)((\phi_2^2(x,\kappa_j(t)), \phi_1^2(x,\kappa_j(t)))^T, (B,C)^T), \tag{70}$$

And then it has:

$$\zeta(t,\kappa_j(t)) = -c_j^2(t)((\phi_2^2(x,\kappa_j(t)), \phi_1^2(x,\kappa_j(t)))^T, (B,C)^T) = \frac{1}{2}\beta(t), \tag{71}$$

Here, the following results have been used:

$$\int_{-\infty}^{\infty} [q(x)\phi_2^2(x,\kappa_j(t)) + r(x)\phi_1^2(x,\kappa_j(t))]\mathrm{d}x = \int_{-\infty}^{\infty} [\phi_1(x,\kappa_j(t))\phi_2(x,\kappa_j(t))]_x \mathrm{d}x = 0, \tag{72}$$

$$\begin{pmatrix} B \\ C \end{pmatrix} = \beta(t)\begin{pmatrix} xq \\ xr \end{pmatrix} + \alpha(t)\sum_{l=2}^{4} L^{l-2}\begin{pmatrix} q \\ r \end{pmatrix}(2\mathrm{i}k)^{4-l}, \tag{73}$$

$$\left((\phi_2^2(x,\kappa_j(t)), \phi_1^2(x,\kappa_j(t)))^T, \begin{pmatrix} xq \\ xr \end{pmatrix}\right) = \int_{-\infty}^{\infty} x(\phi_1(x,\kappa_j(t))\phi_2(x,\kappa_j(t)))_x\mathrm{d}x = -\frac{1}{2c_j^2(t)}. \tag{74}$$

Thus, Equation (66) reads:

$$\phi_t(x,\kappa_j(t)) - N\phi(x,\kappa_j(t)) = \frac{1}{2}\beta(t)\phi(x,\kappa_j(t)). \tag{75}$$

Note the asymptotic properties when $x \to +\infty$:

$$N \to \begin{pmatrix} n_{11} & 0 \\ 0 & -n_{11} \end{pmatrix},\ n_{11} = \frac{1}{2}[\delta(t) + \beta(t)2\mathrm{i}\kappa_j(t)]x + \frac{1}{2}\alpha(t)(2\mathrm{i}\kappa_j(t))^3 + \frac{1}{2}\gamma(t), \tag{76}$$

$$\phi(x,\kappa_j(t)) \to c_j(t)\begin{pmatrix} 0 \\ 1 \end{pmatrix}\mathrm{e}^{\mathrm{i}\kappa_j(t)x}, \tag{77}$$

$$\phi_t \to \frac{\mathrm{d}c_j(t)}{\mathrm{d}t}\begin{pmatrix} 0 \\ 1 \end{pmatrix}\mathrm{e}^{\mathrm{i}\kappa_j(t)x} + \mathrm{i}\frac{\mathrm{d}\kappa_j(t)}{\mathrm{d}t}c_j(t)x\begin{pmatrix} 0 \\ 1 \end{pmatrix}\mathrm{e}^{\mathrm{i}\kappa_j(t)x}, \tag{78}$$

from Equation (75) we reach:

$$\frac{\mathrm{d}\kappa_j(t)}{\mathrm{d}t} = -\frac{\mathrm{i}}{2}[\delta(t) + 2\mathrm{i}\kappa_j(t)\beta(t)], \tag{79}$$

$$\frac{\mathrm{d}c_j(t)}{\mathrm{d}t} = \left\{\left[\frac{1}{2}\alpha(t)(2\mathrm{i}\kappa_j(t))^3 + \frac{1}{2}\gamma(t)\right] + \frac{1}{2}\beta(t)\right\}c_j(t). \tag{80}$$

Directly solving Equations (79) and (80) yields Equations (59) and (60). By a similar way, we also obtain:

$$\frac{\mathrm{d}\overline{\kappa}_j(t)}{\mathrm{d}t} = -\frac{\mathrm{i}}{2}[\delta(t) + 2\mathrm{i}\overline{\kappa}_j(t)\beta(t)], \tag{81}$$

$$\frac{\mathrm{d}\overline{c}_j(t)}{\mathrm{d}t} = -\left\{\left[\frac{1}{2}\alpha(t)(2\mathrm{i}\overline{\kappa}_j(t))^3 + \frac{1}{2}\gamma(t)\right] + \frac{1}{2}\beta(t)\right\}\overline{c}_j(t), \tag{82}$$

and hence reach Equations (62) and (63).

Secondly, we deal with the real continuous spectrum k. Taking a solution $\varphi(x,k)$ of Equation (58), then we can see that $Q(x,k) = \varphi_t(x,k) - N\varphi(x,k)$ solves Equation (58). Therefore, there are two linearly independent fundamental solutions $\varphi(x,k)$ and $\widetilde{\varphi}(x,k)$ of Equation (58), so that:

$$\varphi_t(x,k) - N\varphi(x,k) = \varpi(t,k)\varphi(x,k) + \theta(t,k)\overline{\varphi}(x,k), \tag{83}$$

where $\varpi(t,k)$ and $\theta(t,k)$ are two functions to be determined. Setting $x \to -\infty$ and using the asymptotic properties:

$$\varphi_t(x,k) \to -\mathrm{i}\frac{\mathrm{d}k}{\mathrm{d}t}x\begin{pmatrix} 1 \\ 0 \end{pmatrix}\mathrm{e}^{-\mathrm{i}kx},\ \varphi(x,k) \to \begin{pmatrix} 1 \\ 0 \end{pmatrix}\mathrm{e}^{-\mathrm{i}kx},\ \overline{\varphi}(x,k) \to \begin{pmatrix} 0 \\ -1 \end{pmatrix}\mathrm{e}^{\mathrm{i}kx}, \tag{84}$$

we have:

$$\mathrm{i}\frac{\mathrm{d}k(t)}{\mathrm{d}t} = \frac{1}{2}[\delta(t) + 2\mathrm{i}\beta(t)k(t)],\ \theta(t,k) = 0,\ \varpi(t,k) = \frac{1}{2}\alpha(t)(2\mathrm{i}k(t))^3 + \frac{1}{2}\gamma(t). \tag{85}$$

Substituting the Jost relationship:

$$\varphi(x,k) = a(t,k)\overline{\phi}(x,k) + b(t,k)\phi(x,k) \tag{86}$$

into Equation (83) and using asymptotic properties:

$$\phi(x,k) \to \begin{pmatrix} 0 \\ 1 \end{pmatrix}\mathrm{e}^{\mathrm{i}kx},\ \overline{\phi}(x,k) \to \begin{pmatrix} 1 \\ 0 \end{pmatrix}\mathrm{e}^{-\mathrm{i}kx},\ (x \to +\infty), \tag{87}$$

we easily derive from Equation (83):

$$\frac{\mathrm{d}a(t,k)}{\mathrm{d}t} = 0, \ \frac{\mathrm{d}b(t,k)}{\mathrm{d}t} = [\alpha(t)(2\mathrm{i}k(t))^3 + \gamma(t)]b(t,k). \tag{88}$$

Similarly, we can also have:

$$\frac{\mathrm{d}\bar{a}(t,k)}{\mathrm{d}t} = 0, \ \frac{\mathrm{d}\bar{b}(t,k)}{\mathrm{d}t} = -[\alpha(t)(2\mathrm{i}k(t))^3 + \gamma(t)]b(t,k). \tag{89}$$

Solving Equations (88) and (89) arrives at Equations (61) and (64). The proof is completed. □

Theorem 2. *Based on the time-dependences of scattering data in Equations (56) and (57) corresponding to the mixed spectral AKNS matrix problem (58), the implicit solutions of Equation (11) can be expressed by:*

$$q = -2K_1(t,x,x), \tag{90}$$

$$r = \frac{K_{2x}(t,x,x)}{K_1(t,x,x)}, \tag{91}$$

where $K(t,x,y) = (K_1(t,x,y), K_2(t,x,y))^T$ *satisfies the Gel'fand-Levitan-Marchenko (GLM) integral equation:*

$$K(t,x,y) - \begin{pmatrix}1\\0\end{pmatrix}\overline{F}(t,x+y) + \begin{pmatrix}0\\1\end{pmatrix}\int_x^{\infty} F(t,z+x)\overline{F}(t,z+y)\mathrm{d}z + \int_x^{\infty} K(t,x,s)\int_x^{\infty} F(t,z+s)\overline{F}(t,z+y)\mathrm{d}z\mathrm{d}s = 0 \tag{92}$$

with:

$$F(t,x) = \frac{1}{2\pi}\int_{-\infty}^{\infty} R(t,k)\mathrm{e}^{\mathrm{i}kx}\mathrm{d}k + \sum_{j=1}^{n} c_j^2(t)\mathrm{e}^{\mathrm{i}\kappa_j(t)x}, \tag{93}$$

$$\overline{F}(t,x) = \frac{1}{2\pi}\int_{-\infty}^{\infty} \overline{R}(t,k)\mathrm{e}^{-\mathrm{i}kx}\mathrm{d}k - \sum_{j=1}^{\overline{n}} \bar{c}_j^2(t)\mathrm{e}^{-\mathrm{i}\bar{\kappa}_j(t)x}, \tag{94}$$

where $R(t,k) = b(t,k)/a(t,k)$, $\overline{R}(t,k) = \bar{b}(t,k)/\bar{a}(t,k)$, $\kappa_j(t)$, $\bar{\kappa}_j(t)$, $c_j(t)$ *and* $\bar{c}_j(t)$ *are determined by Equations (59)–(64).*

Proof of Theorem 2. Since the proof is similar to that in [5], we omit it here. However, it is worth noting that the scattering data in Equations (93) and (94) are different. The proof is finished. □

5. Reflectiveless Potential Solutions of Equation (11)

Theorem 3. *In the case of the reflection potentials* $R(t,k) = \overline{R}(t,k) = 0$, *explicit solutions of Equation (11) can be formulated as follows:*

$$q = 2\mathrm{tr}(W^{-1}(x,t)\overline{\Lambda}(x,t)\overline{\Lambda}^T(x,t)), \tag{95}$$

$$r = -\frac{\frac{\partial}{\partial x}\mathrm{tr}(W^{-1}(x,t)P(x,t)\frac{\partial}{\partial x}P^T(x,t))}{\mathrm{tr}(W^{-1}(x,t)\overline{\Lambda}(x,t)\overline{\Lambda}^T(x,t))}, \tag{96}$$

with:

$$W(x,t) = I + P(x,t)P^T(x,t), \ P(x,t) = \left(\frac{c_j(t)\bar{c}_j(t)}{\kappa_j(t)-\bar{\kappa}_j(t)}\mathrm{e}^{\mathrm{i}(\kappa_j(t)-\bar{\kappa}_j(t))x}\right)_{\overline{n}\times n}, \tag{97}$$

$$\overline{\Lambda} = (\bar{c}_1(t)\mathrm{e}^{-\mathrm{i}\bar{\kappa}_1(t)x}, \bar{c}_2(t)\mathrm{e}^{-\mathrm{i}\bar{\kappa}_2(t)x}, \cdots, \bar{c}_n(t)\mathrm{e}^{-\mathrm{i}\bar{\kappa}_n(t)x})^T, \tag{98}$$

where tr(·) represents the trace of matrix, I is the $\overline{n}\times\overline{n}$ identity matrix, while $\kappa_j(t)$, $\overline{\kappa}_j(t)$, $c_j(t)$ and $\overline{c}_j(t)$ are determined by Equations (59), (60), (62) and (63).

Proof of Theorem 3. We use $K(t,x,y)=(K_1(t,x,y),K_2(t,x,y))^T$ to rewrite Equation (92) as:

$$K_1(t,x,y)-\overline{F}_d(t,x+y)+\int_x^{\infty}K_1(t,x,s)\int_x^{\infty}F_d(t,z+s)\overline{F}_d(t,z+y)\mathrm{d}z\mathrm{d}s=0, \tag{99}$$

$$K_2(t,x,y)+\int_x^{\infty}F_d(t,z+x)\overline{F}_d(t,z+y)\mathrm{d}z+\int_x^{\infty}K_2(t,x,s)\int_x^{\infty}F_d(t,z+s)\overline{F}_d(t,z+y)\mathrm{d}z\mathrm{d}s=0. \tag{100}$$

Considering $R(t,k)=\overline{R}(t,k)=0$, we simplify Equations (93) and (94) as:

$$\int_x^{\infty}F_d(t,s+z)\overline{F}_d(t,z+y)\mathrm{d}z=-\sum_{j=1}^{n}\sum_{m=1}^{\overline{n}}\frac{\mathrm{i}c_j^2(t)\overline{c}_m^2(t)}{k_j-\overline{k}_m}\mathrm{e}^{\mathrm{i}k_j(x+s)-\mathrm{i}\overline{k}_m(x+y)}. \tag{101}$$

We suppose that:

$$K_1(x,y,t)=\sum_{p=1}^{\overline{n}}\overline{c}_p(t)g_p(x,t)\mathrm{e}^{-\mathrm{i}\overline{k}_py}, \tag{102}$$

$$K_2(x,y,t)=\sum_{p=1}^{\overline{n}}\overline{c}_p(t)h_p(x,t)\mathrm{e}^{-\mathrm{i}\overline{k}_py}, \tag{103}$$

and substitute them into Equations (99) and (100), then the following equations are derived for $m=1,2,\cdots,\overline{n}$:

$$g_m(x,t)+\overline{c}_m(t)\mathrm{e}^{-\mathrm{i}\overline{k}_mx}+\sum_{j=1}^{n}\sum_{p=1}^{\overline{n}}\frac{c_j^2(t)\overline{c}_m(t)\overline{c}_p(t)}{(k_j-\overline{k}_m)(k_j-\overline{k}_p)}\mathrm{e}^{\mathrm{i}(2k_j-\overline{k}_m-\overline{k}_p)x}g_p(x,t)=0, \tag{104}$$

$$h_m(x,t)-\sum_{j=1}^{n}\frac{1}{(k_j-\overline{k}_m)}c_j^2(t)\overline{c}_m(t)\mathrm{e}^{\mathrm{i}(2k_j-\overline{k}_m)x}+\sum_{j=1}^{n}\sum_{p=1}^{\overline{n}}\frac{c_j^2(t)\overline{c}_m(t)\overline{c}_p(t)}{(k_j-\overline{k}_m)(k_j-\overline{k}_p)}\mathrm{e}^{\mathrm{i}(2k_j-\overline{k}_m-\overline{k}_p)x}h_p(x,t)=0. \tag{105}$$

Using the notations:

$$g(x,t)=(g_1(x,t),g_2(x,t),\cdots,g_{\overline{n}}(x,t))^T, \tag{106}$$

$$h(x,t)=(h_1(x,t),h_2(x,t),\cdots,h_{\overline{n}}(x,t))^T, \tag{107}$$

$$\Lambda=(c_1(t)\mathrm{e}^{-\mathrm{i}\kappa_1(t)x},c_2(t)\mathrm{e}^{-\mathrm{i}\kappa_2(t)x},\cdots,c_n(t)\mathrm{e}^{-\mathrm{i}\kappa_n(t)x})^T, \tag{108}$$

we can rewrite Equations (104) and (105) as:

$$W(x,t)g(x,t)=-\overline{\Lambda}(x,t), \tag{109}$$

$$W(x,t)h(x,t)=\mathrm{i}P(x,t)\Lambda(x,t). \tag{110}$$

When $W^{-1}(x,t)$ exists, Equations (109) and (110) give:

$$g(x,t)=-W^{-1}(x,t)\overline{\Lambda}(x,t), \tag{111}$$

$$h(x,t)=\mathrm{i}W^{-1}(x,t)P(x,t)\Lambda(x,t). \tag{112}$$

Substituting Equations (111) and (112) into Equations (102) and (103), we have:

$$K_1(x,y,t)=-\mathrm{tr}(W^{-1}(x,t)\overline{\Lambda}(x,t)\overline{\Lambda}^T(y,t)), \tag{113}$$

$$K_2(x,y,t)=\mathrm{itr}(W^{-1}(x,t)E(x,t)\Lambda(x,t)\overline{\Lambda}^T(y,t)). \tag{114}$$

We finally obtain Equations (95) and (96) by the substitution of Equations (113) and (114) into Equations (90) and (91). The proof is finished. □

As two special cases of Equations (95) and (96), we first consider $n = \overline{n} = 1$, then Equations (95) and (96) become:

$$q = \frac{2\overline{c}_1^2(t)\mathrm{e}^{-2\mathrm{i}\overline{\kappa}_1(t)x}}{1 + \frac{c_1^2(t)\overline{c}_1^2(t)}{(\kappa_1(t)-\overline{\kappa}_1(t))^2}\mathrm{e}^{2\mathrm{i}(\kappa_1(t)-\overline{\kappa}_1(t))x}}, \tag{115}$$

$$r = \frac{2c_1^2(t)\mathrm{e}^{2\mathrm{i}\kappa_1 x}}{1 + \frac{c_1^2(t)\overline{c}_1^2(t)}{(\kappa_1(t)-\overline{\kappa}_1(t))^2}\mathrm{e}^{2\mathrm{i}(\kappa_1(t)-\overline{\kappa}_1(t))x}}, \tag{116}$$

where:

$$\kappa_1(t) = \mathrm{e}^{\int_0^t \beta(\tau)\mathrm{d}\tau}\left[\kappa_1(0) - \frac{\mathrm{i}}{2}\int_0^t \mathrm{e}^{\int_0^\tau \beta(w)\mathrm{d}w}\delta(\tau)\mathrm{d}\tau\right], \tag{117}$$

$$c_1^2(t) = c_1^2(0)\mathrm{e}^{\int_0^t [\alpha(t)(2\mathrm{i}k_j(\tau))^3+\beta(\tau)+\gamma(\tau)]\mathrm{d}t}, \tag{118}$$

$$\overline{\kappa}_1(t) = \mathrm{e}^{\int_0^t \beta(\tau)\mathrm{d}\tau}\left[\overline{\kappa}_1(0) - \frac{\mathrm{i}}{2}\int_0^t \mathrm{e}^{\int_0^\tau \beta(w)\mathrm{d}w}\delta(\tau)\mathrm{d}\tau\right], \tag{119}$$

$$\overline{c}_1^2(t) = \overline{c}_1^2(0)\mathrm{e}^{-\int_0^t [\alpha(t)(2\mathrm{i}k_j(\tau))^3+\beta(\tau)+\gamma(\tau)]\mathrm{d}t}. \tag{120}$$

Selecting $\kappa_1(0) = 0.5$, $\alpha(t) = t-1$, $\beta(t) = t$, $\gamma(t) = t^2+1$ and $\delta(t) = i$, from Equation (109) we have:

$$\kappa_1(t) = \mathrm{e}^{\frac{t^2}{2}}\left[0.5 + \frac{1}{2}\sqrt{\frac{\pi}{2}}\mathrm{Erf}\left(\frac{t}{\sqrt{2}}\right)\right], \tag{121}$$

where $\mathrm{Erf}(\cdot)$ is the error function. We depict in Figure 1 the dynamical evolution of the spectrum κ_1. It can be seen from Figure 1 that the dynamical evolution of κ_1 presents nonlinear characteristics.

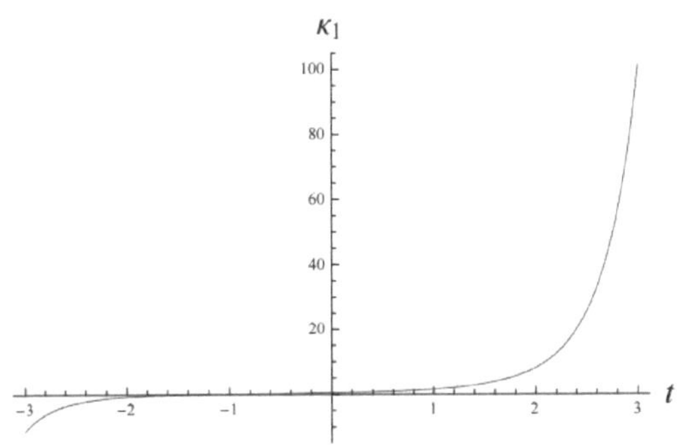

Figure 1. Nonlinear dynamical evolution of the spectrum κ_1 in Equation (121).

In Figures 2 and 3, the space–time dynamical evolutions of solutions (115) and (116) are shown by setting $\kappa_1(0) = 0.5$, $\overline{\kappa}_1(0) = 0.3$, $c_1(0) = 1$, $\overline{c}_1(0) = -2 \times 10^{-15}$, $\alpha(t) = t-1$, $\beta(t) = t$, $\gamma(t) = t^2+1$ and $\delta(t) = i$. We can see from Figure 2 that the space–time dynamical evolution of solution (115) has the characteristics of a bell-shaped soliton. However, Figure 3 shows that the space–time dynamical evolution of solution (116) does not have the characteristics of a soliton, but its amplitude increases infinitely with the increase in time.

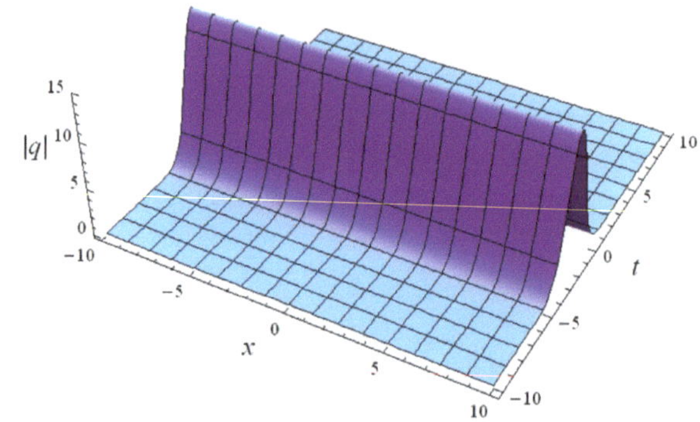

Figure 2. Space–time dynamical evolution of solution (115) with $\kappa_1(0) = 0.5$, $\overline{\kappa}_1(0) = 0.3$, $c_1(0) = 1$, $\overline{c}_1(0) = -2 \times 10^{-15}$, $\alpha(t) = t - 1$, $\beta(t) = t$, $\gamma(t) = t^2 + 1$ and $\delta(t) = i$.

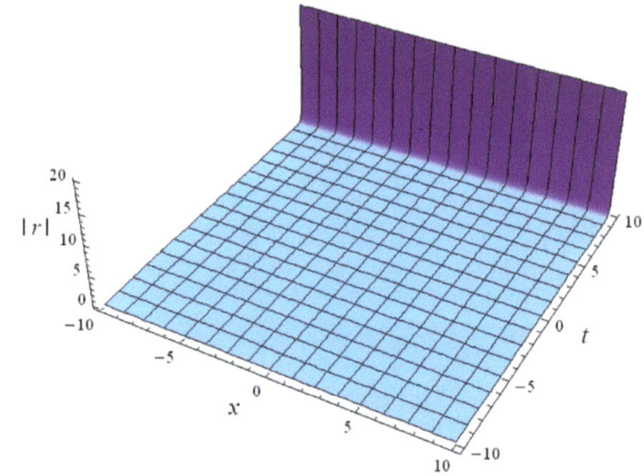

Figure 3. Space–time dynamical evolution of solution (116) with $\kappa_1(0) = 0.5$, $\overline{\kappa}_1(0) = 0.3$, $c_1(0) = 1$, $\overline{c}_1(0) = -2 \times 10^{-15}$, $\alpha(t) = t - 1$, $\beta(t) = t$, $\gamma(t) = t^2 + 1$ and $\delta(t) = i$.

For $n = \overline{n} = 2$, we select $\kappa_1(0) = 0.5$, $\kappa_2(0) = -1$, $\alpha = t - 1$, $\beta = 1$, $\gamma = t + 1$ and $\delta(t) = i$, then two cases of Equation (59) for $j = 1$ and $j = 2$ give:

$$\kappa_1(t) = e^t \left[0.5 + \frac{1}{2}(1 - \cosh t + \sinh t)\right], \tag{122}$$

$$\kappa_2(t) = e^t \left[-1 + \frac{1}{2}(1 - \cosh t + \sinh t)\right]. \tag{123}$$

In Figures 4 and 5, we depict the dynamical evolution of the spectrum κ_1 in Equation (122) and κ_2 in Equation (123), respectively. It can be seen from Figures 4 and 5 that the dynamical evolution of κ_1 and κ_2 presents nonlinear characteristics.

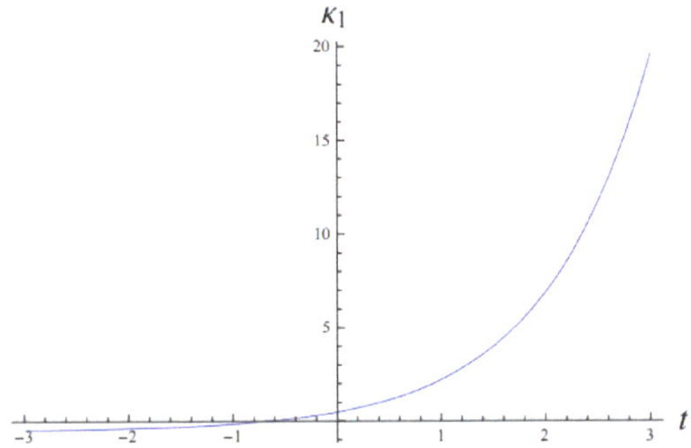

Figure 4. Nonlinear dynamical evolution of the spectrum κ_1 in Equation (122).

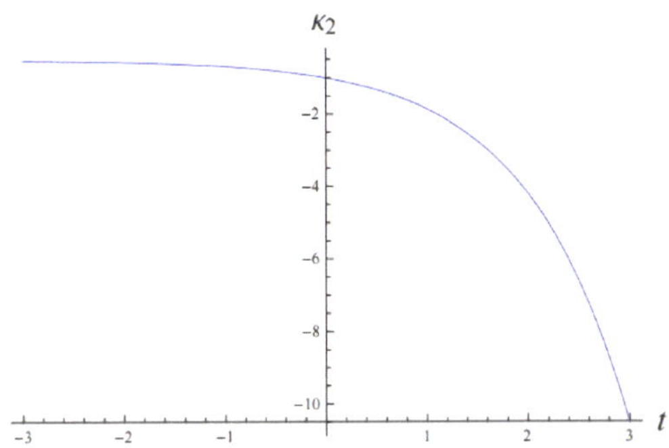

Figure 5. Nonlinear dynamical evolution of the spectrum κ_2 in Equation (123).

It can be seen from Figures 6 and 7 that the space–time dynamical evolution of solution determined by Equation (95) shows a multipoint feature. However, Figures 8 and 9 show that in addition to the multipoint feature of the space–time dynamical evolution of the solution determined by Equation (96), its amplitude also shows a feature of increase with time.

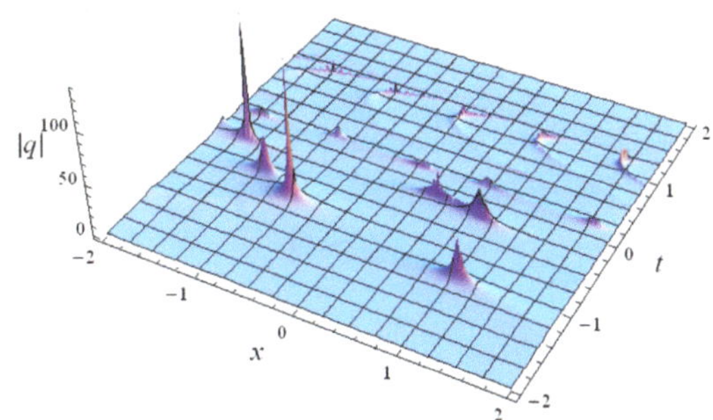

Figure 6. Space–time dynamical evolution of the solution determined by Equation (95) with $\kappa_1(0) = 0.5$, $\kappa_2(0) = -1$, $\alpha = t - 1$, $\beta = 1$, $\gamma = t + 1$ and $\delta(t) = i$.

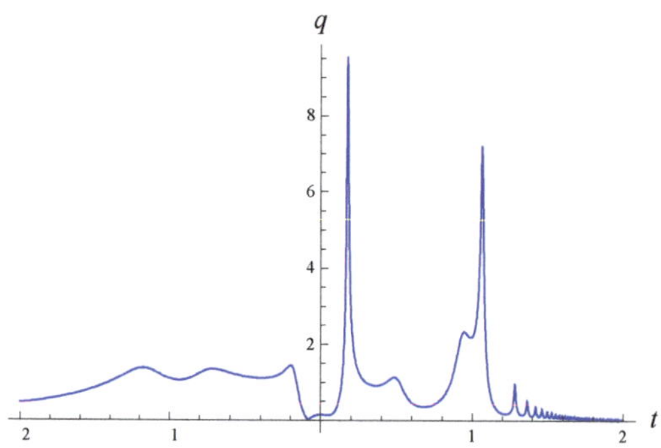

Figure 7. Profile at the position $x = 0$ of space–time of dynamical evolution of solution determined by Equation (95) with $\kappa_1(0) = 0.5$, $\kappa_2(0) = -1$, $\alpha = t - 1$, $\beta = 1$, $\gamma = t + 1$ and $\delta(t) = i$.

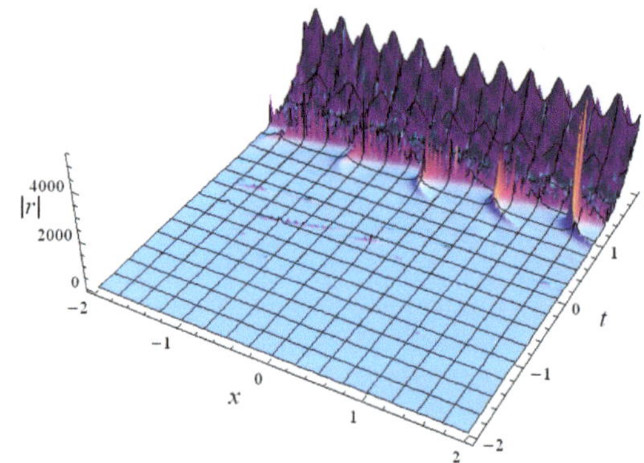

Figure 8. Space–time dynamical evolution of solution determined by solution (96) with $\kappa_1(0) = 0.5$, $\kappa_2(0) = -1$, $\alpha = t - 1$, $\beta = 1$, $\gamma = t + 1$ and $\delta(t) = i$.

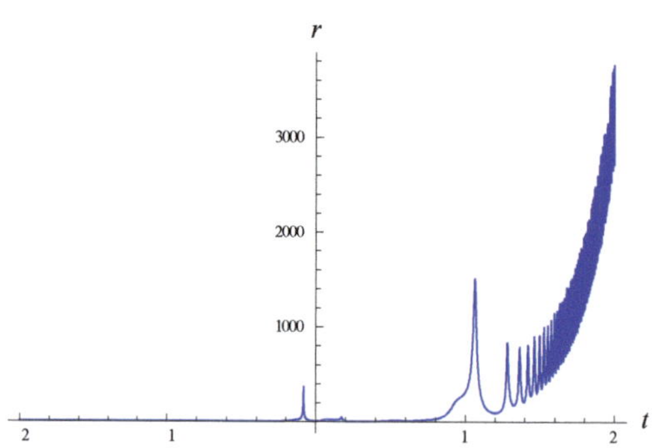

Figure 9. Profile at the position $x = 0$ of space–time of dynamical evolution of solution determined by Equation (96) with $\kappa_1(0) = 0.5$, $\kappa_2(0) = -1$, $\alpha = t - 1$, $\beta = 1$, $\gamma = t + 1$ and $\delta(t) = i$.

6. Conclusions

In short, we have derived the mixed spectral integrable Equation (11), time-dependences of scattering data (59)–(64), implicit solutions (90) and (91), and explicit reflectionless potential solutions (95) and (96). As far as we know, these obtained results are novel. Especially, the spectra with error function and hyperbolic functions in Equations (113)–(115) are new, by which the solutions (95) and (96) with $n = \overline{n} = 1$ and $n = \overline{n} = 2$ depicted in

Figures 2, 3 and 6–9 are obtained. Compared with the mixed spectral AKNS hierarchy [7] mentioned earlier in Equation (28) and the other results in [15–18], the work of this paper has some differences. Specifically, Equation (11) or its operator form (25) are different from the following equations [15–18]:

$$\begin{pmatrix} q \\ r \end{pmatrix}_t = L^3\begin{pmatrix} -q \\ r \end{pmatrix} + \sum_{n=0}^{5} L^n\begin{pmatrix} -xq \\ xr \end{pmatrix}, \tag{124}$$

associated with [15]:

$$\mathrm{i}\frac{\mathrm{d}k}{\mathrm{d}t} = \frac{1}{2}\sum_{n=0}^{5}(2\mathrm{i}k)^5,\ A = \partial^{-1}(r,q)\begin{pmatrix} -B \\ C \end{pmatrix} - \frac{1}{2}(2\mathrm{i}k)^3 - \frac{1}{2}\left[\sum_{n=0}^{5}(2\mathrm{i}k)^n\right]x; \tag{125}$$

$$\begin{pmatrix} q \\ r \end{pmatrix}_t = L\begin{pmatrix} -q \\ r \end{pmatrix} + \sum_{n=0}^{3} L^n\begin{pmatrix} -xq \\ xr \end{pmatrix}, \tag{126}$$

associated with [16]:

$$\mathrm{i}\frac{\mathrm{d}k}{\mathrm{d}t} = \frac{1}{2}\sum_{n=0}^{3}(2\mathrm{i}k)^n,\ A = \partial^{-1}(r,q)\begin{pmatrix} -B \\ C \end{pmatrix} - \frac{1}{2}(2\mathrm{i}k)^3 - \frac{1}{2}\left[\sum_{n=0}^{3}(2\mathrm{i}k)^n\right]x; \tag{127}$$

$$\begin{pmatrix} q \\ r \end{pmatrix}_t = t\begin{pmatrix} -q \\ r \end{pmatrix} + \sum_{n=0}^{2} L^n\begin{pmatrix} -xq \\ xr \end{pmatrix}, \tag{128}$$

associated with [17]:

$$\mathrm{i}\frac{\mathrm{d}k}{\mathrm{d}t} = \frac{1}{2}\sum_{n=0}^{2}(2\mathrm{i}k)^n,\ A = \partial^{-1}(r,q)\begin{pmatrix} -B \\ C \end{pmatrix} - \frac{1}{2}t - \frac{1}{2}\left[\sum_{n=0}^{2}(2\mathrm{i}k)^n\right]x; \tag{129}$$

$$\begin{pmatrix} q \\ r \end{pmatrix}_t = \alpha(t)L\begin{pmatrix} -q \\ r \end{pmatrix} + \sum_{n=0}^{2} \beta_n(t)L^n\begin{pmatrix} -xq \\ xr \end{pmatrix}, \tag{130}$$

associated with [18]:

$$\mathrm{i}\frac{\mathrm{d}k}{\mathrm{d}t} = \frac{1}{2}\sum_{n=0}^{2}\beta_n(t)(2\mathrm{i}k)^n,\ A = \partial^{-1}(r,q)\begin{pmatrix} -B \\ C \end{pmatrix} - \mathrm{i}k - \frac{1}{2}\left[\sum_{n=0}^{2}\beta_n(t)(2\mathrm{i}k)^n\right]x. \tag{131}$$

The construction of meaningful integrable evolution equations based on the AKNS matrix problem (4) with some other different spectra and their exact solutions are worth studying. This paper gives the feasibility of constructing mixed spectral integrable evolution equations which are solvable in the framework of the inverse scattering method with time-varying spectrum. Therefore, we also conclude that Equation (11) constructed in this paper is also integrable in the sense of inverse scattering.

Author Contributions: Methodology, S.Z.; software, B.X.; writing—original draft preparation, J.G.; writing—review and editing, S.Z. All authors have read and agreed to the published version of the manuscript.

Funding: This research was supported by the Liaoning BaiQianWan Talents Program of China (LRS2020[78]), the Natural Science Foundation of Education Department of Liaoning Province of China (LJ2020002), the National Natural Science Foundation of China (11547005) and the Natural Science Foundation of Xinjiang Autonomous Region of China (2020D01B01).

Institutional Review Board Statement: Not applicable.

Informed Consent Statement: Not applicable.

Data Availability Statement: The data in the manuscript are available from the corresponding author upon request.

Conflicts of Interest: The authors declare no conflict of interest.

References

1. Lax, P.D. Integrals of nonlinear equations of evolution and solitary waves. *Commun. Pure Appl. Math.* **1968**, *21*, 467–490. [CrossRef]
2. Ablowitz, M.J.; Kaup, D.J.; Newell, A.C.; Segur, H. The inverse scattering transform-Fourier analysis for nonlinear problems. *Stud. Appl. Math.* **1974**, *53*, 249–315. [CrossRef]
3. Ablowitz, M.J.; Clarkson, P.A. *Solitons, Nonlinear Evolution Equations and Inverse Scattering*; Cambridge University Press: Cambridge, MA, USA, 1991.
4. Tu, G.Z. The trace identity—A powerful tool for constructing the Hamiltonian structure of integrable systems. *J. Math. Phys.* **1989**, *30*, 330–338. [CrossRef]
5. Chen, D.Y. *Introduction of Soliton*; Science Press: Beijing, China, 2006.
6. Ning, T.K.; Chen, D.Y.; Zhang, D.J. Soliton-like solutions for a nonisospectral KdV hierarchy. *Chaos Soliton. Fract.* **2004**, *21*, 395–401. [CrossRef]
7. Ning, T.K.; Chen, D.Y.; Zhang, D.J. The exact solutions for the nonisospectral AKNS hierarchy through the inverse scattering transform. *Physica A* **2004**, *339*, 248–266. [CrossRef]
8. Zhang, S.; Xu, B. *Constructive Methods for Nonlinear Integrable Systems*; Science Press: Beijing, China, 2022.
9. Gardner, C.S.; Greene, J.M.; Kruskal, M.D.; Miura, R.M. Method for solving the Korteweg-de Vries equation. *Phys. Rev. Lett.* **1967**, *19*, 1095–1197. [CrossRef]
10. Hirota, R. Exact envelope-soliton solutions of a nonlinear wave equation. *J. Math. Phys.* **1973**, *14*, 805–809. [CrossRef]
11. Matveev, V.B.; Salle, M.A. *Darboux Transformation and Soliton*; Springer: Berlin/Heidelberg, Germany, 1991.
12. Wang, M.L. Exact solutions for a compound KdV—Burgers equation. *Phys. Lett. A* **1996**, *213*, 279–287. [CrossRef]
13. Fan, E.G. Travelling wave solutions in terms of special functions for nonlinear coupled evolution systems. *Phys. Lett. A* **2002**, *300*, 243–249. [CrossRef]
14. He, J.H.; Wu, X.H. Exp-function method for nonlinear wave equations. *Chaos Soliton. Fract.* **2006**, *30*, 700–708. [CrossRef]
15. Zhang, S.; You, C.H. Inverse scattering transform for new mixed spectral Ablowitz–Kaup–Newell–Segur equations. *Therm. Sci.* **2020**, *24*, 2437–2444. [CrossRef]
16. Xu, B.; Zhang, S. Integrability, exact solutions and nonlinear dynamics of a nonisospectral integral-differential system. *Open Phys.* **2019**, *17*, 299–306. [CrossRef]
17. Zhang, S.; Hong, S.Y. Lax integrability and soliton solutions for a nonisospectral integro-differential system. *Complexity* **2017**, *2017*, 9457078. [CrossRef]
18. Xu, B.; Zhang, S. Derivation and soliton dynamics of a new non-isospectral and variable-coefficient system. *Therm. Sci.* **2019**, *23*, S639–S646. [CrossRef]

mathematics

MDPI

Article

Application of the Exp$(-\varphi(\xi))$-Expansion Method to Find the Soliton Solutions in Biomembranes and Nerves

Attia Rani [1,†], Muhammad Shakeel [1], Mohammed Kbiri Alaoui [2], Ahmed M. Zidan [2], Nehad Ali Shah [3,†] and Prem Junsawang [4,*]

1 Department of Mathematics, University of Wah, Wah Cantt 47040, Pakistan
2 Department of Mathematics, College of Science, King Khalid University, P.O. Box 9004, Abha 61413, Saudi Arabia
3 Department of Mechanical Engineering, Sejong University, Seoul 05006, Korea
4 Department of Statistics, Faculty of Science, Khon Kaen University, Khon Kaen 40002, Thailand
* Correspondence: prem@kku.ac.th
† These authors contributed equally to this work and are co-first authors.

Abstract: Heimburg and Jackson devised a mathematical model known as the Heimburg model to describe the transmission of electromechanical pulses in nerves, which is a significant step forward. The major objective of this paper was to examine the dynamics of the Heimburg model by extracting closed-form wave solutions. The proposed model was not studied by using analytical techniques. For the first time, innovative analytical solutions were investigated using the exp$(-\varphi(\xi))$-expansion method to illustrate the dynamic behavior of the electromechanical pulse in a nerve. This approach generates a wide range of general and broad-spectral solutions with unknown parameters. For the definitive value of these constraints, the well-known periodic- and kink-shaped solitons were recovered. By giving different values to the parameters, the 3D, 2D, and contour forms that constantly modulate in the form of an electromechanical pulse traveling through the axon in the nerve were created. The discovered solutions are innovative, distinct, and useful and might be crucial in medicine and biosciences.

Keywords: nonlinear partial differential equations; exp$(-\varphi(\xi))$-expansion method; Heimburg model; traveling wave solutions

MSC: 83C15; 35A20; 35C05; 35C07; 35C08

Citation: Rani, A.; Shakeel, M.; Kbiri Alaoui, M.; Zidan, A.M.; Shah, N.A.; Junsawang, P. Application of the Exp$(-\varphi(\xi))$-Expansion Method to Find the Soliton Solutions in Biomembranes and Nerves. *Mathematics* **2022**, *10*, 3372. https://doi.org/10.3390/math10183372

Academic Editors: Almudena del Pilar Marquez Lozano and Vladimir Iosifovich Semenov

Received: 10 August 2022
Accepted: 14 September 2022
Published: 16 September 2022

1. Introduction

Nonlinear partial differential equations (NLPDEs) have recently proven to be a powerful tool in multidisciplinary studies [1–11]. Exact solutions to these equations are crucial in a variety of physical phenomena, including fluid mechanics, control theory, hydrodynamics, geochemistry, optics, plasma, and so on. So far, a number of innovative techniques for obtaining traveling wave solutions of these equations have recently been developed. The modified Jacobian elliptic function expansion technique was implemented to extract soliton solutions for the modified Liouville equation and for the system of shallow water wave equations by Zahran et al. [12]. The extended simple equation method was implemented to obtain soliton solutions of a modified Benjamin–Bona–Mahony equation, shallow water wave equations, and the nonlinear microtubules model by Khater [13]. Nonlinear evolution equations (NLEEs) were examined using the tanh method by Wazwaz [14]. An extended tanh method was applied to extract the exact soliton solutions of NLEEs by El-Wakil and Abdou [15]. The KdV equation was examined using the sine–cosine method [16]. The homogeneous balance method was implemented to obtain the exact solutions of the Gardner equation and the burger equation by Radhal and Duraisamy [17]. Ren and Zhang [18] investigated the (2 + 1)-dimensional Nizhnik–Novikov–Veselov model using the F-expansion method. The kink soliton solutions of the B-type Kadomtsev–Petviashvili equation were explored via the multiple exp-function method by Darvishi et al. [19]. Using the exp-function method, the exact solutions of the (2 + 1)-dimensional nonlinear system of Schrödinger equations were explored by Khani et al. [20] and so on [21–26].

Aside from these models, the Heimburg model of the nerve impulse is another important one. The soliton model is a mathematical model that represents mechanical processes in biomembranes. The model assumes that the nerve axon, which is modeled as a cylinder-shaped biomembrane, transits from the fluid to a gel structure at a suitable temperature below normal temperature [27]. Lautrup et al. [28] analyzed the Heimburg–Jackson model numerically, while Peets et al. [29] reported the solitonic solutions of the modified Heimburg–Jackson model.

The main goal of this work was to use the $\exp(-\varphi(\xi))$-expansion method to find some exact traveling wave solutions of the Heimburg model. For the first time, innovative analytical solutions were investigated using the $\exp(-\varphi(\xi))$-expansion method to demonstrate the dynamic behavior of the electromechanical pulse in a nerve. This method is commonly used to find the various types of soliton solutions of nonlinear differential equations (NLDEs). For example, the $\exp(-\varphi(\xi))$-expansion method was implemented to explore the exact solutions of the nonlinear double-chain model of DNA and a diffusive predator–prey model by Mahmoud et al. [30], the $\exp(-\phi(\zeta))$-expansion technique was used for soliton solutions of the nonlinear Schrödinger system by Pankaj et al. [31].

The following is the structure of the paper: In Section 2, we summarize the nonlinear Heimburg model. The Section 3 is about the methodology. In the Section 4, we analyze the nonlinear Heimburg model using the $\exp(-\varphi(\xi))$-expansion technique. The results are discussed with the help of graphs in the Section 5. Finally, we draw some conclusions.

2. Heimburg Model Equation

The voltage variation across the nerve membrane is most frequently described as a propagating version of the action potential [32–35]. This voltage difference, which manifests as an electrical pulse going up the nerve axon, is caused by unequal distributions of positive and negative ions on each side of the membrane. The nerve axon is viewed as an electrical circuit in the Hodgkin–Huxley model [32–34], in which proteins are represented as resistors and the membrane as capacitors. The membrane's ion currents produce a voltage pulse that travels along the nerve axon. Consider the nerve axon as a one-dimensional cylinder that experiences lateral density excitations. The following equation governs sound propagation in the absence of dispersion:

$$\frac{\partial^2 \Delta\rho^A}{\partial\tau^2} = \frac{\partial}{\partial z}\left(c^2 \frac{\partial \Delta\rho^A}{\partial z}\right), \tag{1}$$

where τ is the time, z is the position along the nerve axon, $\Delta\rho^A = \rho^A - \rho_0^A$ is the difference in nerve axon area density between the density of the gel state (ρ^A) and the density of the fluid state (ρ_0^A), and $c = \sqrt{1/\kappa_s^A\rho^A}$ is the sound velocity which depends on density. We did not attempt to derive the aforementioned equation here because it is connected to the hydrodynamic Euler equation.

The phases of gel and liquid are essentially incompressible. A minor increase in pressure can lead to a considerable rise in density by changing liquid into gel at densities close to the phase transition where the two phases coexist. The compression modulus is significantly smaller close to this phase transition. As a result, we can approximate the sound speed, c, as

$$c^2 = \frac{1}{\rho^A \kappa_S^A} = c_0^2 + \alpha\Delta\rho^A + \beta\left(\Delta\rho^A\right)^2, \tag{2}$$

with $\alpha < 0$ and $\beta > 0$. Additionally, the velocity of sound is frequency dependent [36]. This indicates that the system is dispersive, which is required for the formation of solitons. For unilamellar dipalmitoyl phosphatidylcholine (DPPC) vesicles, one gets $c_0 = 176.6$ m/s, $\alpha = -16.6c_0^2/\rho_0^A$ and $\beta = 79.5c_0^2/(\rho_0^A)^2$ with $\rho_0^A = 4.035 \times 10^{-3}$ g/m^2, assuming a bulk temperature of T = 45 °C [37]. By introducing a dispersive term, we are able to approximate the dispersive effects outlined above, $-h\frac{\partial^4 \Delta\rho^A}{\partial z^4}$ with $h > 0$, in Equation (1), and we obtain

$$\frac{\partial^2 \Delta\rho^A}{\partial \tau^2} = \frac{\partial}{\partial z}\left(\left[c_0^2 + \alpha\Delta\rho^A + \beta\left(\Delta\rho^A\right)^2\right]\frac{\partial \Delta\rho^A}{\partial z}\right) + v\frac{\partial^2}{\partial z^2}\left(\frac{\partial \Delta\rho^A}{\partial \tau}\right) - h\frac{\partial^4 \Delta\rho^A}{\partial z^4}. \tag{3}$$

Equation (3) is known as the Heimburg model [27] with a damping term added to the system. According to Heimburg and Jackson [37], the density change that causes the nerve impulse and the mechanical responses that accompany it might be characterized by Equation (3). It describes how an area density pulse $\Delta\rho^A$ propagates through the nerve axon when damping is taken into consideration. The equation implies that nerve impulses propagate through a nerve axon via contraction and viscous dissipation of lipid molecules, with z being the position of the nerve impulse at time τ, and v and h denoting the friction of the nerve axon and dispersion, respectively.

The axon's lateral compressibility is accounted for by K_s^A, while $c_0^2 = \frac{1}{K_s^A \rho_0^A}$, $\alpha = -\frac{1}{K_s^A (\rho_0^A)^2}$, and $\beta = \frac{1}{K_s^A (\rho_0^A)^3}$. Take the following dimensionless variables u, x, and t which are given below:

$$u = \frac{\Delta\rho^A}{\rho_0^A},\ x = \frac{c_0 z}{\sqrt{h}},\ t = \frac{c_0^2 \tau}{\sqrt{h}}. \tag{4}$$

We obtain the following dimensionless density–wave equation Equation (3) with these new variables:

$$\frac{\partial^2 u}{\partial t^2} = \frac{\partial}{\partial x}\left(\left(1 + pu + qu^2\right)\frac{\partial u}{\partial x}\right) - \frac{\partial^4 u}{\partial x^4} + \mu\frac{\partial^3 u}{\partial x^2 \partial t}, \tag{5}$$

where $\mu = \frac{v}{\sqrt{h}}$, $q = \frac{(\rho_0^A)^2}{c_0^2}\beta$, and $p = \frac{\rho_0^A}{c_0^2}\alpha$.

3. Analysis of method

Consider the general form of the NLPDE

$$Y(u, u_x,\ u_t,\ u_{xx},\ u_{xt},\ \ldots) = 0, \tag{6}$$

Here, Y is polynomial in $u(x,t)$. The main steps of this method are outlined below:

Step1: Consider the transformation:

$$u(x,t) = v(\xi),\ \xi = x - \omega t, \tag{7}$$

where ω is the velocity of the density pulse. Equation (7) transforms Equation (6) into the following form:

$$Z(v, v', v'', v''', \ldots) = 0. \tag{8}$$

Step 2: Assume that the solution of Equation (8) can be written as follows by a polynomial in $\exp(-\varphi(\xi))$.

$$v(\xi) = A_m(\exp(-\varphi(\xi)))^m + A_{m-1}(\exp(-\varphi(\xi)))^{m-1} + \ldots. \tag{9}$$

In the above equation, A_m and A_{m-1} are the constants such that $A_m \neq 0$, and $\varphi(\xi)$ satisfies the following ODE:

$$\varphi'(\xi) = \exp(-\varphi(\xi)) + Q\exp(\varphi(\xi)) + P, \tag{10}$$

where Q and P are arbitrary constants.

Step 3: To obtain integer m, we apply the homogeneous principle in Equation (8).

There are the following five cases:
Case 1: When $P^2 - 4Q > 0$ and $Q \neq 0$,

$$\varphi(\xi) = ln\left\{\frac{1}{2Q}\left(-\sqrt{P^2 - 4Q}tanh\left(\frac{\sqrt{P^2 - 4Q}}{2}(\xi + a_1)\right) - Q\right)\right\}. \tag{11}$$

Case 2: When $P^2 - 4Q < 0$ and $Q \neq 0$,

$$\varphi(\xi) = ln\left\{\frac{1}{2Q}\left(-P + \sqrt{4Q - P^2}tan\left(\frac{\sqrt{4Q - P^2}}{2}(\xi + a_1)\right)\right)\right\}. \tag{12}$$

Case 3: When $P \neq 0$ and $Q = 0$,

$$\varphi(\xi) = -ln\left\{\frac{P}{(\exp(P(\xi + a_1)) - 1)}\right\}. \tag{13}$$

Case 4: When $P^2 - 4Q = 0$ and $P \neq 0, Q \neq 0$,

$$\varphi(\xi) = ln\left\{\frac{2(P(\xi + a_1)) + 2}{P^2(\xi + a_1)}\right\}. \tag{14}$$

Case 5: When $P = 0$ and $Q = 0$,

$$\varphi(\xi) = ln(\xi + a_1), \tag{15}$$

where a_1 is the constant of integration.

Step 4: By inserting Equation (9) into (8) along with (10), Equation (8) converts into a polynomial in $\exp(-\varphi(\xi))$. We obtain a series of equations for A_m, ω, P, and Q by setting each coefficient of this polynomial to 0. From these equations, the unknown constants A_m, ω, P, and Q can be obtained using computational tools such as Maple, and the novel soliton solutions of Equation (6) can be generated by utilizing these values in Equation (9).

4. Application of the Method

Utilizing the $\exp(-\varphi(\xi))$-expansion method, we created exact traveling wave solutions to the Heimburg model. By using Equation (7) in (5), we obtain:

$$\omega^2 v'' - \alpha_1 (v')^2 - 2\beta_1 v (v')^2 - v'' - \alpha_1 v v'' - \beta_1 v^2 v'' + v'^{v} + \mu\omega v''' = 0, \tag{16}$$

where $\alpha_1 = p$ and $\beta_1 = q$. Balancing between the terms v'^{v} and $v^2 v''$ in Equation (14) yields $m = 1$ as shown in Appendix A.

Hence, from Equation (9), we obtain:

$$v(\xi) = A_0 + A_1 e^{-\varphi(\xi)}, \tag{17}$$

where A_0 and A_1 are arbitrary constants. Putting Equation (17) into (16) with (10), Equation (16) converts into the polynomial in $\exp(-\varphi(\xi))$. By setting the coefficients of the polynomial equal to 0, a set of equations for A_0, A_1, P, and Q is obtained as shown in Appendix A. By solving these equations using computational software Maple 18, we obtain:

1st Solution Set:

$$P = \frac{1}{\sqrt{6}\sqrt{\beta_1}}\sqrt{-2\beta_1\omega^2(6 - \mu^2) + 3\alpha_1^2 + 12\beta_1(2Q - 1)},$$

$$A_0 = \frac{\mu\omega}{\sqrt{6}\sqrt{\beta_1}} - \frac{\alpha_1}{2\beta_1} - \frac{1}{2\beta_1}\sqrt{2\beta_1\omega^2(6 - \mu^2) + 3\alpha_1^2 + 12\beta_1(2Q - 1)},$$

$$A_1 = -\frac{\sqrt{6}}{\sqrt{\beta_1}}. \tag{18}$$

By using these results in Equation (17), we obtain:

$$v(\xi) = \frac{\mu\omega}{\sqrt{6}\sqrt{\beta_1}} - \frac{\alpha_1}{2\beta_1} - \frac{1}{2\beta_1}\sqrt{2\beta_1\omega^2(6-\mu^2) + 3\alpha_1^2 + 12\beta_1(2Q-1)} - \frac{\sqrt{6}}{\sqrt{\beta_1}}e^{-\varphi(\xi)}, \tag{19}$$

Case 1: For $P^2 - 4Q > 0$ and $Q \neq 0$, we obtain:

$$\begin{aligned} v(\xi) = {} & \tfrac{\mu\omega}{\sqrt{6}\sqrt{\beta_1}} - \tfrac{\alpha_1}{2\beta_1} - \tfrac{1}{2\beta_1}\sqrt{2\beta_1\omega^2(6-\mu^2) + 3\alpha_1^2 + 12\beta_1(2Q-1)} \\ & + \frac{2\sqrt{6}Q}{\sqrt{\beta_1}\left(\sqrt{\mathrm{P}^2-4Q}tanh\left(\frac{\sqrt{\mathrm{P}^2-4Q}}{2}(\xi+a_1)\right)+\mathrm{P}\right)}. \end{aligned} \tag{20}$$

Case 2: For $P^2 - 4Q < 0$ and $Q \neq 0$, we obtain:

$$\begin{aligned} v(\xi) = {} & \tfrac{\mu\omega}{\sqrt{6}\sqrt{\beta_1}} - \tfrac{\alpha_1}{2\beta_1} - \tfrac{1}{2\beta_1}\sqrt{2\beta_1\omega^2(6-\mu^2) + 3\alpha_1^2 + 12\beta_1(2Q-1)} - \\ & \frac{2\sqrt{6}Q}{\sqrt{\beta_1}\left(-\mathrm{P}+\sqrt{4Q-\mathrm{P}^2}tan\left(\frac{\sqrt{4Q-\mathrm{P}^2}}{2}(\xi+a_1)\right)\right)}. \end{aligned} \tag{21}$$

Case 3: For $P \neq 0$ and $Q = 0$, we obtain:

$$\begin{aligned} v(\xi) = {} & \tfrac{\mu\omega}{\sqrt{6}\sqrt{\beta_1}} - \tfrac{\alpha_1}{2\beta_1} - \tfrac{1}{2\beta_1}\sqrt{2\beta_1\omega^2(6-\mu^2) + 3\alpha_1^2 + 12\beta_1(2Q-1)} - \\ & \frac{P\sqrt{6}}{\sqrt{\beta_1}(\exp(\mathrm{P}(\xi+a_1))-1)}. \end{aligned} \tag{22}$$

Case 4: For $P^2 - 4Q = 0$ and $P \neq 0, Q \neq 0$, we obtain:

$$\begin{aligned} v(\xi) = {} & \tfrac{\mu\omega}{\sqrt{6}\sqrt{\beta_1}} - \tfrac{\alpha_1}{2\beta_1} - \tfrac{1}{2\beta_1}\sqrt{2\beta_1\omega^2(6-\mu^2) + 3\alpha_1^2 + 12\beta_1(2Q-1)} - \\ & \frac{\sqrt{6}\,p^2}{\sqrt{\beta_1}}\frac{\mathrm{P}^2(\xi+a_1)}{(2\mathrm{P}(\xi+a_1)+2\ )}. \end{aligned} \tag{23}$$

Case 5: For $P = 0$ and $Q = 0$, we obtain:

$$\begin{aligned} v(\xi) = {} & \tfrac{\mu\omega}{\sqrt{6}\sqrt{\beta_1}} - \tfrac{\alpha_1}{2\beta_1} - \tfrac{1}{2\beta_1}\sqrt{2\beta_1\omega^2(6-\mu^2) + 3\alpha_1^2 + 12\beta_1(2Q-1)} - \\ & \frac{\sqrt{6}}{\sqrt{\beta_1(\xi+a_1)}}. \end{aligned} \tag{24}$$

2nd Solution Set:

$$\begin{gathered} Q = \frac{1}{24\beta_1}\left(2\beta_1\omega^2\left(\mu^2-6\right) + 6\beta_1\left(P^2+2\right) - 3\alpha_1^2\right), \\ A_0 = \frac{1}{6\beta_1}\left(\sqrt{6\beta_1}(\mu\omega - 3\mathrm{P}) - 3\alpha_1\right),\ A_1 = -\frac{\sqrt{6}}{\sqrt{\beta_1}}. \end{gathered} \tag{25}$$

By using these results in Equation (17), we obtain:

$$v(\xi) = \frac{1}{6\beta_1}\left(\sqrt{6\beta_1}(\mu\omega - 3\mathrm{P}) - 3\alpha_1\right) - \frac{\sqrt{6}}{\sqrt{\beta_1}}e^{-\varphi(\xi)}, \tag{26}$$

Case 1: For $P^2 - 4Q > 0$ and $Q \neq 0$, we obtain:

$$v(\xi) = \frac{1}{6\beta_1}\left(\sqrt{6\beta_1}(\mu\omega - 3\mathrm{P}) - 3\alpha_1\right) + \frac{2\sqrt{6}Q}{\sqrt{\beta_1}\left(\sqrt{\mathrm{P}^2-4\overline{Q}}tanh\left(\frac{\sqrt{\mathrm{P}^2-4\overline{Q}}}{2}(\xi+a_1)\right)+\mathrm{P}\right)}. \tag{27}$$

Case 2: For $P^2 - 4Q < 0$ and $Q \neq 0$, we obtain:

$$v(\xi) = \frac{1}{6\beta_1}\left(\sqrt{6\beta_1}(\mu\omega - 3\mathrm{P}) - 3\alpha_1\right) + \frac{2\sqrt{6}Q}{\sqrt{\beta_1}\left(-\sqrt{4Q - \mathrm{P}^2}tan\left(\frac{\sqrt{4Q - \mathrm{P}^2}}{2}(\xi + a_1)\right) + P\right)}. \tag{28}$$

Case 3: For $P \neq 0$ and $Q = 0$, we obtain:

$$v(\xi) = \frac{1}{6\beta_1}\left(\sqrt{6\beta_1}(\mu\omega - 3\mathrm{P}) - 3\alpha_1\right) - \frac{\mathrm{P}\sqrt{6}}{\sqrt{\beta_1}(\exp(\mathrm{P}(\xi + a_1)) - 1)}. \tag{29}$$

Case 4: For $P^2 - 4Q = 0$ and $P \neq 0, Q \neq 0$, we obtain:

$$v(\xi) = \frac{1}{6\beta_1}\left(\sqrt{6\beta_1}(\mu\omega - 3\mathrm{P}) - 3\alpha_1\right) - \frac{\sqrt{6}\,\mathrm{P}^2}{\sqrt{\beta_1}}\frac{\mathrm{P}^2(\xi + a_1)}{(2\mathrm{P}(\xi + a_1) + 2\,)}. \tag{30}$$

Case 5: For $P = 0$ and $Q = 0$, we obtain:

$$v(\xi) = \frac{1}{6\beta_1}\left(\sqrt{6\beta_1}(\mu\omega - 3\mathrm{P}) - 3\alpha_1\right) - \frac{\sqrt{6}}{\sqrt{\beta_1}(\xi + a_1)}. \tag{31}$$

3rd Solution Set:

$$\begin{gathered} P = \tfrac{1}{6\sqrt{\beta_1}}\left(-2\sqrt{6}\beta_1 A_0 + 2Q\omega\sqrt{\beta_1} - \alpha_1\sqrt{6}\right), \\ Q = \tfrac{1}{36\beta_1}\left(-2\sqrt{6}\beta_1^{\frac{3}{2}}\omega\mu A_0 + 4\beta_1\omega^2\mu^2 - \sqrt{6\beta_1}\alpha_1\omega\mu + 6\beta_1^2 A_0^2 - 18\beta_1\omega^2 + 6\alpha_1\beta_1 A_0 - 3\alpha_1^2 + 18\beta_1\right),\ A_1 = -\tfrac{\sqrt{6}}{\sqrt{\beta_1}}. \end{gathered} \tag{32}$$

By using these results in Equation (17), we obtain:

$$v(\xi) = A_0 - \frac{\sqrt{6}}{\sqrt{\beta_1}}e^{-\varphi(\xi)}, \tag{33}$$

Case 1: For $P^2 - 4Q > 0$ and $Q \neq 0$, we obtain:

$$v(\xi) = A_0 + \frac{2\sqrt{6}Q}{\sqrt{\beta_1}\left(\sqrt{\mathrm{P}^2 - 4Q}tanh\left(\frac{\sqrt{\mathrm{P}^2 - 4Q}}{2}(\xi + a_1)\right) + \mathrm{P}\right)}. \tag{34}$$

Case 2: For $P^2 - 4Q < 0$ and $Q \neq 0$, we obtain:

$$v(\xi) = A_0 - \frac{2\sqrt{6}Q}{\sqrt{\beta_1}\left(-\mathrm{P} + \sqrt{4Q - \mathrm{P}^2}tan\left(\frac{\sqrt{4Q - \mathrm{P}^2}}{2}(\xi + a_1)\right)\right)}. \tag{35}$$

Case 3: For $P \neq 0$ and $Q = 0$, we obtain:

$$v(\xi) = A_0 - \frac{P\sqrt{6}}{\sqrt{\beta_1}(\exp(\mathrm{P}(\xi + a_1)) - 1)}. \tag{36}$$

Case 4: For $P^2 - 4Q = 0$ and $P \neq 0, Q \neq 0$, we obtain:

$$v(\xi) = A_0 - \frac{\sqrt{6}}{\sqrt{\beta_1}}\frac{\mathrm{P}^2(\xi + a_1)}{(2\mathrm{P}(\xi + a_1) + 2\,)}. \tag{37}$$

Case 5: For $P = 0$ and $Q = 0$, we obtain:

$$v(\xi) = A_0 - \frac{\sqrt{6}}{\sqrt{\beta_1}(\xi + a_1)}. \tag{38}$$

In all the above cases, $\xi = x - \omega t$.

It is important to note that the acquired traveling wave solutions of the stated model are diversified and that for certain values of the free parameters, new and more general solutions are found. The accuracy of the obtained findings is also ensured by plugging the obtained solutions into the given equation with the Maple 18 software. The key benefit of the suggested approach is that, when we vary P and Q with some free parameters, it provides a number of new exact traveling wave solutions that are more general. The exact solutions are crucial for understanding the underlying internal dynamics of natural phenomena. The explicit solutions representing several forms of solitary wave solutions are regulated in the typical nerve impulse shape based on the variation in the physical parameters.

5. Results and Discussion

The 2D, 3D, and contour shapes of some of the collected results are revealed with the help of Wolfram Mathematica. We discovered that set-1 comprises solutions (20)–(24). These solutions have a large number of parameters. Because the parameters influence the shape of the solution, we can generate a wide range of graphs by inputting arbitrary values for the parameters. Using the graphs shown, we can determine the nature of solitons. Furthermore, set-2 provides adequate new solutions (27)–(30), and set-3 comprises solutions (34)–(38). Figures 1–4 show the 2D, 3D, and contour conspiracies of some of the obtained findings. For the sake of clarity, the graphs of some of the discovered solutions are provided here.

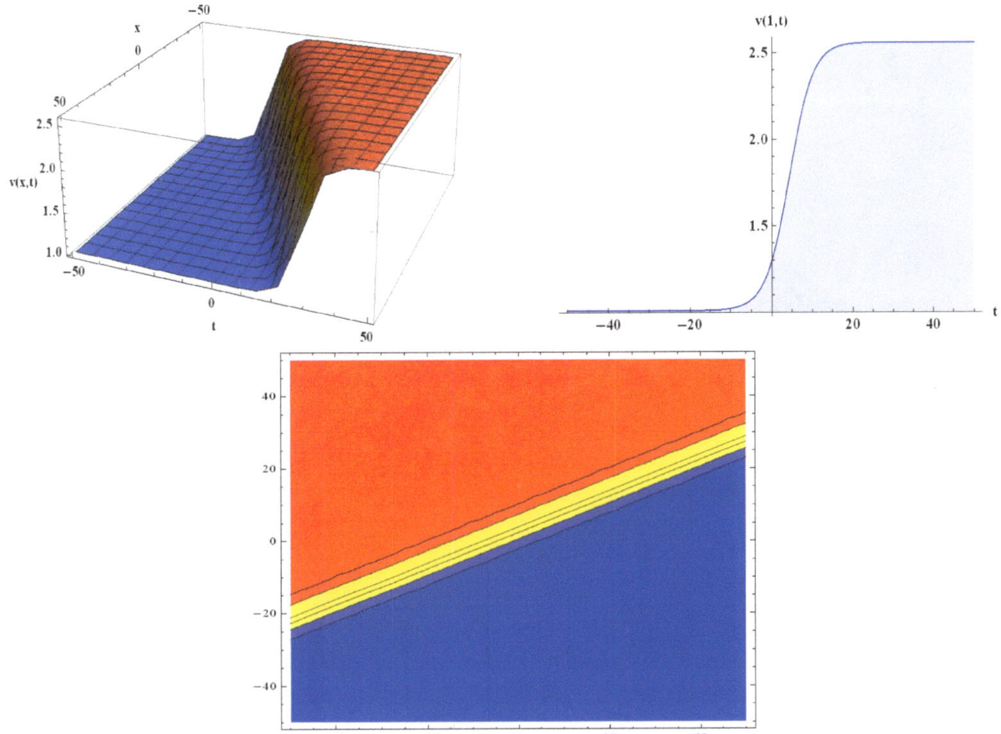

Figure 1. Three-dimensional, two-dimensional, and contour conspiracies for solution (20) for $\alpha_1 = -0.9, \beta_1 = 2, \mu = 1, P = 0.8, \omega = 2, Q = 0.8, a_1 = 5$.

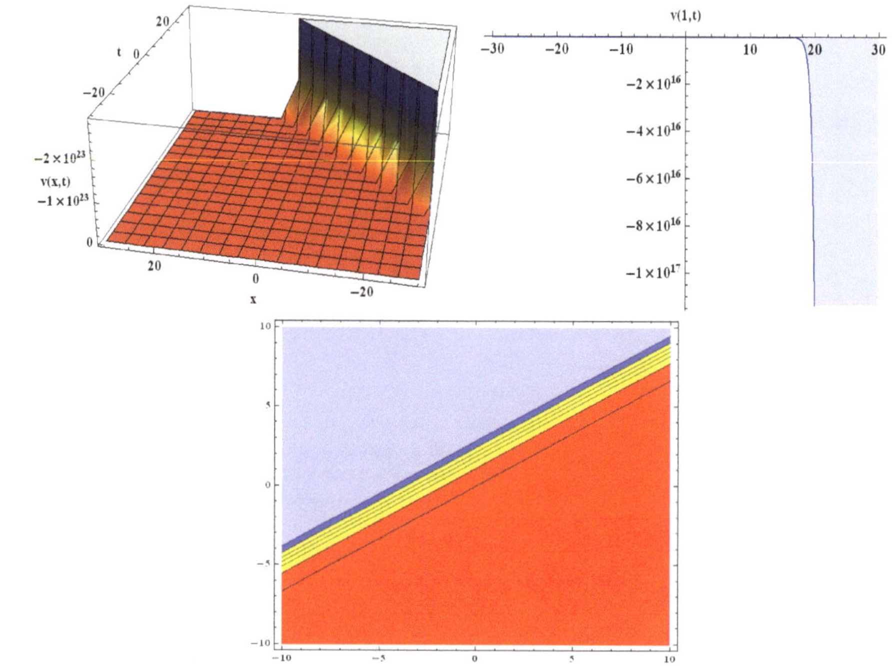

Figure 2. Three-dimensional, two-dimensional, and contour conspiracies for solution (22) for $\alpha_1 = -1.5, \beta_1 = 2, \mu = 1, \omega = 1, Q = 1, a_1 = 1$.

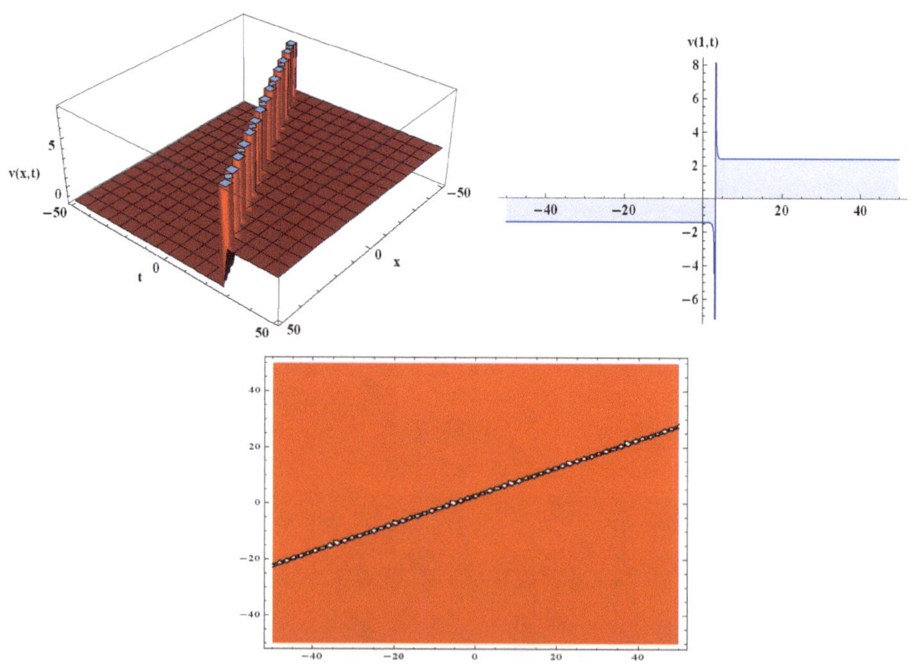

Figure 3. Three-dimensional, two-dimensional, and contour conspiracies for solution (27) $\alpha_1 = -0.9, \beta_1 = 2, \omega = 1, P = 0.8, a_1 = 5$.

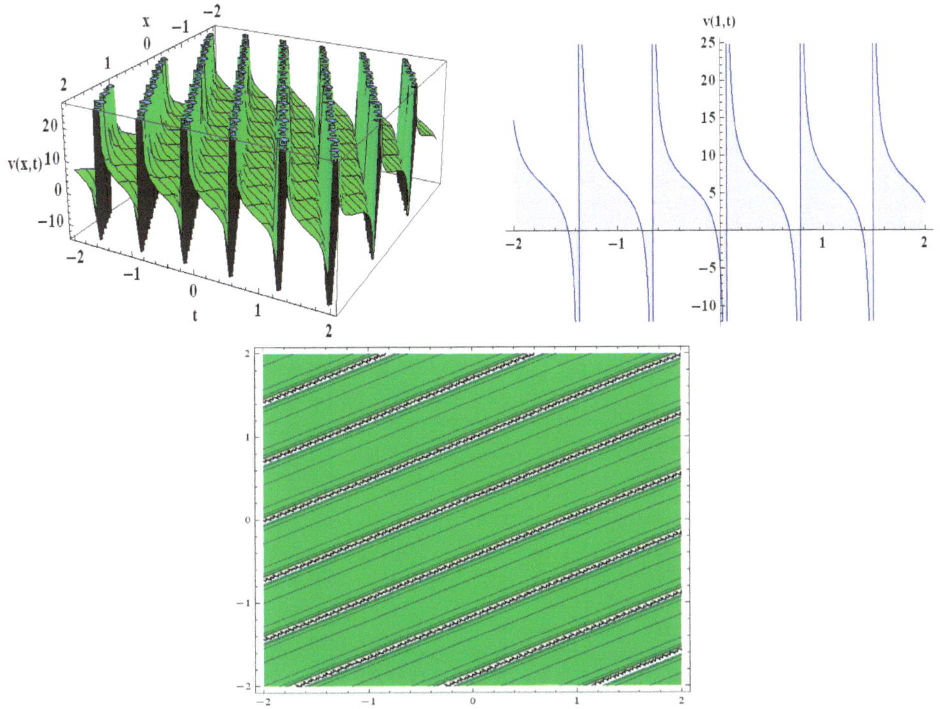

Figure 4. Three-dimensional, two-dimensional, and contour conspiracies for solution (35) for $\alpha_1 = -4, \beta_1 = 3, \mu = 1, Q = 5.6, a_1 = 5$.

The many types of graphs are created using the wave solution. When the free parameters associated with the solution are altered, the shape of the traveling wave changes. From the Heimburg model equation, we acquire the number of exact solutions along with unknown parameters.

The attained solutions (20) and (22) involve the parameters $\alpha_1, \beta_1, \mu, \omega$, P, and a_1. For the values of $\alpha_1 = -0.9, \beta_1 = 2, \mu = 1, \omega = 2, P = 0.8$, and $a_1 = 5$, in solution (20), the kink-shaped input is regulated and permanently stabilized in the typical pulse shape along the nerve axon (Figure 1). Similarly, for $\alpha_1 = -1.5, \beta_1 = 2, \mu = 1, \omega = 1, Q = 1$, and $a_1 = 1$ in solution (22), the kink-shaped input is regulated and permanently stabilized in the typical pulse shape along the nerve axon (Figure 2). For $\alpha_1 = -0.9, \beta_1 = 2, \omega = 2, P = 0.8$, and $a_1 = 5$, in Equation (27), the kink-shaped input is obtained (Figure 3). For $\alpha_1 = -4, \beta_1 = 3, Q = 5.6$, and $a_1 = 5$, in Equation (35), the periodic-shaped input is regulated in the typical pulse shape (Figure 4). The 3D and contour plots are shown for $-50 \leq x, t \leq 50$, and the 2D conspiracy is shown for $-50 \leq t \leq 50$, x = 1, in Figures 1 and 3; the 3D and contour plots are shown for $-30 \leq x, t \leq 3$, and the 2D conspiracy is shown for $-30 \leq t \leq 30$, x = 1, in Figure 2; the 3D and contour plots are shown for $-2 \leq x, t \leq 2$, and the 2D conspiracy is shown for $-2 \leq t \leq 2$, x = 1, in Figure 4.

The Heimburg model's nonlinear dynamic nature is shown in Figures 1–4. Different varieties of traveling waves are described in the inferred graphical renderings. Numerous novel exact solutions, including periodic kink, and singular-kink soliton solutions are discovered. The graphical presentation shows that the four distinct profiles constantly modulate in the form of an electromechanical pulse traveling through the axon in the nerve [27]. The findings demonstrate that the implemented technique is reliable, proficient, and dominant when it comes to analyzing different kinds of NLPDEs.

6. Conclusions

Not just in neurophysiology but also in mathematical physics, the process by which the nerve impulse is generated and propagated across the axon has been a critical challenge. We discovered the exact traveling wave solutions of the Heimburg model of neuroscience which is one of the most intriguing topics in modern bio-physics since the nerve is the foundation of life. The $\exp(-\varphi(\xi))$-expansion method was utilized to analyze the Heimburg model in this research article. Traveling wave solutions were explored using the above-mentioned model. This method yields traveling wave solutions with arbitrary parameters expressed as kink, singular-kink, and periodic-wave solutions. The graphical presentation shows that the four distinct profiles constantly modulate into the pulse pattern as they travel through the axon. It is worth noting that the findings of this study are revealed for the first time, in comparison to earlier investigations. The accuracy of the results was tested using Maple 18 and putting the obtained findings into the original equation. The solutions provided are novel, distinctive, and practical and might be essential in the fields of medicine and biosciences. In other words, the analytical expression of solitary solutions may be useful for the precise determination of the control pulse's magnitude. Additional research is required on the fascinating challenge of wave propagation in biomembranes. A thorough analysis of the dissipative effects and coupling with the action potential will be discussed in the next work.

Author Contributions: Data curation, A.R.; Formal analysis, P.J.; Funding acquisition, P.J.; Software, M.K.A.; Validation, A.M.Z.; Writing – original draft, M.S.; Writing – review & editing, N.A.S. All authors have read and agreed to the published version of the manuscript.

Funding: This research received no external funding.

Data Availability Statement: No data were used to support this study.

Acknowledgments: The authors extends their appreciation to the Deanship of Scientific Research at King Khalid University, Abha 61413, Saudi Arabia, for funding this work through a research group program under grant number R.G.P.-2/65/43. This research received funding support from the NSRF via the Program Management Unit for Human Resources & Institutional Development, Research and Innovation, (grant number B05F650018).

Conflicts of Interest: The authors declare no conflict of interest.

Appendix A

Balancing between the terms v'^{v} and v^2v'' in Equation (14) yields

$$m + 4 = 2m + m + 2,$$

$$m = 1.$$

Putting Equation (15) into (14) with (8), Equation (14) converts into the polynomial in $\exp(-\varphi(\xi))$. By setting the coefficients of the polynomial equal to 0, a set of equations for A_0, A_1, P, and Q is obtained as follows:

$$2\beta_1 A_0 A_1 Q^2 + Q^2\omega P^2 + \beta_1 PQA_0^2 + 2Q^3\omega + \alpha_1 A_1 Q^2 + \alpha_1 PQA_0 - QP^3 - \omega^2 PQ - 8Q^2P + PQ = 0,$$

$$2\beta_1 A_1^2 Q^2 + 6\beta_1 A_0 A_1 PQ + Q\omega P^3 + \beta_1 P^2 A_0^2 + 8Q^2\omega P + 3\alpha_1 A_1 PQ + 2\beta_1 QA_0^2 + \alpha_1 P^2 A_0 - P^4 - \omega^2 P^2 + 2\alpha_1 QA_0 - 22QP^2 - 2\omega^2 Q - 16Q^2 + P^2 + 2Q = 0,$$

$$5\beta_1 A_1^2 PQ + 4\beta_1 A_0 A_1 P^2 + 8\beta_1 A_0 A_1 Q + 7Q\omega P^2 + 2\alpha_1 A_1 P^2 + 3\beta_1 PA_0^2 + 8Q^2\omega + 4\alpha_1 A_1 Q + 3\alpha_1 PA_0 - 15P^3 - 3P\omega^2 - 60PQ + 3P = 0,$$

$$3\beta_1 A_1^2 P^2 + 62\beta_1 A_1^2 Q + 10\beta_1 A_0 A_1 P + 12PQ\omega + 5\alpha_1 A_1 P + 2\beta_1 A_0^2 + 2\alpha_1 A_0 - 50P^2 - 2\omega^2 - 40Q + 2 = 0,$$
$$7\beta_1 A_1^2 P + 6\beta_1 A_0 A_1 + 6Q\omega + 3\alpha_1 A_1 - 60P = 0,$$
$$4\beta_1 A_1^2 - 24 = 0. \quad (A1)$$

References

1. Miah, M.M.; Seadawy, A.R.; Ali, H.M.S.; Akbar, M.A. Abundant closed form wave solutions to some nonlinear evolution equations in mathematical physics. *J. Ocean Eng. Sci.* **2020**, *5*, 269–278. [CrossRef]
2. Barman, H.K.; Seadawy, A.R.; Akbar, M.A.; Baleanu, D. Competent closed form soliton solutions to the Riemann wave equation and the Novikov-Veselov equation. *Results Phys.* **2020**, *17*, 103131. [CrossRef]
3. Akbar, M.A.; Akinyemi, L.; Yao, S.-W.; Jhangeer, A.; Rezazadeh, H.; Khater, M.M.; Ahmad, H.; Inc, M. Soliton solutions to the Boussinesq equation through sine-Gordon method and Kudryashov method. *Results Phys.* **2021**, *25*, 104228. [CrossRef]
4. Younis, M.; Seadawy, A.R.; Baber, M.; Husain, S.; Iqbal, M.; Rizvi, S.R.; Baleanu, D. Analytical optical soliton solutions of the Schrödinger-Poisson dynamical system. *Results Phys.* **2021**, *27*, 104369. [CrossRef]
5. Akbar, M.A.; Kayum, M.A.; Osman, M.S.; Abdel-Aty, A.H.; Eleuch, H. Analysis of voltage and current flow of electrical transmission lines through mZK equation. *Results Phys.* **2021**, *20*, 103696. [CrossRef]
6. Shah, N.A.; El-Zahar, E.R.; Chung, J.D. Fractional Analysis of Coupled Burgers Equations within Yang Caputo-Fabrizio Operator. *J. Funct. Spaces* **2022**, *2022*, 6231921. [CrossRef]
7. Chang, Y.F. Neural synergetics, lorenz model of brain, soliton-chaos double solutions and physical neurobiology. *NeuroQuantology* **2013**, *11*, 56–62. [CrossRef]
8. Shah, N.A.; Hamed, Y.S.; Abualnaja, K.M.; Chung, J.-D.; Shah, R.; Khan, A. A Comparative Analysis of Fractional-Order Kaup–Kupershmidt Equation within Different Operators. *Symmetry* **2022**, *14*, 986. [CrossRef]
9. Alquran, M.; Krishnan, E.V. Applications of sine-gordon expansion method for a reliable treatment of some nonlinear wave equations. *Nonlinear Stud.* **2016**, *23*, 639–649.
10. Rupp, D.E.; Selker, J.S. On the use of the Boussinesq equation for interpreting recession hydrographs from sloping aquifers. *Water Resour. Res.* **2006**, *42*, 1–15. [CrossRef]
11. Rani, A.; Khan, N.; Ayub, K.; Khan, M.Y.; Mahmood-Ul-Hassan, Q.; Ahmed, B.; Ashraf, M. Solitary Wave Solution of Nonlinear PDEs Arising in Mathematical Physics. *Open Phys.* **2019**, *17*, 381–389. [CrossRef]
12. Zahran, E.H.M.; Khater, M.M.A. Exact Traveling Wave Solutions for the System of Shallow Water Wave Equations and Modified Liouville Equation Using Extended Jacobian Elliptic Function Expansion Method. *Am. J. Comput. Math.* **2014**, *04*, 455–463. [CrossRef]
13. Khater, M.M.A. The Modified Simple Equation Method and its Applications in Mathematical Physics and Biolog. *Glob. J. Sci. Front. Res. Math. Decis. Sci.* **2015**, *15*, 69–86.
14. Wazwaz, A.M. The tanh method for traveling wave solutions of nonlinear equations. *Appl. Math. Comput.* **2004**, *154*, 713–723. [CrossRef]
15. El-Wakil, S.A.; Abdou, M.A. New exact travelling wave solutions using modified extended tanh-function method. *Chaos Solitons Fractals* **2007**, *31*, 840–852. [CrossRef]
16. Wazwaz, A.M. A sine-cosine method for handling nonlinear wave equations. *Math. Comput. Model.* **2004**, *40*, 499–508. [CrossRef]
17. Radha, B.; Duraisamy, C. The homogeneous balance method and its applications for finding the exact solutions for nonlinear equations. *J. Ambient Intell. Humaniz. Comput.* **2021**, *12*, 6591–6597. [CrossRef]
18. Ren, Y.J.; Zhang, H.Q. A generalized F-expansion method to find abundant families of Jacobi Elliptic Function solutions of the (2 + 1)-dimensional Nizhnik-Novikov-Veselov equation. *Chaos Solitons Fractals* **2006**, *27*, 959–979. [CrossRef]
19. Darvishi, M.T.; Najafi, M.; Arbabi, S.; Kavitha, L. Exact propagating multi-anti-kink soliton solutions of a (3+1)-dimensional B-type Kadomtsev–Petviashvili equation. *Nonlinear Dyn.* **2016**, *83*, 1453–1462. [CrossRef]
20. Khani, F.; Darvishi, M.T.; Farmany, A.; Kavitha, L. New exact solutions of coupled (2+1)-dimensional nonlinear systems of Schrödinger equations. *ANZIAM J.* **2011**, *52*, 110–121. [CrossRef]
21. Ilie, M.; Biazar, J.; Ayati, Z. The first integral method for solving some conformable fractional differential equations. *Opt. Quantum Electron.* **2018**, *50*, 1–11. [CrossRef]
22. He, W.; Chen, N.; Dassios, I.; Shah, N.A.; Chung, J.D. Fractional System of Korteweg-De Vries Equations via Elzaki Transform. *Mathematics* **2021**, *9*, 673. [CrossRef]
23. Manafian, J.; Ilhan, O.A.; Mohammed, S.A. Forming localized waves of the nonlinearity of the dna dynamics arising in oscillator-chain of peyrard-bishop model. *AIMS Math.* **2020**, *5*, 2461–2483. [CrossRef]
24. Ilhan, O.A.; Manafian, J.; Alizadeh, A.; Baskonus, H.M. New exact solutions for nematicons in liquid crystals by the tan (ϕ/ 2) -expansion method arising in fluid mechanics. *Eur. Phys. J. Plus* **2020**, *135*, 1–19. [CrossRef]
25. Shah, N.A.; Dassios, I.; El-Zahar, E.R.; Chung, J.D. An Efficient Technique of Fractional-Order Physical Models Involving ρ-Laplace Transform. *Mathematics* **2022**, *10*, 816. [CrossRef]

26. Rani, A.; Ashraf, M.; Ahmad, J.; Ul-Hassan, Q.M. Soliton solutions of the Caudrey–Dodd–Gibbon equation using three expansion methods and applications. *Opt. Quantum Electron.* **2022**, *54*, 1–19. [CrossRef]
27. GAchu, F.; Kakmeni, F.M.M.; Dikande, A.M. Breathing pulses in the damped-soliton model for nerves. *Phys. Rev. E* **2018**, *97*, 1–9. [CrossRef]
28. Lautrup, B.; Appali, R.; Jackson, A.D.; Heimburg, T. The stability of solitons in biomembranes and nerves. *Eur. Phys. J. E. Soft Matter.* **2011**, *34*, 1–9. [CrossRef] [PubMed]
29. Peets, T.; Tamm, K.; Engelbrecht, J. On the role of nonlinearities in the Boussinesq-type wave equations. *Wave Motion* **2017**, *71*, 113–119. [CrossRef]
30. Abdelrahman, M.A.E.; Zahran, E.H.M.; Khater, M.M.A. The Exp (-φ(ξ))-Expansion Method and Its Application for Solving Nonlinear Evolution Equations. *Int. J. Mod. Nonlinear Theory Appl.* **2015**, *4*, 37–47. [CrossRef]
31. Pankaj, R.D.; Kumar, A.; Singh, B.; Meena, M.L. Exp $(-\varphi(\xi))$ expansion method for soliton solution of nonlinear Schrödinger system. *J. Interdiscip. Math.* **2022**, *25*, 89–97. [CrossRef]
32. Hodgkin, A.L.; Huxley, A.F. Propagation of electrical signals along giant nerve fibers. *Proc. R. Soc. Lond. B. Biol. Sci.* **1952**, *140*, 177–183. [CrossRef]
33. Hodgkin, A.L.; Huxley, A.F. Resting and action potentials in single nerve fibres. *J. Physiol.* **1945**, *104*, 176–195. [CrossRef]
34. Hodgkin, A.L.; Huxley, A.F. Currents carried by sodium and potassium ions through the membrane of the giant axon of loligo. *J. Physiol.* **1952**, *116*, 449–472. [CrossRef]
35. FitzHugh, R. Impulses and Physiological States in Theoretical Models of Nerve Membrane. *Biophys. J.* **1961**, *1*, 445–466. [CrossRef]
36. Mitaku, S.; Date, T. Anomalies of nanosecond ultrasonic relaxation in the lipid bilayer transition. *BBA—Biomembr.* **1982**, *688*, 411–421. [CrossRef]
37. Heimburg, T.; Jackson, A.D. On soliton propagation in biomembranes and nerves. *Proc. Natl. Acad. Sci. USA* **2005**, *102*, 9790–9795. [CrossRef]

mathematics

Article

Analytical Method for Generalized Nonlinear Schrödinger Equation with Time-Varying Coefficients: Lax Representation, Riemann-Hilbert Problem Solutions

Bo Xu [1,2,*] and Sheng Zhang [3,*]

1 School of Mathematics, China University of Mining and Technology, Xuzhou 221116, China
2 School of Educational Sciences, Bohai University, Jinzhou 121013, China
3 School of Mathematical Sciences, Bohai University, Jinzhou 121013, China
* Correspondence: bxu@bhu.edu.cn (B.X.); szhangchina@126.com or szhang@bhu.edu.cn (S.Z.)

Abstract: In this paper, a generalized nonlinear Schrödinger (gNLS) equation with time-varying coefficients is analytically studied using its Lax representation and the associated Riemann-Hilbert (RH) problem equipped with a symmetric scattering matrix in the Hermitian sense. First, Lax representation and the associated RH problem of the considered gNLS equation are established so that solution of the gNLS equation can be transformed into the associated RH problem. Secondly, using the solvability of unique solution of the established RH problem, time evolution laws of the scattering data reconstructing potential of the gNLS equation are determined. Finally, based on the determined time evolution laws of scattering data, the long-time asymptotic solution and N-soliton solution of the gNLS equation are obtained. In addition, some local spatial structures of the obtained one-soliton solution and two-soliton solution are shown in the figures. This paper shows that the RH method can be extended to nonlinear evolution models with variable coefficients, and the curve propagation of the obtained N-soliton solution in inhomogeneous media is controlled by the selection of variable–coefficient functions contained in the models.

Keywords: gNLS equation with time-varying coefficients; Lax representation; RH problem; scattering data; long-time asymptotic solution; N-soliton solution

MSC: 37K40; 37K10; 35Q15; 35C08

Citation: Xu, B.; Zhang, S. Analytical Method for Generalized Nonlinear Schrödinger Equation with Time-Varying Coefficients: Lax Representation, Riemann-Hilbert Problem Solutions. *Mathematics* **2022**, *10*, 1043. https://doi.org/10.3390/math10071043

Academic Editors: Almudena del Pilar Marquez Lozano and Vladimir Iosifovich Semenov

Received: 7 February 2022
Accepted: 22 March 2022
Published: 24 March 2022

1. Introduction

Nonlinear problems are full of challenges, and these have attracted the extensive attention of researchers. One of the important achievements of nonlinear mathematical physics in recent decades is the discovery of certain nonlinear partial differential equations (PDEs) with important applications and analytical solutions. For example, the classical NLS equation has practical applications in many fields [1], including optics, oceanography, biology, economics and so on. There are many effective methods for solving nonlinear PDEs analytically, such as inverse scattering method [2], Darboux transformation [3], Hirota bilinear method [4] and other methods [5–14].

When an inhomogeneous medium is considered, the variable–coefficient model is usually closer to the essence of the phenomenon. Generally, solving variable–coefficient equations is more difficult than solving constant-coefficient ones. In most cases, it is necessary to embed appropriate coefficient functions in the solution process of the existing analytical methods, see [15] for an ingenious work extending inverse scattering method to deal with a variable–coefficient NLS equation. Owing to the fact that Schrödinger-type equations are widely used in many fields and differential equations with variable–coefficient functions often model dynamic processes in non-uniform media, this paper considers a model in nonlinear fiber optics, namely the following gNLS equation with gain [16]:

$$\mathrm{i}\psi_z = \frac{\beta(z)}{2}\phi_{\tau\tau} - \gamma(z)|\psi|^2\psi + \mathrm{i}\frac{g(z)}{2}\psi, \tag{1}$$

where $\psi = \psi(z,t)$; the three functions $\beta(z)$, $\gamma(z)$ and $g(z)$ of propagation distance z represent the group velocity dispersion parameter, nonlinearity parameter and distributed gain function, respectively; $|\psi|$ denotes the module of ψ; and i is the imaginary unit. For convenience, we take the transformations:

$$\psi(z,\tau) = u(x,t),\ \beta(z) = -\beta(t),\ \gamma(z) = \beta(t),\ g(z) = 2\mathrm{i}\alpha(t). \tag{2}$$

Then, Equation (1) is converted to the gNLS equation with time-varying coefficients:

$$\mathrm{i}u_t + \frac{\beta(t)}{2}u_{xx} + \beta(t)|u|^2u + \alpha(t)u = 0. \tag{3}$$

Here, $\alpha(t)$ and $\beta(t)$ are assumed to be real integrable functions, while u and all its partial derivatives with respect to x and t approach zero quickly enough as $|x| \to \infty$.

The analytical method adopted in this paper for Equation (3) is the RH method [17], which was developed based on the IST [2]. The RH method is an analytical method that does not need to solve the Gel'fand-Levitan-Marchenko integral equation and can also analyze the long-time asymptotic behavior of the obtained implicit analytical solutions. In recent years, the RH method has achieved many applications, such as [17–28]. One of the important developments of RH method is Deift-Zhou's nonlinear steepest descent method [18].

The basic idea of the RH method is to establish the relationship between the solution of nonlinear PDE to be solved and the solution of associated solvable RH problem using the eigenfunction, then to solve the RH problem, and finally obtain the solution of nonlinear PDE. In the literature, there are some results, such as [8,16,29–35], that have been obtained for the gNLS Equation (3). However, as far as we know, there is still no research on the RH problem of Equation (3), and the relevant work is worth exploring. Equation (3) is integrable; the Lax presentation, which provides a basis of the study of the associated RH problem is given in Section 2.

With the help of the given Lax presentation, the associated RH problem is established in Section 3 to connect the solution of Equation (3) and that of the established RH problem, and then the time evolution laws of scattering data in the RH problem are determined. In Section 4, the long-time asymptotic solution and N-soliton solution of Equation (3) are obtained. At the same time, some spatial structures of the obtained one-soliton solution and two-soliton solution are shown by selecting several special cases of the time-varying functions.

2. Lax Presentation and RH Problem

We introduce, in this section, the linear spectral problem in the matrix forms:

$$F_x + \mathrm{i}\xi\sigma_3 F = \phi F, \tag{4}$$

$$F_t + \mathrm{i}[\xi^2\beta(t) - \frac{1}{2}\alpha(t)]\sigma_3 F = \varphi F, \tag{5}$$

where ξ is the complex spectral parameter; $F = F(x,t,\xi)$ is the eigenfunction in matrix form; the notations σ_3, ϕ and φ stand for

$$\sigma_3 = \begin{pmatrix} 1 & 0 \\ 0 & -1 \end{pmatrix}, \ \phi = \begin{pmatrix} 0 & u \\ -u^* & 0 \end{pmatrix} \text{ and}$$
$$\varphi = \begin{pmatrix} \frac{\mathrm{i}}{2}\beta(t)\left|u\right|^2 & \frac{\mathrm{i}}{2}\beta(t)u_x \\ \frac{\mathrm{i}}{2}\beta(t)u_x^* & -\frac{\mathrm{i}}{2}\beta(t)\left|u\right|^2 \end{pmatrix} + \xi\beta(t)\phi; \tag{6}$$

and the symbol * is complex conjugate.

It is easy to check that the compatibility condition $F_{xt} = F_{tx}$ is equivalent to Equation (3). Therefore, we say that the gNLS Equation (3) has Lax integrability, and its Lax representations are Equations (4) and (5).

Considering the asymptotic condition of the previously assumed boundary value that u and all its partial derivatives, with respect to x and t, approach zeros quickly as $|x| \to \infty$, we have the asymptotic Jost solution of Equations (4) and (5):

$$F \to \mathrm{e}^{-\mathrm{i}\vartheta(x,t,\xi)\sigma_3}, \ |x| \to \infty, \tag{7}$$

with

$$\vartheta(x,t,\xi) = \xi x + \int_0^t [\xi^2\beta(\tau) - \frac{1}{2}\alpha(\tau)]\mathrm{d}\tau. \tag{8}$$

By the transformation:

$$K(x,t,\xi) \to F\mathrm{e}^{\mathrm{i}\vartheta(x,t,\xi)\sigma_3}, \tag{9}$$

we transform Equations (4) and (5) into the following forms:

$$K_x + \mathrm{i}\xi[\sigma_3, K] = \phi K, \tag{10}$$

$$K_t + \mathrm{i}[\xi^2\beta(t) - \frac{1}{2}\alpha(t)][\sigma_3, \varphi] = \varphi K, \tag{11}$$

so that the eigenfunction K has the boundary condition:

$$K_\pm \to I, \ x \to \pm\infty, \tag{12}$$

where $K_\pm$ means the boundary conditions of K at the positive infinity and negative infinity respectively, and I denotes the second-order identity matrix. In the case where the boundary conditions (12) hold, the x-part of the Lax representation, that is, Equation (10) has the solutions [17]:

$$K_- = I + \int_{-\infty}^x \mathrm{e}^{-\mathrm{i}\xi(x-y)\sigma_3}\phi(y)K_-(y,\xi)\mathrm{e}^{\mathrm{i}\xi(x-y)\sigma_3}\mathrm{d}y, \tag{13}$$

$$K_+ = I - \int_x^{\infty} \mathrm{e}^{-\mathrm{i}\xi(x-y)\sigma_3}\phi(y)K_+(y,\xi)\mathrm{e}^{\mathrm{i}\xi(x-y)\sigma_3}\mathrm{d}y, \tag{14}$$

which enable the following relationships to be established:

$$K_- = K_+\mathrm{e}^{-\mathrm{i}\xi\sigma_3}M(\xi)\mathrm{e}^{\mathrm{i}\xi\sigma_3}, \ \xi \in R, \tag{15}$$

by means of the scattering matrix:

$$M(\xi) = \begin{pmatrix} m_{11}(\xi) & m_{12}(\xi) \\ m_{21}(\xi) & m_{22}(\xi) \end{pmatrix}. \tag{16}$$

Since the determinant $\det K_{\pm} = 1$ [17], which shows that the matrix $K_{\pm}$ is reversible, we can see from Equation (15) that $\det M(\xi) = 1$ and then obtain the inverse matrix of the scattering matrix $M(\xi)$:

$$M^{-1}(\xi) = \begin{pmatrix} \hat{m}_{11}(\xi) & \hat{m}_{12}(\xi) \\ \hat{m}_{21}(\xi) & \hat{m}_{22}(\xi) \end{pmatrix} = \begin{pmatrix} m_{22}(\xi) & -m_{12}(\xi) \\ -m_{21}(\xi) & m_{22}(\xi) \end{pmatrix}. \tag{17}$$

Due to $K_{\pm}^{H}(x, \xi^*) = K_{\pm}^{-1}(x, \xi)$, with H standing for the Hermitian conjugate, one knows that the symmetric relation $M^H(\xi^*) = M^{-1}(\xi)$ leads to the equalities $m_{11}^*(\xi^*) = m_{22}(\xi)$ and $m_{12}^*(\xi^*) = -m_{21}(\xi)$.

With the help of notations $K_{\pm} = ((K_{\pm})_1, (K_{\pm})_2)$ and $K_{\pm}^{-1} = ((K_{\pm}^{-1})_1, (K_{\pm}^{-1})_2)^T$, we introduce the matrices:

$$\phi^{+} = K_{-}H_1 + K_{+}H_2 = ((K_{-})_1, (K_{+})_2), \tag{18}$$

$$\phi^{-} = H_1 K_{-}^{-1} + H_2 K_{+}^{-1} = \begin{pmatrix} (K_{-}^{-1})^1 \\ (K_{+}^{-1})^2 \end{pmatrix}, \tag{19}$$

where $(K_{\pm})_s$ and $(K_{\pm}^{-1})^s$ denote the vector in the s-th row and that in the s-th column of $K_{\pm}$, respectively, and $H_1 = \mathrm{diag}(1,0)$ and $H_2 = \mathrm{diag}(0,1)$ are two special diagonal matrices. Clearly, ϕ^+ and ϕ^- enable Equation (10) and its adjoint equation to be true, that is to say:

$$\phi_x^{+} + \mathrm{i}\xi[\sigma_3, \phi^{+}] = \phi\phi^{+}, \tag{20}$$

$$\phi_x^{-} + \mathrm{i}\xi[\sigma_3, \phi^{-}] = \phi^{-}\phi. \tag{21}$$

The Taylor series of $\phi^{\pm}$ gives:

$$\phi^{\pm} = I + \frac{\phi_1^{\pm}}{\xi} + O(\xi^{-2}). \tag{22}$$

We insert ϕ^+ and ϕ^- into Equations (20) and (21) and compare the coefficients of ξ^{-1}, and then one has

$$\phi = \mathrm{i}[\sigma_3, K_1^{+}] = -\mathrm{i}[\sigma_3, K_1^{-}]. \tag{23}$$

Thus, solution u of the gNLS Equation (3) is converted to $\phi^{\pm}$ by the following formula:

$$u = \pm 2\mathrm{i}(\phi_1^{\pm})_{12} = \pm 2\mathrm{i} \lim_{\lambda \to \infty} (\xi\phi^{\pm})_{12}, \tag{24}$$

with $(\phi_1^{\pm})_{12}$ representing the element locations at the intersection of the first row and the second column of $\phi_1^{\pm}$. Here, $\phi^{\pm}$ will be determined by the matrix RH problem established by Equations (18) and (19):

$$\begin{array}{ll} \text{(i)} & \phi^{\pm}(x,\xi) \text{ are analytic in } \xi \in C_{\pm}; \\ \text{(ii)} & \phi^{-}(x,\xi)\phi^{+}(x,\xi) = \Omega(x,\xi) \text{ for } \xi \in R; \\ \text{(iii)} & \phi^{\pm}(x,\xi) \to I \text{ for } \xi \in C_{\pm} \to \infty \end{array} \tag{25}$$

where C_+ and C_- are the upper and lower half complex planes, respectively; $\mathbb{R}$ is the set of real numbers; and $\Omega(x, \lambda)$ is the jump matrix:

$$\Omega(x,\xi) = \mathrm{e}^{-\mathrm{i}\xi\sigma_3} \begin{pmatrix} 1 & \hat{s}_{12}(\xi) \\ s_{21}(\xi) & 1 \end{pmatrix} \mathrm{e}^{\mathrm{i}\xi\sigma_3}. \tag{26}$$

3. Solvability of RH Problem and Time Evolution Laws for Scattering Data

The RH Problem (25) established above is solvable and always has a unique solution. More detailed proof can be found in [17]; the difference is because the time evolution laws of the scattering data involved are different. In fact, from Equations (15), (18) and (19), we can see that

$$\det\phi^{+} = \hat{m}_{22}(\xi) = m_{11}(\xi),\ \det\phi^{-} = m_{22}(\xi) = \hat{m}_{11}(\xi), \tag{27}$$

where the symmetry relation $M^{H}(\xi^{*}) = M^{-1}(\xi)$ has been used.

When $\det\phi^{\pm}(\lambda) \neq 0$, the RH problem (30) is regular. Then, Plemelj formula [36] can be used to obtain a unique solution of Equation (25):

$$(\phi^{+})^{-1}(\xi) = I + \frac{1}{2\mathrm{i}\pi}\int_{-\infty}^{\infty} \frac{\hat{\Omega}(\xi)(\phi^{+})^{-1}(\xi)}{s-\xi}\mathrm{d}s,\ \xi \in \mathbb{C}_{+}, \tag{28}$$

with

$$\hat{\Omega}(\xi) = I - \Omega(\xi) = -\mathrm{e}^{-\mathrm{i}\xi\sigma_3}\begin{pmatrix} 0 & \hat{m}_{12}(\xi) \\ m_{21}(\xi) & 0 \end{pmatrix}\mathrm{e}^{\mathrm{i}\xi\sigma_3}. \tag{29}$$

In the case of $\det\phi^{\pm}(\xi) = 0$, the relation $M^{H}(\xi^{*}) = M^{-1}(\xi)$ makes the numbers of the conjugate zeros of $\det\phi^{+}(\xi) = 0$ and $\det\phi^{-}(\xi) = 0$ must be equal. Thus, we suppose that $\det\phi^{+}(\xi) = 0$ has conjugate zeros $\xi_j, \xi_2, \cdots, \xi_N \in C_{+}$ and denote the conjugate zeros of $\det\phi^{-}(\xi) = 0$ as $\overline{\xi}_j = \xi_j^{*} \in C_{-}(j = 1, 2, \cdots, N)$. For the irregular case of the RH Problem (25), we consider the systems of linear equations:

$$\phi^{+}(\xi_j)v_j(\xi_j) = 0,\ (j = 1, 2, \cdots, N), \tag{30}$$

$$\overline{v}_j(\overline{\xi}_j)\phi^{-}(\overline{\xi}_j) = 0,\ (j = 1, 2, \cdots, N), \tag{31}$$

where non-zero row vector $v_j(\xi_j)$ and non-zero column vector $\overline{v}_j(\overline{\xi}_j)$ are solutions of Equations (30) and (31), respectively. The Hermitian conjugate of Equation (30), together with the symmetry relation $(\phi^{+})^{H}(\xi_j^{*}) = \phi^{-}(\overline{\xi}_j)$, gives

$$v_j^{H}(\xi_j)\phi^{-}(\overline{\xi}_j) = 0. \tag{32}$$

Then, Equations (31) and (32) lead to the symmetry relation $\overline{v}_j(\overline{\xi}_j) = v_j^{H}(\xi_j)$. Based on these preparations and theorem [37], the irregular RH Problem (25) with $\det\phi^{\pm}(\xi) = 0$ can be transformed into a regular one. Thus, we indirectly arrive at the proof that the irregular RH Problem (25) has a unique solution, and therefore the solution of Equation (24) can be determined as follows:

$$\phi_1^{+}(\xi) = \sum_{k=1}^{N}\sum_{j=1}^{N} v_k(P^{-1})_{kj}\overline{v}_j + \frac{1}{2\mathrm{i}\pi}\int_{-\infty}^{\infty} Q(s)\hat{\Omega}(s)Q^{-1}(s)(\hat{\phi}^{+})^{-1}(s)\mathrm{d}s, \tag{33}$$

with

$$(\hat{\phi}^{+})^{-1}(\xi) = I + \frac{1}{2\mathrm{i}\pi}\int_{-\infty}^{\infty} \frac{Q(s)\hat{\Omega}(s)Q^{-1}(s)(\hat{\phi}^{+})^{-1}(s)}{s-\xi}\mathrm{d}s,\ \xi \in C_{+}, \tag{34}$$

$$Q(\xi) = I + \sum_{k=1}^{N}\sum_{j=1}^{N} \frac{v_k(P^{-1})_{kj}\overline{v}_j}{\xi - \overline{\xi}_j},\ Q^{-1}(\xi) = I - \sum_{k=1}^{N}\sum_{j=1}^{N} \frac{v_k(P^{-1})_{kj}\overline{v}_j}{\xi - \xi_k}, \tag{35}$$

$$P = (p_{kj})_{N\times N},\ p_{kj} = \frac{\overline{v}_k v_j}{\overline{\xi}_k - \xi_j},\ (1 \leq k, j \leq N). \tag{36}$$

The solvability of RH Problem (25) lays a theoretical foundation for the determination of the corresponding scattering data.

Theorem 1. *Let $u(x,t)$ solve the gNLS Equation (3). Then, the scattering data:*

$$\{m_{21}(\xi), m_{21}(\xi), \hat{m}_{12}(\xi), (\xi \in \mathbb{R}); \xi_j, \overline{\xi}_j, v_j, \overline{v}_j, (j = 1, 2, \cdots, N)\}, \tag{37}$$

determined by the regular RH problem (30) have the time evolution laws:

$$m_{21}(t,\xi) = m_{21}(0,\xi)e^{2i\int_0^t[\xi^2\beta(\tau)-\frac{1}{2}\alpha(\tau)]d\tau}, \tag{38}$$

$$\hat{m}_{12}(t,\xi) = \hat{m}_{12}(0,\xi)e^{-2i\int_0^t[\xi^2\beta(\tau)-\frac{1}{2}\alpha(\tau)]d\tau}, \tag{39}$$

$$\xi_j(t) = \xi_j(0),\ \overline{\xi}_j(t) = \overline{\xi}_j(0), \tag{40}$$

$$v_j(x,t,\xi_j) = e^{-i\{\xi_j(0)x+\int_0^t[\xi_j^2(0)\beta(\tau)-\frac{1}{2}\alpha(\tau)]d\tau\}\sigma_3}v_j(0,0,\xi_j(0)), \tag{41}$$

$$\overline{v}_j(x,t,\overline{\xi}_j) = e^{i\{\overline{\xi}_j(0)x+\int_0^t[\overline{\xi}_j^2(0)\beta(\tau)-\frac{1}{2}\alpha(\tau)]d\tau\}\sigma_3}\overline{v}_j(0,0,\xi_j(0)). \tag{42}$$

Proof of Theorem 1. It is necessary to rewrite Equation (15) as:

$$K_-e^{-i\xi\sigma_3} = K_+e^{-i\xi\sigma_3}M(\xi),\ \xi \in R. \tag{43}$$

Differentiating the left side of Equation (48) with respect to t, we arrive at

$$K_{-,t}e^{-i\xi\sigma_3} = -i[\xi^2\beta(t) - \frac{1}{2}\alpha(t)][\sigma_3, K_-]e^{-i\xi\sigma_3} + \varphi K_-e^{-i\xi\sigma_3}, \tag{44}$$

by employing Equation (15). It is easy to see from Equation (44) that the left side of Equation (43) solves Equation (11). We, therefore, know that the right side of Equation (43) is a solution of Equation (11). Then, the substitution of the right side of Equation (43) into Equation (11) together with the boundary condition (12) yields

$$\frac{dM(t,\xi)}{dt} + i[\xi^2\beta(t) - \frac{1}{2}\alpha(t)]K_+e^{-i\xi\sigma_3}[\sigma_3, M(t,\xi)] = 0. \tag{45}$$

Similarly, we easily see that $K_+e^{-i\xi\sigma_3} = K_-e^{-i\xi\sigma_3}M^{-1}(t,\xi)$ is also a solution of Equation (11). Putting $K_-e^{-i\xi\sigma_3}M^{-1}(t,\xi)$ into Equation (11) and using the boundary condition (12) yields:

Considering Equations (16) and (17) and comparing the elements of Equations (45) and (46), we gain

$$\frac{dM^{-1}(t,\xi)}{dt} + i[\xi^2\beta(t) - \frac{1}{2}\alpha(t)]K_+e^{-i\xi\sigma_3}[\sigma_3, M^{-1}(t,\xi)] = 0 \tag{46}$$

$$\frac{dm_{21}(t,\xi)}{dt} = 2i[\xi^2\beta(t) - \frac{1}{2}\alpha(t)]m_{21}(t,\xi), \tag{47}$$

$$\frac{d\hat{m}_{12}(t,\xi)}{dt} = -2i[\xi^2\beta(t) - \frac{1}{2}\alpha(t)]\hat{m}_{12}(t,\xi), \tag{48}$$

$$\frac{d\hat{m}_{22}(t,\xi)}{dt} = 0,\ \frac{dm_{22}(t,\xi)}{dt} = 0. \tag{49}$$

Solving Equations (47) and (48), we reach Equations (38) and (39). Equation (27) indicates that, if $\xi_j(t)$ and $\overline{\xi}_j(t)$ are the zeros of $\det\phi^+(t,\xi)$ and $\det\phi^-(t,\xi)$, they are also the zeros of $\hat{m}_{22}(t,\xi)$ and $m_{22}(t,\xi)$. In view of Equation (49), one can see that $\xi_j(t)$ and $\overline{\xi}_j(t)$ are independent from t. This means that Equation (40) is true.

To prove Equations (41) and (42), it is necessary to differentiate Equation (30) with respect to x and t, and then one has

$$\phi_x^+(x,t,\xi_j)v_j(x,t,\xi_j) + \phi^+(x,t,\xi_j)v_{j,x}(x,t,\xi_j) = 0,\ (j = 1,2,\cdots,N), \tag{50}$$

$$\phi_t^+(x,t,\xi_j)v_j(x,t,\xi_j) + \phi^+(x,t,\xi_j)v_{j,t}(x,t,\xi_j) = 0,\ (j = 1,2,\cdots,N). \tag{51}$$

Using Equations (11) and (18) yields

$$\phi_t^+(x,t,\xi_j)v_j(x,t,\xi_j) = -\mathrm{i}[\xi^2\beta(t) - \frac{1}{2}\alpha(t)][\sigma_3, \phi^+(x,t,\xi_j)] + \varphi\phi^+. \tag{52}$$

Substituting Equations (20) and (52) into Equations (50) and (51), we gain

$$\phi^+(x,t,\xi_j)(v_{j,x}(x,t,\xi_j) + \mathrm{i}\xi_j\sigma_3 v_j(x,t,\xi_j)) = 0,\ (j=1,2,\cdots,N), \tag{53}$$

$$\phi^+(x,t,\xi_j)\left\{v_{j,t}(x,t,\xi_j) + \mathrm{i}[\xi^2\beta(t) - \frac{1}{2}\alpha(t)]\sigma_3 v_j(x,t,\xi_j)\right\} = 0,\ (j=1,2,\cdots,N), \tag{54}$$

by the usage of Equation (30). Solving Equations (53) and (54), one can obtain Equation (41). In a similar way, Equation (42) can be obtained using Equations (11), (21) and (31). □

4. Long-Time Asymptotic Solution and *N*-Soliton Solution

Based on Equations (38) and (39), the time evolution laws of the Jump matrix $\hat{\Omega}(x,t,\xi)$ can be determined as follows:

$$\hat{\Omega}(x,t,\xi) = \begin{pmatrix} 0 & -\hat{m}_{12}(0,\xi)\mathrm{e}^{-2\mathrm{i}\xi\vartheta(x,t,\xi)\sigma_3} \\ m_{21}(0,\xi)\mathrm{e}^{2\mathrm{i}\xi\vartheta(x,t,\xi)\sigma_3} & 0 \end{pmatrix}, \tag{55}$$

where $\vartheta(x,t,\xi)$ is determined by Equation (8). Generally, with the above scattering data in Equations (38)–(42), one can obtain solution of the gNLS Equation (3) theoretically. However, we still have difficulty in calculating the integral in Equation (33) for $\hat{\Omega}(x,t,\xi) \neq 0$. In this case, the asymptotic solution of the gNLS Equation (3) when $t \to \infty$ can be derived from Equation (24). For instance, if we let $\hat{\xi} = \xi t^{1/2\gamma}$ and $\beta(t) = t^{1/\gamma - 1}$ for any $1 \leq \gamma \in R$, the integral contained in Equation (38) tends to zero at a rate of $t^{-1/\gamma}$. We, therefore, obtain the following long-time asymptotic solution of the gNLS Equation (3):

$$u(x,t) \to 2\mathrm{i}\left(\sum_{k=1}^{N}\sum_{j=1}^{N} v_k(P^{-1})_{kj}\overline{v}_j\right)_{12},\ t \to \infty, \tag{56}$$

where P and v_k are calculated using Equations (36) and (41), while the calculation of $\overline{v}_k$ can restore to Equation (42) or the symmetry relation $\overline{v}_j = v_j^H$.

In the reflectionless case, we next construct an N-soliton solution of the NLS Equation (3). Setting $\hat{m}_{12}(0,\xi) = 0$ and $m_{21}(0,\xi) = 0$, and then one has $\hat{\Omega}(x,t,\xi) = 0$. In this case, Equation (33) is simplified as

$$\phi_1^+(x,t) = \sum_{k=1}^{N}\sum_{j=1}^{N} v_k(P^{-1})_{kj}\overline{v}_j. \tag{57}$$

To determine P^{-1} in Equation (57), we further select the complex number c_j and let $v_j(0,0,\xi_j(0)) = (c_j, 1)$. Then, Equations (41) and (42) give

$$v_j(x,t,\xi_j) = \begin{pmatrix} c_j\mathrm{e}^{\theta_j} \\ \mathrm{e}^{-\theta_j} \end{pmatrix}, \tag{58}$$

$$\overline{v}_j(x,t,\overline{\xi}_j) = v_j^H(x,t,\xi_j^*) = (c_j^*\mathrm{e}^{\theta_j^*}, \mathrm{e}^{-\theta_j^*}). \tag{59}$$

where

$$\theta_j = -\mathrm{i}\xi_j(0)x - \mathrm{i}\int_0^t [\xi_j^2(0)\beta(\tau) - \frac{1}{2}\alpha(\tau)]\mathrm{d}\tau,\ \xi_j(0) \in C_+, \tag{60}$$

Finally, with the help of Equations (24) and (58)–(60), one obtains the N-soliton solution of NLS Equation (3):

$$u(x,t) = 2\mathrm{i}\left(\sum_{k=1}^{N}\sum_{j=1}^{N} c_k \mathrm{e}^{\theta_k-\theta_j^*}(P^{-1})_{kj}\right)_{12} = -2\mathrm{i}\frac{\det S}{\det P}. \tag{61}$$

where θ_k and θ_k^* can be determined by Equation (60),

$$S = \begin{pmatrix} 0 & c_1\mathrm{e}^{\theta_1} & \cdots & c_N\mathrm{e}^{\theta_N} \\ \mathrm{e}^{-\theta_1^*} & p_{11} & \cdots & p_{1N} \\ \cdots & \cdots & \cdots & \cdots \\ \mathrm{e}^{-\theta_N^*} & p_{N1} & \cdots & p_{NN} \end{pmatrix}, \; P = (p_{kj})_{N\times N}, \; p_{kj} = \frac{c_k^* c_j \mathrm{e}^{\theta_j+\theta_k^*} + \mathrm{e}^{-\theta_j-\theta_k^*}}{\bar{\xi}_k(0) - \xi_j(0)}. \tag{62}$$

As a special case of Equation (61), $N = 1$ is selected, and then one has:

$$u(x,t) = -2\mathrm{i}\frac{-c_1\mathrm{e}^{\theta_1-\theta_1^*}}{\frac{c_1 c_1^* \mathrm{e}^{\theta_1+\theta_1^*} + \mathrm{e}^{-\theta_1-\theta_1^*}}{\bar{\xi}_1(0)-\xi_1(0)}}. \tag{63}$$

Further letting $\xi_1(0) = a + \mathrm{i}b(a, b > 0 \in R)$ and $c_1 = \mathrm{e}^{-2b\delta_0+\mathrm{i}w_0}(d_0, w_0 \in R)$ yields $\bar{\xi}_1(0) = \xi_1^*(0) = a - \mathrm{i}b$ and $c_1 c_1^* = \mathrm{e}^{-2b\delta_0}$. Thus, Equation (63) becomes

$$u(x,t) = 4b\frac{\mathrm{e}^{-2b\delta_0+\mathrm{i}w_0}\mathrm{e}^{-2\mathrm{i}ax-2\mathrm{i}\int_0^t [(a^2-b^2)\beta(\tau)-\frac{1}{2}\alpha(\tau)]\mathrm{d}\tau}}{\mathrm{e}^{-4b\delta_0}\mathrm{e}^{-2bx-4ab\int_0^t \beta(\tau)\mathrm{d}\tau-2bd_0} + \mathrm{e}^{2bx+4ab\int_0^t \beta(\tau)\mathrm{d}\tau+2bd_0}}, \tag{64}$$

which can be rewritten as:

$$u(x,t) = 4b\frac{\mathrm{e}^{-2\mathrm{i}ax-2\mathrm{i}\int_0^t [(a^2-b^2)\beta(\tau)-\frac{1}{2}\alpha(\tau)]\mathrm{d}\tau+\mathrm{i}w_0}}{\mathrm{e}^{-2bx-4ab\int_0^t \beta(\tau)\mathrm{d}\tau-2b\delta_0} + \mathrm{e}^{2bx+4ab\int_0^t \beta(\tau)\mathrm{d}\tau+2b\delta_0}}. \tag{65}$$

Finally, the one-soliton solution of the gNLS Equation (3) can be obtained as follows:

$$u(x,t) = 2b\mathrm{e}^{-2\mathrm{i}\eta}\mathrm{sech}[2b(x + 2a\int_0^t \beta(\tau)\mathrm{d}\tau - \delta_0)], \tag{66}$$

where

$$\eta = ax + \int_0^t [(a^2-b^2)\beta(\tau) - \frac{1}{2}\alpha(\tau)]\mathrm{d}\tau - \frac{1}{2}w_0. \tag{67}$$

In Figures 1–4, four spatial structures of the one-soliton solution (66) are shown by selecting the same parameters $a = 1$, $b = 0.1$, $\delta_0 = 6$ and $w_0 = 0.5$, however, with different time-varying coefficients: $\alpha(t) = \sin(t^2)$ and $\beta(t) = 1 + \mathrm{sech}(t)$ in Figure 1; $\alpha(t) = t^2$ and $\beta(t) = 1 + \sin(1 + 0.4t)$ in Figure 2; $\alpha(t) = \tanh(t)$ and $\beta(t) = 1 + \cos(t)$ in Figure 3; and $\alpha(t) = \tanh(t)$ and $\beta(t) = 1$ in Figure 4. Figures 1–4 show that the four bell one-solitons propagating along the negative x-axis have different velocities: variable velocities in Figures 1–3 and uniform velocity in Figure 4. Form Equation (67), we can see that $\beta(t)$ and $\alpha(t)$ determine the frequency of the soliton vibration.

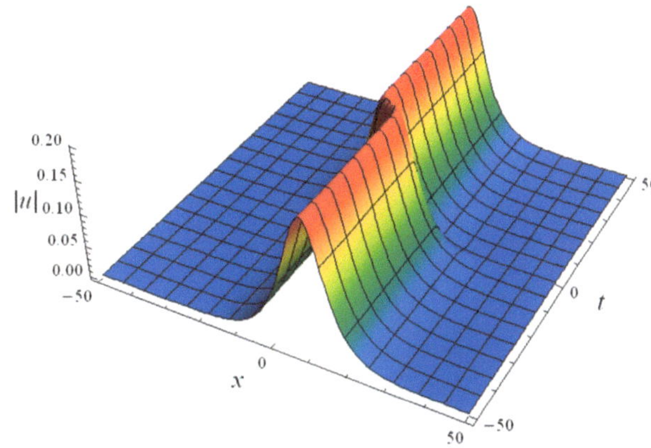

Figure 1. Spatial structure of the one-soliton solution (66) with $\alpha(t) = \sin(t^2)$ and $\beta(t) = 1 + \mathrm{sech}(t)$.

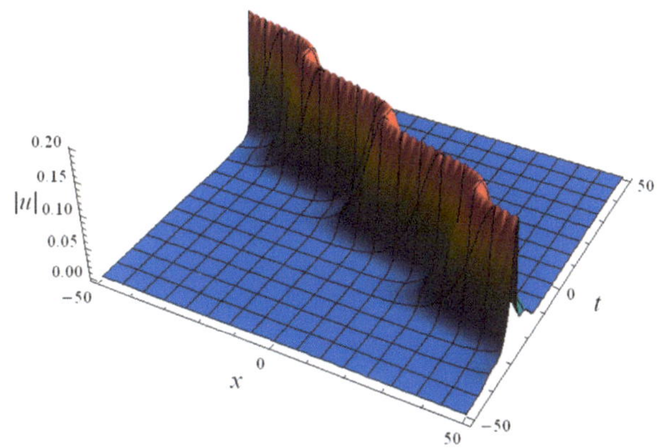

Figure 2. Spatial structure of the one-soliton solution (66) with $\alpha(t) = t^2$ and $\beta(t) = 1 + \sin(1 + 0.4t)$.

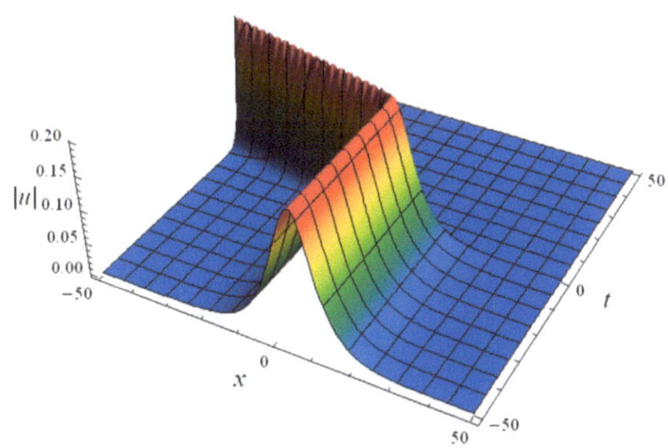

Figure 3. Spatial structure of the one-soliton solution (66) with $\alpha(t) = \tanh(t)$ and $\beta(t) = 1 + \cos(t)$.

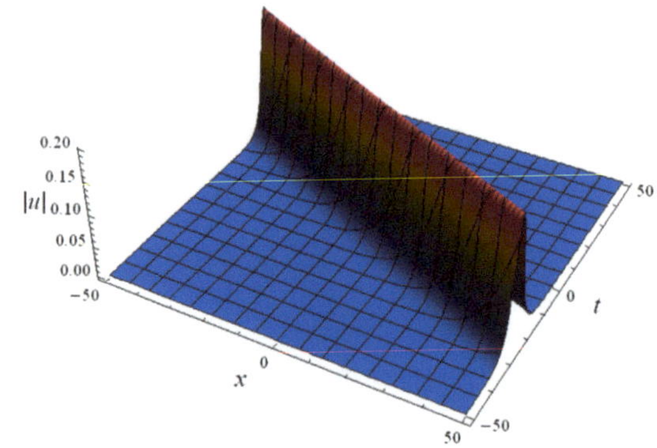

Figure 4. Spatial structure of the one-soliton solution (66) with $\alpha(t) = \tanh(t)$ and $\beta(t) = 1$.

When $N \geq 2$, solution (61) cannot be written as a hyperbolic function like Equation (66). For the selection of $N = 2$, Equation (61) gives

$$u(x,t) = -2\mathrm{i}\frac{c_1 \mathrm{e}^{\theta_1-\theta_2^*} p_{12} + c_2 \mathrm{e}^{\theta_2-\theta_1^*} p_{21} - c_2 \mathrm{e}^{\theta_2-\theta_2^*} p_{11} - c_1 \mathrm{e}^{\theta_1-\theta_1^*} p_{22}}{p_{11}p_{22} - p_{12}p_{21}}, \tag{68}$$

with

$$p_{11} = \frac{c_1^* c_1 \mathrm{e}^{\theta_1^*+\theta_1} + \mathrm{e}^{-\theta_1^*-\theta_1}}{\bar{\xi}_1(0) - \xi_1(0)}, \; p_{12} = \frac{c_1^* c_2 \mathrm{e}^{\theta_1^*+\theta_2} + \mathrm{e}^{-\theta_1^*-\theta_2}}{\bar{\xi}_1(0) - \xi_2(0)}, \tag{69}$$

$$p_{21} = \frac{c_2^* c_1 \mathrm{e}^{\theta_2^*+\theta_1} + \mathrm{e}^{-\theta_2^*-\theta_1}}{\bar{\xi}_2(0) - \xi_1(0)}, \; p_{22} = \frac{c_2^* c_2 \mathrm{e}^{\theta_2^*+\theta_2} + \mathrm{e}^{-\theta_2^*-\theta_2}}{\bar{\xi}_2(0) - \xi_2(0)}, \tag{70}$$

where θ_1 and θ_2 are determined by Equation (65), $\bar{\xi}_1(0) = \xi_1^*(0)$ and $\bar{\xi}_2(0) = \xi_2^*(0)$. In Figures 5–7, a collision between bell two-solitons determined by solution (68) is shown by setting the parameters $c_1 = 1$, $c_2 = 1$, $\xi_1(0) = 0.3 + 0.3\mathrm{i}$, $\xi_2(0) = 0.4 + 0.4\mathrm{i}$, $\alpha(t) = t$ and $\beta(t) = \tanh(0.2t)$. It can be seen from Figures 5–7 that, after interaction, two solitons moving in the opposite directions along the x-axis move away from each other in the original opposite direction. This is different from the interaction between two solitons with the variable coefficient $\alpha(t) = t$ and the constant coefficient $\beta(t) = 1$, which continue to move forward after passing through each other as shown in Figures 8–10.

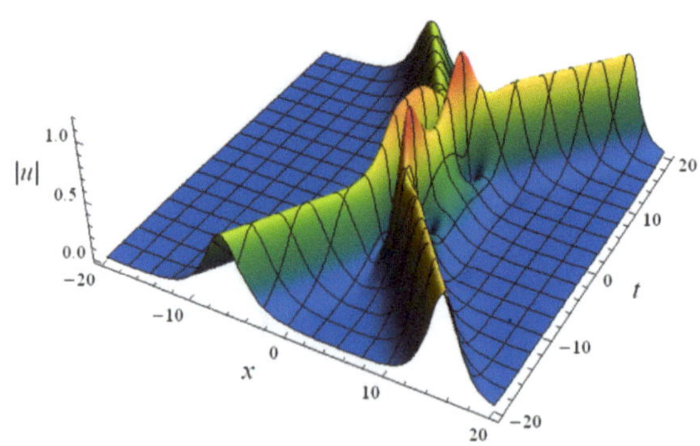

Figure 5. Spatial structure of the two-soliton solution (68) with $\alpha(t) = t$ and $\beta(t) = \tanh(0.3t)$.

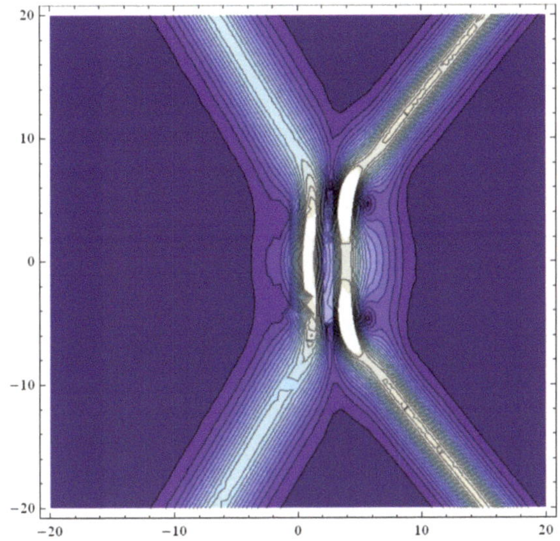

Figure 6. Contour of the two-soliton solution (68) with $\alpha(t) = t$ and $\beta(t) = \tanh(0.3t)$.

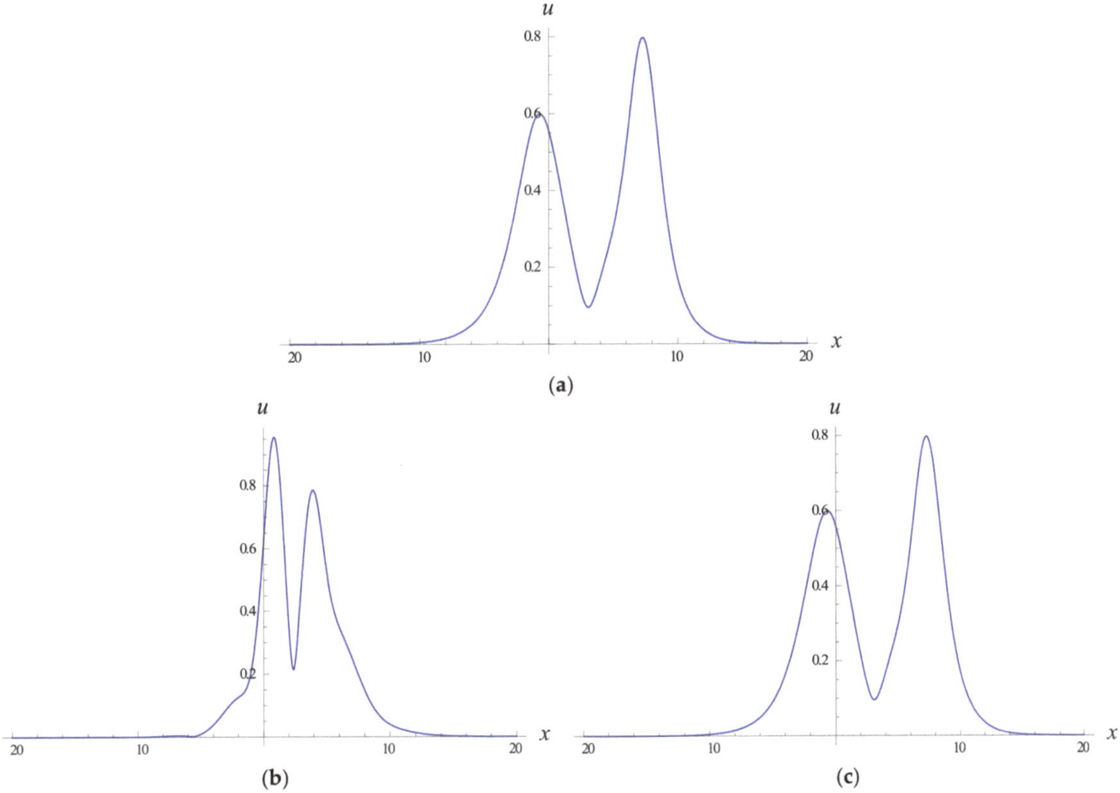

Figure 7. Interaction of the two-soliton solution (68) with $\alpha(t) = t$ and $\beta(t) = \tanh(0.3t)$: (**a**) $t = -10$, (**b**) $t = 0$ and (**c**) $t = 10$.

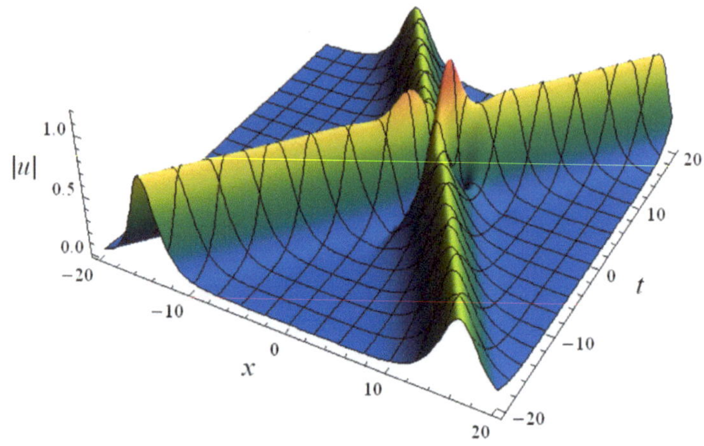

Figure 8. Spatial structure of the two-soliton solution (68) with $\alpha(t) = t$ and $\beta(t) = 1$.

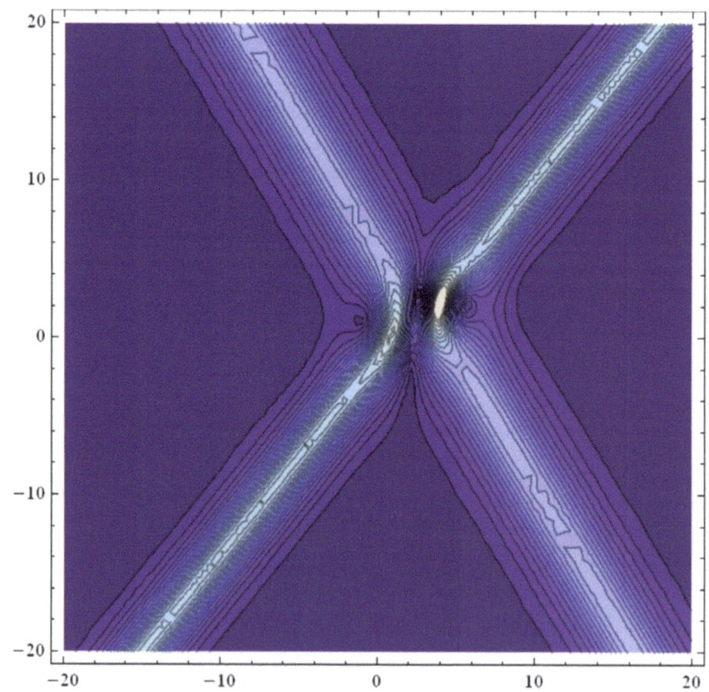

Figure 9. Contour of the two-soliton solution (68) with $\alpha(t) = t$ and $\beta(t) = 1$.

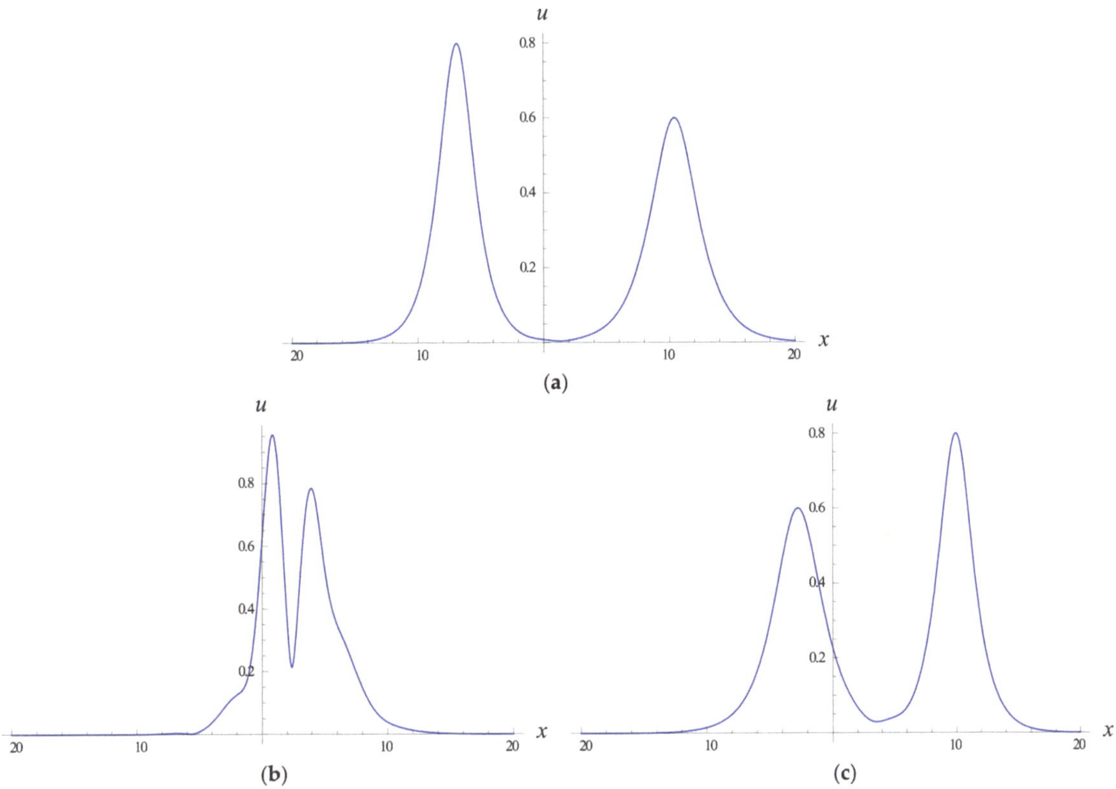

Figure 10. Interaction of the two-soliton solution (68) with $\alpha(t) = t$ and $\beta(t) = 1$: (**a**) $t = -10$, (**b**) $t = 0$ and (**c**) $t = 10$.

5. Conclusions

Taking the gNLS Equation (3) as an example, this paper presented a positive answer to the feasibility of extending the RH method [17] to nonlinear evolution equations with variable coefficients. Due to the derived Lax representation in Equations (4) and (5) and their transformation forms (10) and (11) with unit boundary values at infinity of spatial independent variables, the solution of the gNLS Equation (3) is transformed into the associated RH problem (30) via Equation (29).

Based on the solvability of the RH Problem (25), we determined the time evolution laws (38)–(42) of the corresponding scattering data, recovered the potential function using the RH method [17] and, finally, obtained the solution (56) with the long-time asymptotic behavior and the N-soliton solution (61). It can be seen from Figures 1–4 that four bell one-solitons propagating from the positive x-axis to the negative x-axis possess different velocities, which make their peaks form different motion trajectories, including the kink trajectory in Figure 1, periodic kink trajectory in Figure 2, straight turning trajectory in Figure 3 and straight-line trajectory in Figure 4. This is due to the different selections of the time-varying coefficient function $\beta(t)$.

Whether the propagation trajectory of the bell soliton peak determined by the one-soliton solution (66) shows a straight line or curve depends on the time-varying coefficient $\beta(t)$. For the multiple soliton solution (61) with $N > 1$, there will be similar peak curve trajectory characteristics. In fact, for the one-soliton solution (66), this point can be verified mathematically. Specifically, from Equation (66), we determined the modulus of the one-soliton solution (66):

$$|u| = 2b\mathrm{sech}[2b(x + 2a\int_0^t \beta(\tau)\mathrm{d}\tau - d_0)], \tag{71}$$

which is a bell soliton solution. The peak coordinates (x,t) of the bell soliton determined by Equation (71) satisfy the equation:

$$x + 2a\int_0^t \beta(\tau)\mathrm{d}\tau - d_0 = 0. \tag{72}$$

Clearly, the parameter controlling the peak trajectory of the above bell one-soliton is the propagation velocity $\dot{x} = -2a\beta(t)$. Therefore, selecting $\beta(t) = 1$ as a constant is the reason why the peak trajectory of the bell one-soliton in Figure 4 is a straight line. In addition, it should be pointed out that, when $\alpha(t) = 0$ and $\beta(t) = 2$, the gNLS Equation (3) becomes the classical NLS equation, and the results obtained in this paper can degenerate into the known ones [17]. Recently, some novel solutions [33–35] of NLS-type equations with variable coefficients have been obtained. A comparison shows that both the long-time asymptotic solution (63) and the N-soliton solution (61) are different from those in [8,16,29–35].

Author Contributions: Methodology, B.X.; software, B.X.; writing—original draft preparation, B.X.; writing—review and editing, S.Z. All authors have read and agreed to the published version of the manuscript.

Funding: This research was supported by the Liaoning BaiQianWan Talents Program of China (2019), the Natural Science Foundation of Education Department of Liaoning Province of China (LJ2020002) and the Natural Science Foundation of Xinjiang Autonomous Region of China (2020D01B01).

Institutional Review Board Statement: Not applicable.

Informed Consent Statement: Not applicable.

Data Availability Statement: The data in the manuscript are available from the corresponding author upon request.

Acknowledgments: The authors would like to express their thanks to the anonymous referees for their valuable comments.

Conflicts of Interest: The authors declare no conflict of interest.

References

1. Liu, W.M.; Kengne, E. *Schrödinger Equations in Nonlinear Systems*; Springer Nature Singapore Pte Ltd.: Singapore, 2019.
2. Gardner, C.S.; Greene, J.M.; Kruskal, M.D.; Miura, R.M. Method for solving the Korteweg-de Vries equation. *Phys. Rev. Lett.* **1967**, *19*, 1095–1197. [CrossRef]
3. Matveev, V.B.; Salle, M.A. *Darboux Transformation and Soliton*; Springer: Berlin, Germany, 1991.
4. Hirota, R. Exact envelope-soliton solutions of a nonlinear wave equation. *J. Math. Phys.* **1973**, *14*, 805–809. [CrossRef]
5. Wang, M.L. Exact solutions for a compound KdV-Burgers equation. *Phys. Lett. A* **1996**, *213*, 279–287. [CrossRef]
6. Fan, E.G. Travelling wave solutions in terms of special functions for nonlinear coupled evolution systems. *Phys. Lett. A* **2002**, *300*, 243–249. [CrossRef]
7. He, J.H.; Wu, X.H. Exp-function method for nonlinear wave equations. *Chaos Soliton. Fract.* **2006**, *30*, 700–708. [CrossRef]
8. Zhang, S.; Xia, T.C. A generalized auxiliary equation method and its application to (2 + 1)-dimensional asymmetric Nizhnik-Novikov-Vesselov equations. *Phys. A Math. Theor.* **2007**, *40*, 227–248. [CrossRef]
9. Ma, W.X.; Lee, J.H. A transformed rational function method and exact solutions to 3 + 1 dimensional Jimbo-Miwa equation. *Chaos Soliton. Fract.* **2009**, *42*, 1356–1363. [CrossRef]
10. Tian, S.F. Initial-boundary value problems for the general coupled nonlinear Schrödinger equation on the interval via the Fokas method. *J. Diff. Equ.* **2016**, *262*, 506–588. [CrossRef]
11. Xu, B.; Zhang, S. A novel approach to time-dependent-coefficient WBK system: Doubly periodic waves and singular nonlinear dynamics. *Complexity* **2018**, *2018*, 3158126. [CrossRef]

12. Xu, B.; Zhang, S. Exact solutions with arbitrary functions of the (4 + 1)-dimensional Fokas equation. *Therm. Sci.* **2019**, *23*, 2403–2411. [CrossRef]
13. Xu, B.; Zhang, S. Exact solutions of nonlinear equations in mathematical physics via negative power expansion method. *J. Math. Phys. Anal. Geo.* **2021**, *17*, 369–387. [CrossRef]
14. Zhang, S.; Xu, B. Painlevé test and exact solutions for (1 + 1)-dimensional generalized Broer-Kaup equations. *Mathematics* **2022**, *10*, 486. [CrossRef]
15. Serkin, V.N.; Hasegawa, A.; Belyaeva, T.L. Nonautonomous solitons in external potentials. *Phys. Rev. Lett.* **2007**, *98*, 074102. [CrossRef]
16. Kruglov, V.I.; Peacock, A.C.; Harvey, J.D. Exact self-similar solutions of the generalized nonlinear Schrödinger equation with distributed coefficients. *Phys. Rev. Lett.* **2003**, *90*, 113902. [CrossRef]
17. Yang, J.K. *Nonlinear Waves in Integrable and Nonintegrable Systems*; SIAM: Philadelphia, PA, USA, 2010.
18. Deift, P.; Zhou, X. A steepest descent method for oscillatory Riemann-Hilbert problems. *Ann. Math.* **1993**, *137*, 295–368. [CrossRef]
19. Xu, J.; Fan, E.G.; Chen, Y. Long-time asymptotic for the derivative nonlinear Schrödinger equation with step-like initial value. *Math. Phys. Anal. Geom.* **2013**, *16*, 253–288. [CrossRef]
20. Ma, W.X. Riemann-Hilbert problems and *N*-soliton solutions for a coupled mKdV system. *J. Geom. Phys.* **2018**, *132*, 45–54. [CrossRef]
21. Hu, B.B.; Xia, T.C.; Ma, W.X. Riemann-Hilbert approach for an initial-boundary value problem of the two-component modified Korteweg-de Vries equation on the half-line. *Appl. Math. Comput.* **2018**, *332*, 148–159. [CrossRef]
22. Wang, D.S.; Guo, B.; Wang, X.L. Long-time asymptotics of the focusing Kundu-Eckhaus equation with nonzero boundary conditions. *J. Differ. Equ.* **2019**, *266*, 5209–5253. [CrossRef]
23. Chen, S.Y.; Yan, Z.Y.; Guo, B.L. Long-time asymptotics for the focusing Hirota equation with non-zero boundary conditions at infinity via the Deift-Zhou approach. *Math. Phys. Anal. Geom.* **2021**, *24*, 17. [CrossRef]
24. Guo, H.D.; Xia, T.C. Multi-soliton solutions for a higher-order coupled nonlinear Schrödinger system in an optical fiber via Riemann-Hilbert approach. *Nonlinear Dyn.* **2021**, *103*, 1805–1816. [CrossRef]
25. Li, Z.Q.; Tian, S.F.; Zhang, T.T.; Yang, J.J. Riemann-Hilbert approach and multi-soliton solutions of a variable-coefficient fifth-order nonlinear Schrödinger equation with N distinct arbitrary-order poles. *Mod. Phys. Lett. B* **2021**, *35*, 2150194. [CrossRef]
26. Zhang, B.; Fan, E.G. Riemann-Hilbert approach for a Schrödinger-type equation with nonzero boundary conditions. *Mod. Phys. Lett. B* **2021**, *35*, 2150208. [CrossRef]
27. Wei, H.Y.; Fan, E.G.; Guo, H.D. Riemann-Hilbert approach and nonlinear dynamics of the coupled higher-order nonlinear Schrödinger equation in the birefringent or two-mode fiber. *Nonlinear Dyn.* **2021**, *104*, 649–660. [CrossRef]
28. Xu, B.; Zhang, S. Riemann-Hilbert approach for constructing analytical solutions and conservation laws of a local time-fractional nonlinear Schrödinger equation. *Symmetry* **2021**, *13*, 13091593. [CrossRef]
29. Serkin, V.N.; Hasegawa, A. Novel soliton solutions of the nonlinear Schrödinger equation model. *Phys. Rev. Lett.* **2000**, *85*, 4502–4505. [CrossRef]
30. Kruglov, V.I.; Harvey, J.D. Asymptotically exact parabolic solutions of the generalized nonlinear Schrödinger equation with varying parameters. *J. Opt. Soc. Am. B* **2006**, *23*, 2541–2550. [CrossRef]
31. Serkin, V.N.; Belyaeva, T.L. Optimal control of dark solitons. *Optik* **2018**, *168*, 827–838. [CrossRef]
32. Zhang, S.; Zhang, L.J.; Xu, B. Rational waves and complex dynamics: Analytical insights into a generalized nonlinear Schrödinger equation with distributed coefficients. *Complexity* **2019**, *2019*, 3206503. [CrossRef]
33. Plemelj, J. Riemannsche Funktionenscharen mit gegebener Monodromiegruppe. *Monatsch. Math. Phys.* **1908**, *19*, 211–246. [CrossRef]
34. Zakharov, V.E.; Shabat, A.B. Integration of nonlinear equations of mathematical physics by the method of inverse scattering. II. *Funct. Anal. Appl.* **1979**, *13*, 166–174. [CrossRef]
35. Mai, X.L.; Li, W.; Dong, S.H. Exact solution to the nonlinear Schrödinger equation with time-dependent coefficients. *Adv. High Energy Phys.* **2021**, *2021*, 6694980. [CrossRef]
36. Guo, Q.; Liu, J. New exact solution to the nonlinear Schrödinger equation with variable coefficients. *Results Phys.* **2020**, *16*, 102857. [CrossRef]
37. Kosti, A.A.; Colreavy-Donnelly, S.; Caraffini, F.; Anastassi, Z.A. Efficient computation of the nonlinear Schrödinger equation with time-dependent coefficients. *Mathematics* **2020**, *8*, 374. [CrossRef]

mathematics

Article

Painlevé Test and Exact Solutions for (1 + 1)-Dimensional Generalized Broer–Kaup Equations

Sheng Zhang [1,*] and Bo Xu [2,3,*]

1 School of Mathematical Sciences, Bohai University, Jinzhou 121013, China
2 School of Mathematics, China University of Mining and Technology, Xuzhou 221116, China
3 School of Educational Sciences, Bohai University, Jinzhou 121013, China
* Correspondence: szhangchina@126.com or szhang@bhu.edu.cn (S.Z.); bxu@bhu.edu.cn (B.X.)

Abstract: In this paper, the Painlevé integrable property of the (1 + 1)-dimensional generalized Broer–Kaup (gBK) equations is first proven. Then, the Bäcklund transformations for the gBK equations are derived by using the Painlevé truncation. Based on a special case of the derived Bäcklund transformations, the gBK equations are linearized into the heat conduction equation. Inspired by the derived Bäcklund transformations, the gBK equations are reduced into the Burgers equation. Starting from the linear heat conduction equation, two forms of N-soliton solutions and rational solutions with a singularity condition of the gBK equations are constructed. In addition, the rational solutions with two singularity conditions of the gBK equation are obtained by considering the non-uniqueness and generality of a resonance function embedded into the Painlevé test. In order to understand the nonlinear dynamic evolution dominated by the gBK equations, some of the obtained exact solutions, including one-soliton solutions, two-soliton solutions, three-soliton solutions, and two pairs of rational solutions, are shown by three-dimensional images. This paper shows that when the Painlevé test deals with the coupled nonlinear equations, the highest negative power of the coupled variables should be comprehensively considered in the leading term analysis rather than the formal balance between the highest-order derivative term and the highest-order nonlinear term.

Keywords: Painlevé integrable property; Painlevé test; leading term analysis; (1 + 1)-dimensional gBK equations; Bäcklund transformations; exact solutions

Citation: Zhang, S.; Xu, B. Painlevé Test and Exact Solutions for (1 + 1)-Dimensional Generalized Broer–Kaup Equations. *Mathematics* **2022**, *10*, 486. https://doi.org/10.3390/math10030486

Academic Editors: Almudena del Pilar Marquez Lozano and Vladimir Iosifovich Semenov

Received: 31 December 2021
Accepted: 30 January 2022
Published: 2 February 2022

1. Introduction

Painlevé analysis is an important method for testing Painlevé integrable property [1–7] of nonlinear partial differential equations (PDEs). If a given nonlinear PDE passes through the Painlevé test, then we say it has Painlevé property [1]. More specifically, Painlevé property or Painlevé integrability for nonlinear PDEs means that the solutions of the given PDE must be "single-valued" in the neighborhood of a movable singularity manifold (non-characteristic). The so-called WTC method of Painlevé analysis proposed by Weiss, Tabor, and Carnevale [2] is an effective approach for Painlevé test of nonlinear PDEs. As pointed out in [4], the celebrated BK equations can be used to describe the propagation with double directions of long waves located in shallow water. In 2013, Zhang, Han, and Tam [8] derived the following (1 + 1)-dimensional gBK equations:

$$v_t = v_{xx} - 2vv_x - 4w_x \tag{1}$$

$$w_t = -w_{xx} - 2(wv)_x - 2v_x \tag{2}$$

In soliton theory, besides the Painlevé analysis method [1–3], there are many alternative methods [9–17] for solving nonlinear PDEs. Generally, each of these methods has its advantages and disadvantages. Both the inverse scattering method [9] and the Darboux transformation [11] depends on the Lax pair of the solved equations; however, constructing the Lax pair sometimes is more difficult than solving the equation itself or even impossible. One of the key steps of the Hirota's bilinear method [10] is finding an effective dependent variable transformation, which is often inseparable from attempts or known solutions. Most of other methods, such as [12–16], are relatively direct, but the hypothetical forms of ansatz solution limit the type of the constructing solutions. The Painlevé analysis method [1–3] can give the relative unified form of the solution of the equation to the greatest extent and can construct various formal solutions from this unified form as needed. It is worth mentioning that one of the advantages of the Painlevé analysis method [1–3] is to provide a useful tool for the reduction or linearization of nonlinear PDEs. The Lax integrability and multiple soliton solutions of Equations (1) and (2) were obtained in [8,18,19]. As far as we know, the Painlevé test of Equations (1) and (2) has not been reported. In this paper, we extend the WTC method [2] to prove the Painlevé property of Equations (1) and (2). At the same time, the Bäcklund transformations, two reductions, and some exact solutions of Equations (1) and (2) have been obtained by using the Painlevé truncation technique. This is due to our consideration of balancing $v_{xx} - 4w_x$ and $2vv_x$ rather than the highest-order derivative term v_{xx} and the highest-order nonlinear term $2vv_x$ in form for Equation (1) in the process of using the Painlevé test to deal with the leading term analysis.

2. Painlevé Test and Painlevé Integrability

Employing the WTC method [2] of Painlevé analysis, we suppose that

$$v = \phi^{-\alpha}\sum_{j=0}^{\infty} v_j\phi^j,\ v_0 \neq 0, \tag{3}$$

$$w = \phi^{-\beta}\sum_{j=0}^{\infty} w_j\phi^j,\ w_0 \neq 0, \tag{4}$$

where ϕ, v_j, and w_j are functions of x and t, α, and β are non-negative integers. Considering the leading term analysis, we take

$$v \sim v_0\phi^{-\alpha} \tag{5}$$

$$w \sim w_0\phi^{-\beta} \tag{6}$$

and therefore have

$$v_x \sim -\alpha v_0\phi^{-\alpha-1}\phi_x,\ v_{xx} \sim \alpha(\alpha+1)v_0\phi^{-\alpha-2}\phi_x^2, \tag{7}$$

$$w_x \sim -\beta w_0\phi^{-\beta-1}\phi_x,\ w_{xx} \sim \beta(\beta+1)w_0\phi^{-\beta-2}\phi_x^2. \tag{8}$$

Using Equations (5)–(8) to balance the following terms of Equations (1) and (2),

$$v_{xx} - 2vv_x - 4w_x = 0, \tag{9}$$

$$-w_{xx} - 2(wv)_x = 0, \tag{10}$$

We arrive at

$$\alpha = 1,\ v_0 = \phi_x, \tag{11}$$

$$\beta = 2,\ w_0 = -\frac{1}{2}\phi_x^2. \tag{12}$$

Thus, Equations (3) and (4) can be rewritten as

$$v = \sum_{j=0}^{\infty} v_j\phi^{j-1}, \tag{13}$$

$$w = \sum_{j=0}^{\infty} w_j\phi^{j-2}. \tag{14}$$

Substituting Equations (13) and (14) into Equations (1) and (2), collecting the same power coefficients of ϕ, and then setting all the coefficients as zeros, we have

$$\phi^{-3}: 2\phi_x(v_0^2 + 4w_0 + v_0\phi_x) = 0, \tag{15}$$

$$\phi^{-2}: 4w_1\phi_x - 2\phi_x v_{0,x} - 4w_{0,x} + v_0(\phi_t + 2v_1\phi_x - 2v_{0,x} - \phi_{xx}) = 0, \tag{16}$$

$$\phi^{j-1}: v_{j+2}[j(j+1)\phi_x^2 - 2jv_0\phi_x] - 4jw_{j+2}\phi_x = F_{j+2,1},\ (j = 0, 1, 2, \cdots), \tag{17}$$

$$\phi^{-4}: 6w_0\phi_x(v_0 - \phi_x) = 0, \tag{18}$$

$$\phi^{-3}: 2[v_0(2w_1\phi_x - w_{0,x}) + \phi_x(-w_1\phi_x + 2w_{0,x}) + w_0(\phi_t + 2v_1\phi_x - v_{0,x} + \phi_{xx})] = 0, \tag{19}$$

$$\phi^{j-2}: v_{j+2}[-2(j-1)w_0\phi_x] - 2(j-1)w_{j+2}(j\phi_x^2 + v_0\phi_x) = F_{j+2,2},\ (j = 0, 1, 2, \cdots), \tag{20}$$

where $F_{j+2,1}$ and $F_{j+2,2}$ are functions of v_0, w_0, v_1, w_1, v_2, w_2, $\cdots$, v_{j+1}, w_{j+1}, ϕ and their partial derivatives with respect to x and t. It is easy to see that Equations (15) and (18) give the same v_0 and w_0 as Equations (11) and (12). From Equation (16), one has

$$v_1 = -\frac{\phi_t + \phi_{xx}}{2\phi_x},\ w_1 = \frac{1}{2}\phi_{xx}. \tag{21}$$

In view of Equations (11), (12), (17), and (20), we drive the determinants of the coefficients of v_j and $w_j (j = 2, 3, 4, \cdots)$

$$\begin{vmatrix} (j-2)(j-3)\phi_x^2 & -4(j-2)\phi_x \\ (j-3)\phi_x^3 & -2(j-3)(j-1)\phi_x^2 \end{vmatrix} = -2(j-2)(j-3)(j^2 - 4j + 1)\phi_x^4. \tag{22}$$

Equation (22) hints that $j = 2, 3$ are the resonance points. Accordingly, v_2 and w_3 or v_3 and w_2 may be the corresponding resonance functions. Fortunately, when we select v_2 and w_3 as the resonance functions and set $v_2 = 0$ and $w_3 = 0$, Equation (20) gives

$$w_2 = \frac{\phi_{xx}(\phi_t + \phi_{xx}) - \phi_x(4\phi_x + \phi_{xt} + \phi_{xxx})}{4\phi_x^2}. \tag{23}$$

Further set $v_j = 0 (j \geq 3)$ and $w_j = 0 (j \geq 4)$; then, Equations (3) and (4) can be truncated. This shows that the (1 + 1)-dimensional gBK Equations (1) and (2) pass the Painlevé test and hence possess the Painlevé integrability.

3. Bäcklund Transformations and Two Reductions

For the (1 + 1)-dimensional gBK Equations (1) and (2), the following Bäcklund transformations hold:

Theorem 1. *The (1 + 1)-dimensional gBK Equations (1) and (2) have the Bäcklund transformations:*

$$v = \frac{\phi_x}{\phi} + v_1 \tag{24}$$

$$w = -\frac{\phi_x^2}{2\phi^2} + \frac{\phi_{xx}}{2\phi} + \frac{1}{2}v_{1,x} - 1, \tag{25}$$

where ϕ and v_1 satisfy the following equations:

$$\phi_t + 2v_1\phi_x + \phi_{xx} = 0, \tag{26}$$

$$v_{1,t} + 2v_1 v_{1,x} + v_{1,xx} = 0. \tag{27}$$

Proof of Theorem 1. Setting $v_j = 0(j \geq 2)$ and $w_j = 0(j \geq 3)$, we can truncate Equations (3) and (4) as

$$v = v_0\phi^{-1} + v_1 \tag{28}$$

$$w = w_0\phi^{-2} + w_1\phi^{-1} + w_2 \tag{29}$$

Substituting Equations (28) and (29), together with v_0, w_0, and w_1 in Equations (11), (12), and (21) into Equations (1) and (2), then collecting the same power coefficients of ϕ and setting all the coefficients as zeros, we can arrive at Equation (26). At the same time, with the help of Equation (26), we have

$$v_{1,t} + 2v_1v_{1,x} + 4w_{2,x} - v_{1,xx} = 0 \tag{30}$$

$$w_2 = \frac{1}{2}v_{1,x} - 1. \tag{31}$$

Inserting Equation (31) into Equation (30), we reach Equation (27). Using Equations (11), (12), (21), and (31), we finally convert Equations (28) and (29) into Equations (24) and (25). □

Corollary 1. *Under the transformations:*

$$v = \frac{\phi_x}{\phi} \tag{32}$$

$$w = -\frac{\phi_x^2}{2\phi^2} + \frac{\phi_{xx}}{2\phi} - 1, \tag{33}$$

the (1 + 1)-dimensional gBK Equations (1) and (2) can be reduced to the linear heat conduction equation:

$$\phi_t + \phi_{xx} = 0 \tag{34}$$

Proof of Corollary 1. Obviously, when $v_1 = 0$, Equations (26) and (27) degenerate into Equation (34). Meanwhile, Equations (24) and (25) become Equations (32) and (33). □

Inspired by Equation (27) derived from the substitution of Equations (31) into Equation (30), we get the following Theorem 2.

Theorem 2. *Suppose that*

$$w = \frac{1}{2}v_x - 1 \tag{35}$$

the (1 + 1)-dimensional gBK Equations (1) and (2) can be reduced to the Burgers equation:

$$v_t + 2vv_x + v_{xx} = 0 \tag{36}$$

Proof of Theorem 2. On the one hand, we can reduce Equation (1) into Equation (36) by using Equation (35). On the other hand, the substitution of Equation (35) into Equation (2) gives

$$(v_t + 2vv_x + v_{xx})_x = 0 \tag{37}$$

Since when Equation (36) is true, Equation (37) naturally holds, we then conclude that Equation (35) can transform Equations (1) and (2) into Equation (36). □

4. Soliton Solutions and Rational Solutions

To construct exact solutions of the (1 + 1)-dimensional gBK Equations (1) and (2), we consider Equations (32)–(34) and suppose that

$$\phi = 1 + e^{k_1(x+c_1t+d_1)} \tag{38}$$

where k_1, c_1, and d_1 are constants. Then, Equation (34) dictates

$$c_1 = -k_1 \tag{39}$$

Thus, we obtain one-soliton solutions of Equations (1) and (2):

$$v = \frac{k_1 e^{k_1(x-k_1t+d_1)}}{1+e^{k_1(x-k_1t+d_1)}} \tag{40}$$

$$w = -\frac{k_1^2 e^{2k_1(x-k_1t+d_1)}}{2(1+e^{k_1(x-k_1t+d_1)})^2} + \frac{k_1^2 e^{k_1(x-k_1t+d_1)}}{2(1+e^{k_1(x-k_1t+d_1)})} - 1. \tag{41}$$

Considering the linearity of Equation (34), we know that

$$\phi = 1 + \sum_{i=1}^{N} e^{k_i(x-k_it+d_i)},\ c_i = -k_i (i = 1, 2, \cdots, N) \tag{42}$$

still solves Equation (34). Thus, we obtain N-soliton solutions of Equations (1) and (2):

$$v = \frac{\sum\limits_{i=1}^{N} k_i e^{k_i(x-k_it+d_i)}}{1+\sum\limits_{i=1}^{N} e^{k_i(x-k_it+d_i)}} = \left[\ln\left(1+\sum_{i=1}^{N} e^{k_i(x-k_it+d_i)}\right)\right]_x, \tag{43}$$

$$w = -\frac{\left(\sum\limits_{i=1}^{N} k_i e^{k_i(x-k_it+d_i)}\right)^2}{2\left(1+\sum\limits_{i=1}^{N} e^{k_i(x-k_it+d_i)}\right)^2} + \frac{\sum\limits_{i=1}^{N} k_i^2 e^{k_i(x-k_it+d_i)}}{2\left(1+\sum\limits_{i=1}^{N} e^{k_i(x-k_it+d_i)}\right)} - 1 = \frac{1}{2}\left[\ln\left(1+\sum_{i=1}^{N} e^{k_i(x-k_it+d_i)}\right)\right]_{xx} - 1 \tag{44}$$

We note here that the more general N-soliton solutions,

$$v = \left\{\ln\left[\left(1+\sum_{i=1}^{N} e^{k_ix+k_i(2A-k_i)t+\xi_i^0}\right)/e^{A(x+At)+\eta^0}\right]\right\}_x, \tag{45}$$

$$w = \frac{1}{2}\left\{\ln\left[\left(1+\sum_{i=1}^{N} e^{k_ix+k_i(2A-k_i)t+\xi_i^0}\right)e^{A(x+At)+\eta^0}\right]\right\}_{xx} - 1, \tag{46}$$

which include Equations (43) and (44) as special cases, have been obtained in our previous work [19]. In order to construct other formal N-soliton solutions of Equations (1) and (2), we assume that

$$\phi = \left(1+\sum_{i=1}^{N} e^{k_i(x+c_it+d_i)}\right)e^{p(x+qt+r)}, \tag{47}$$

where k_i, c_i, d_i, p, q, and r are all constants. Then, the substitution of Equation (47) into Equation (34) determines the relations:

$$c_i = -k_i - 2p,\ q = -p. \tag{48}$$

We therefore gain the formal N-soliton solutions of Equations (1) and (2):

$$v = \ln\left[\left(1+\sum_{i=1}^{N} e^{k_i[x-(k_i-2p)t+d_i]}\right)e^{p(x-pt+r)}\right]_x, \tag{49}$$

$$w = \frac{1}{2}\left\{\ln\left[\left(1+\sum_{i=1}^{N} e^{k_i[x-(k_i-2p)t+d_i]}\right)e^{p(x-pt+r)}\right]\right\}_{xx} - 1. \tag{50}$$

Here, it should be noted when Equation (50) is equivalent to Equation (46), then Equation (49) is different from Equation (45). Besides, we easily see that Equations (43) and (44) are special cases of Equations (49) and (50). The equivalent forms of Equations (49) and (50) are also helpful for the comparison, which reads

$$v = \frac{\sum_{i=1}^{N} k_i e^{k_i[x-(k_i-2p)t+d_i]}}{1+\sum_{i=1}^{N} e^{k_i[x-(k_i-2p)t+d_i]}} + p \tag{51}$$

$$w = -\frac{\left[\sum_{i=1}^{N} k_i e^{k_i[x-(k_i-2p)t+d_i]}\right]^2}{2\left[1+\sum_{i=1}^{N} e^{k_i[x-(k_i-2p)t+d_i]}\right]^2} + \frac{\sum_{i=1}^{N} k_i^2 e^{k_i[x-(k_i-2p)t+d_i]}}{2\left[1+\sum_{i=1}^{N} e^{k_i[x-(k_i-2p)t+d_i]}\right]} - 1. \tag{52}$$

In Figure 1, the one-soliton solutions determined by Equations (51) and (52) are shown by setting $k_1 = -1$, $d_1 = 0$, and $p = -0.6$. It can be seen from Figure 1 that the one-soliton solution determined by Equation (51) possesses kink structure, and the one-soliton solution determined by Equation (52) has bell structure.

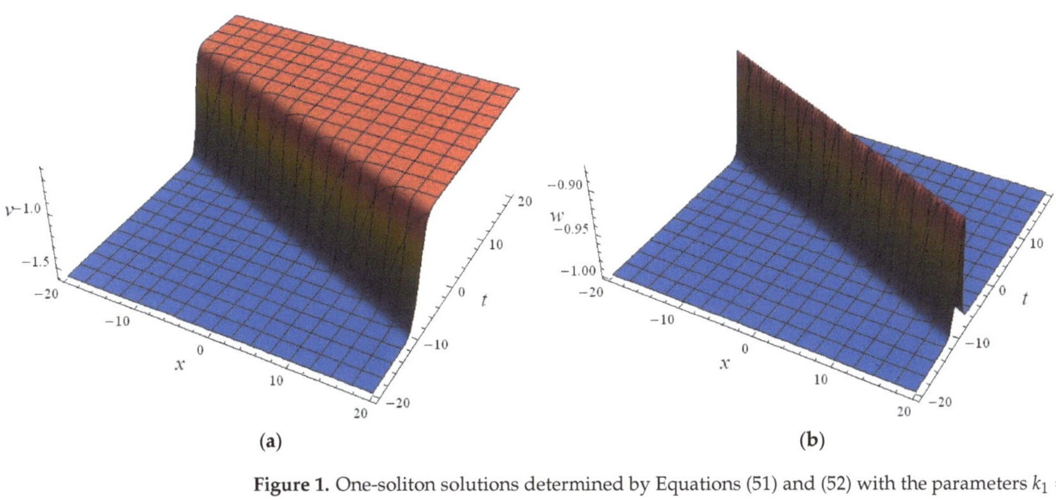

Figure 1. One-soliton solutions determined by Equations (51) and (52) with the parameters $k_1 = -1$, $d_1 = 0$, and $p = -0.6$: (**a**) One kink-soliton solution; (**b**) one bell-soliton solution.

The head-on two kink-soliton solution and the head-on two bell-soliton solution determined by Equations (51) and (52) are shown in Figure 2 by setting $k_1 = 1$, $k_2 = -1.1$, $d_1 = 0$, $d_2 = 0$, and $p = -0.6$.

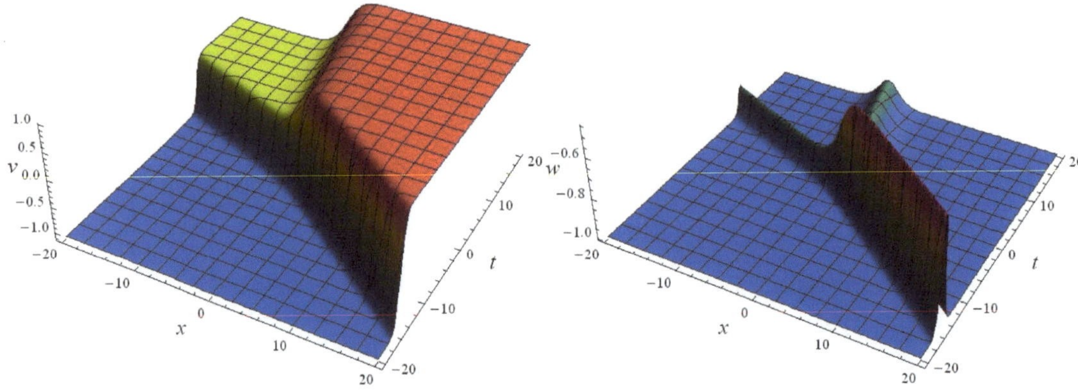

(a) (b)

Figure 2. Two-soliton solutions determined by Equations (51) and (52) with the parameters $k_1 = 1$, $k_2 = -1.1$, $d_1 = 0$, $d_2 = 0$, and $p = -0.6$: (**a**) Two kink-soliton solution; (**b**) two bell-soliton solution.

The interaction of three kink-soliton solution determined by Equation (51) and the interaction of three bell-soliton solution determined by Equation (52) are shown in Figure 3; there, the parameters are selected as $k_1 = -1$, $k_2 = -1.5$, $k_3 = 0.8$, $d_1 = 0$, $d_2 = 0$, $d_3 = 0$, and $p = -0.6$.

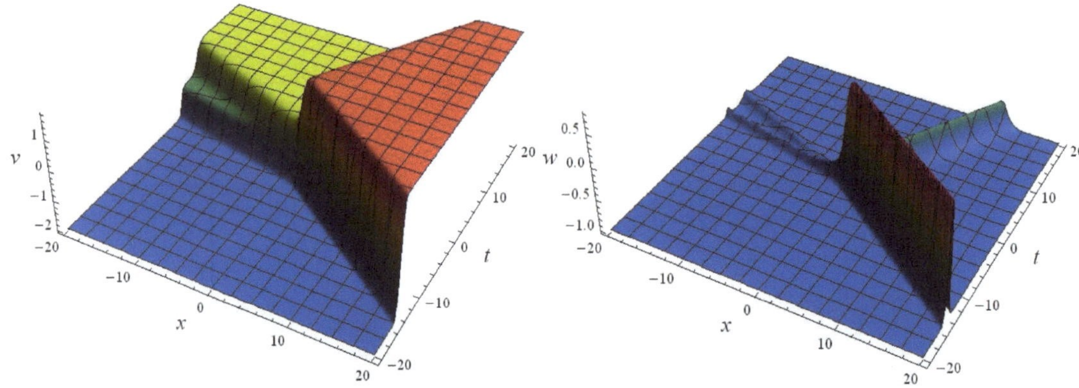

(a) (b)

Figure 3. Three-soliton solutions determined by Equations (51) and (52) with the parameters $k_1 = -1$, $k_2 = -2$, $k_3 = 0.8$, $d_1 = 0$, $d_2 = 0$, $d_3 = 0$, and $p = -0.6$: (**a**) Kink three-soliton solution; (**b**) bell three-soliton solution.

In addition to the soliton solutions obtained above, some other types of exact solutions of Equations (1) and (2) can also be obtained. For example, if we choose Equation (26) in the form

$$\phi = k(x - ct) + d \tag{53}$$

where k, c, and d are all constants, then one has

$$v_0 = k,\ v_1 = \frac{1}{2}c,\ w_0 = -\frac{1}{2}k^2,\ w_1 = 0,\ w_2 = -1. \tag{54}$$

The rational solutions of Equations (1) and (2) are then obtained as follows:

$$v = \frac{k}{k(x - ct) + d} + \frac{1}{2}c, \tag{55}$$

$$w = -\frac{k^2}{2[k(x-ct)+d]^2} - 1. \tag{56}$$

In Figure 4, the rational solutions (55) and (56) are shown by selecting $c = 1, k = -2$, and $d = 3$. A direct computation shows that the singularities in Figure 4 occur at each point on the straight line $2(x-t) - 3 = 0$.

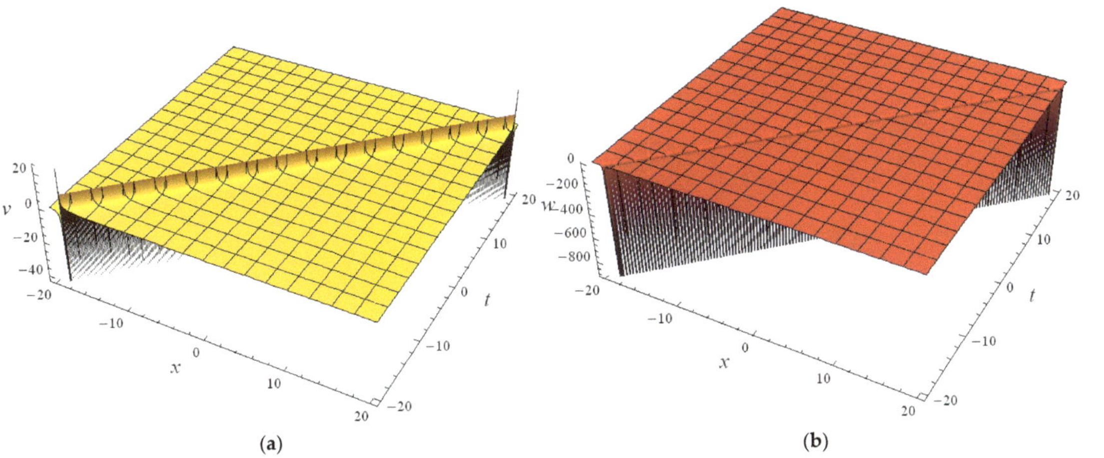

Figure 4. Rational solutions (55) and (56) with the parameters $c = 1, k = -2$, and $d = 3$: (**a**) Rational solution (55); (**b**) rational solution (56).

The above results obtained benefit from the correct selection of the resonance coefficient functions for v_2 and w_3 in Equations (17) and (20). In fact, if we keep the generality of v_2, the equation $v_2 v_{2,x} = 0$ will appear in the operation of the above Painlevé test. When $v_2 = 0$ is selected, then the rational solutions (55) and (56) can be obtained by employing Equation (53). To avoid repetition, we omit them here. However, when we select $v_{2,x} = 0$ together with Equation (53), the similar operations give

$$v_2 = \frac{1}{2kt - s} \tag{57}$$

where s is an arbitrary constant and hence produces the general rational solutions of Equations (1) and (2):

$$v = \frac{k}{k(x-ct)+d} + \frac{k(x-ct)+d}{2kt-s} + \frac{1}{2}c \tag{58}$$

$$w = -\frac{k^2}{2[k(x-ct)+d]^2} + \frac{k}{2(2kt-s)} - 1 \tag{59}$$

which causes singularities to occur in two straight lines, $k(x-ct)+d = 0$ and $2kt - s = 0$, in the case of $k \neq 0$. See Figure 5 for the rational solutions (58) and (59) with the parameters $c = 1, k = -2, d = 3$, and $s = -2$.

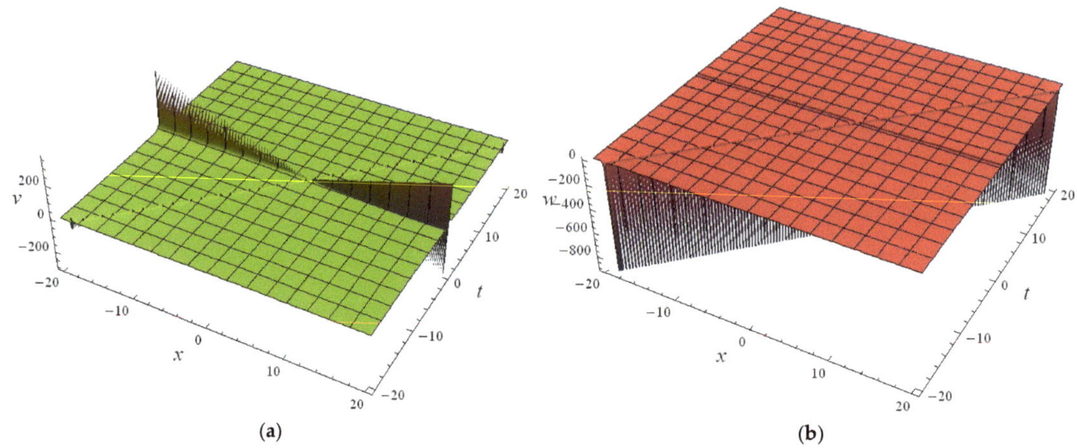

Figure 5. Rational solutions (58) and (59) with the parameters $c = 1$, $k = -2$, $d = 3$, and $s = -2$: (**a**) Rational solution (58); (**b**) rational solution (59).

5. Conclusions

We have proved the Painlevé integrable property of the (1 + 1)-dimensional gBK Equations (1) and (2). This is due to the key step adopted in this paper to balance Equations (9) and (10) rather than the equation

$$v_{xx} - 2vv_x = 0 \tag{60}$$

and Equation (10) in the process of using the Painlevé test to deal with the leading term analysis. If the leading terms of Equations (60) and (10) are balanced, then $\alpha = 1$ and $v_0 = -\phi_x$ are derived from Equation (60) by similar computations using Equations (5) and (6). Substituting $\alpha = 1$ and $v_0 = -\phi_x$ into Equation (10) and balancing the leading terms yields $\beta = -1$ or $\beta = -2$. This contradicts the prior assumption that β is a nonnegative integer. Although the derivative of v_{xx} is one order higher than that of w_x in form, the highest negative power of ϕ is balanced, and $v_{xx} \sim 2v_0\phi_x^2\phi^{-3}$ and $w_x \sim -2w_0\phi_x\phi^{-3}$ are the same power as discussed in this paper. For other coupled nonlinear PDEs, this should be noted in the leading term analysis. Of course, this noteworthy point will not appear in a single model. We think this point is very important and will directly affect whether the Painlevé test can be passed if the equation under consideration has Painlevé property. To the best of our knowledge, it has not been reported in the literature.

Based on the Painlevé truncation, the Bäcklund transformations (24)–(27) and exact solutions of Equations (1) and (2) have been obtained, including the N-soliton solutions (43) and (44), (49) and (50), rational solutions (55) and (56), and (58) and (59). These obtained soliton solutions and rational solutions may provide useful information for explaining some relevant nonlinear physical phenomena. This shows the importance of the Bäcklund transformations (24)–(27) in constructing exact solutions to a great extent. Using the Bäcklund transformations (24)–(27) to construct other types of exact solutions of Equations (1) and (2) is worthy of study. Besides, the gBK Equations (1) and (2) are reduced into two simple forms by the aid of the benefits of the Bäcklund transformations (24)–(27). One reduced form gives the linear heat conduction Equation (34), and the other ones arrive at the Burgers Equation (36). Compared with the bilinear forms (2.8a) and (2.8b) [19], which are nonlinear, of Equations (1) and (2), the linear heat conduction Equation (34) in this paper is much simpler. Based on the bilinear forms (2.8a) and (2.8b) [19], the perturbation truncation technique of the Hirtoa's bilinear method [10] can obtain the truncated expansion with any finite terms consisting of the solutions to construct. However, the Painlevé truncation, as did in this paper for Equations (1) and (2), will generally stop at the resonance point, and the number of expansion terms of the solution is small. It is difficult for the Painlevé analysis method [1–3] to construct the formal solutions of Equations (1) and (2) with any number of expansion terms, and the convergence of the infinite term series expansion solution (3) or (4) is far from being solved.

Author Contributions: Methodology, B.X.; software, S.Z.; writing—original draft preparation, B.X.; writing—review and editing, S.Z. All authors have read and agreed to the published version of the manuscript.

Funding: This research was supported by the Liaoning BaiQianWan Talents Program of China (2019), the Natural Science Foundation of Education Department of Liaoning Province of China (LJ2020002) and the Natural Science Foundation of Xinjiang Autonomous Region of China (2020D01B01).

Institutional Review Board Statement: Not applicable.

Informed Consent Statement: Not applicable.

Data Availability Statement: The data in the manuscript are available from the corresponding author upon request.

Acknowledgments: The authors express their thanks to the three anonymous referees for the valuable and helpful comments.

Conflicts of Interest: The authors declare no conflict of interest.

References

1. Ablowitz, M.J.; Clarkson, P.A. *Solitons, Nonlinear Evolution Equations and Inverse Scattering*; Cambridge University Press: New York, NY, USA, 1991.
2. Weiss, J.; Tabor, M.; Carnvale, C. The Painlevé property for partial differential equations. *J. Math. Phys.* **1983**, *24*, 522–526. [CrossRef]
3. Kruskal, M.D.; Joshi, N.; Halburd, R. Analytic and asymptotic methods for nonlinear singularity analysis: A review and extensions of tests for the Painlevé property. *Lect. Notes Phys.* **1997**, *495*, 171–205.
4. Zhang, S.L.; Wu, B.; Lou, S.Y. Painlevé analysis and special solutions of generalized Broer-Kaup equations. *Phys. Lett. A* **2002**, *300*, 40–48. [CrossRef]
5. Kumar, S.; Singh, K.; Gupta, R.K. Painlevé analysis, Lie symmetries and exact solutions for (2+1)-dimensional variable coefficients Broer-Kaup equations. Commun. *Nonlinear Sci. Numer. Simulat.* **2012**, *17*, 1529–1541. [CrossRef]
6. Zhang, S.; Chen, M.T.; Qian, W.Y. Painleve analysis for a forced Korteveg-de Vries equation arisen in fluid dynamics of internal solitary waves. *Therm. Sci.* **2015**, *19*, 1223–1226. [CrossRef]
7. Zhang, S.; Chen, M.T. Painlevé integrability and new exact solutions of the (4+1)-dimensional Fokas equation. *Math. Probl. Engin.* **2015**, *2015*, 367425. [CrossRef]
8. Zhang, Y.F.; Han, Z.; Tam, H.W. An integrable hierarchy and Darboux transformations, bilinear Bäcklund transformations of a reduced equation. *Appl. Math. Comput.* **2013**, *219*, 5837–5848. [CrossRef]
9. Gardner, C.S.; Greene, J.M.; Kruskal, M.D.; Miura, R.M. Method for solving the Korteweg-de Vries equation. *Phys. Rev. Lett.* **1967**, *19*, 1095–1197. [CrossRef]
10. Hirota, R. Exact envelope-soliton solutions of a nonlinear wave equation. *J. Math. Phys.* **1973**, *14*, 805–809. [CrossRef]
11. Matveev, V.B.; Salle, M.A. *Darboux Transformation and Soliton*; Springer: Berlin/Heidelberg, Germany, 1991.
12. Wang, M.L. Exact solutions for a compound KdV-Burgers equation. *Phys. Lett. A* **1996**, *213*, 279–287. [CrossRef]
13. Fan, E.G. Travelling wave solutions in terms of special functions for nonlinear coupled evolution systems. *Phys. Lett. A* **2002**, *300*, 243–249. [CrossRef]
14. He, J.H.; Wu, X.H. Exp-function method for nonlinear wave equations. *Chaos Soliton. Fract.* **2006**, *30*, 700–708. [CrossRef]
15. Zhang, S.; Xia, T.C. A generalized auxiliary equation method and its application to (2+1)-dimensional asymmetric Nizhnik-Novikov-Vesselov equations. *Phys. A Math. Theor.* **2007**, *40*, 227. [CrossRef]
16. Ma, W.X.; Lee, J.H. A transformed rational function method and exact solutions to 3+1 dimensional Jimbo-Miwa equation. *Chaos Soliton. Fract.* **2009**, *42*, 1356–1363. [CrossRef]
17. Tian, S.F.; Zhang, H.Q. Riemann theta functions periodic wave solutions and rational characteristics for the (1 + 1)-dimensional and (2 + 1)-dimensional Ito equation. *Chaos Soliton. Fract.* **2013**, *47*, 27–41. [CrossRef]
18. Zhang, S.; Liu, D.D. The third kind of Darboux transformation and multisoliton solutions for generalized Broer-Kaup equations. *Turk. J. Phys.* **2015**, *39*, 165–177. [CrossRef]
19. Zhang, S.; Zheng, X.W. N-soliton solutions and nonlinear dynamics for two generalized Broer-Kaup systems. *Nonlinear Dyn.* **2022**, *107*, 1179–1193. [CrossRef]

Σ mathematics

MDPI

Article

Diffusion-Wave Type Solutions to the Second-Order Evolutionary Equation with Power Nonlinearities

Alexander Kazakov * and Anna Lempert

Matrosov Institute for System Dynamics and Control Theory of Siberian Branch of Russian Academy of Sciences, 664033 Irkutsk, Russia; lempert@icc.ru
* Correspondence: kazakov@icc.ru

Abstract: The paper deals with a nonlinear second-order one-dimensional evolutionary equation related to applications and describes various diffusion, filtration, convection, and other processes. The particular cases of this equation are the well-known porous medium equation and its generalizations. We construct solutions that describe perturbations propagating over a zero background with a finite velocity. Such effects are known to be atypical for parabolic equations and appear as a consequence of the degeneration of the equation at the points where the desired function vanishes. Previously, we have constructed it, but here the case of power nonlinearity is considered. It allows for conducting a more detailed analysis. We prove a new theorem for the existence of solutions of this type in the class of piecewise analytical functions, which generalizes and specifies the earlier statements. We find and study exact solutions having the diffusion wave type, the construction of which is reduced to the second-order Cauchy problem for an ordinary differential equation (ODE) that inherits singularities from the original formulation. Statements that ensure the existence of global continuously differentiable solutions are proved for the Cauchy problems. The properties of the constructed solutions are studied by the methods of the qualitative theory of differential equations. Phase portraits are obtained, and quantitative estimates are determined by constructing and analyzing finite difference schemes. The most significant result is that we have shown that all the special cases for incomplete equations take place for the complete equation, and other configurations of diffusion waves do not arise.

Keywords: nonlinear partial differential equation; porous medium equation; diffusion wave; existence theorem; analytical solution; power series; majorant method; exact solution

MSC: 35K57

Citation: Kazakov, A.; Lempert, A. Diffusion-Wave Type Solutions to the Second-Order Evolutionary Equation with Power Nonlinearities. *Mathematics* **2022**, *10*, 232. https://doi.org/10.3390/math10020232

Academic Editors: Almudena del Pilar Marquez Lozano and Vladimir Iosifovich Semenov

Received: 27 December 2021
Accepted: 12 January 2022
Published: 12 January 2022

1. Introduction

This article continues our study of one special class of solutions to a second-order nonlinear evolutionary equation [1]. We consider the equation having the following general form:

$$T_t = (\Phi_1(T))_{xx} + (\Phi_2(T))_x + \Phi_3(T). \tag{1}$$

Here t, x are independent variables: t is time, x is a spatial variable, $T(t,x)$ is an unknown function, and Φ_i, $i = 1,2,3$ are the specified functions. From a physical point of view, the function Φ_1 describes diffusion processes (diffusion term), Φ_2 corresponds to convection processes (convection term), and Φ_3 is a source or a sink.

Equation (1) is parabolic if $\Phi_1'(T) \geq 0$. Solutions that hold the parabolic type of the equation are usually studied. However, for the completeness of the study, negative case can also be considered.

A detailed bibliography overview is given in our first article devoted to the problem considered [1]. Let us briefly recollect some essential points. First, we should mention classical monographs that significantly influenced developing the theory of nonlinear parabolic equations [2–4]. Second, we point out the articles in which, apparently, the authors presented diffusion wave-type solutions for the first time [5,6]. Let us especially note book [7], which presents the mathematical theory of the porous medium equation and thorough state-of-art.

Recall that the case where Φ_1 is a power function, and $\Phi_2 = \Phi_3 \equiv 0$ is the porous medium Equation [7]. It is rich in applications and describes the filtration of an ideal gas in porous formations [6], the radiant (nonlinear) thermal conductivity [5], as well as population dynamics processes [8].

If Φ_1 and Φ_3 are power functions, and $\Phi_2 \equiv 0$, then (1) becomes the generalized porous medium Equation [7] or the nonlinear heat equation with a source [9]. This equation describes the same processes as the porous medium equation, but allows us to consider the inflow or outflow of matter or heat.

Assuming $\Phi_3 \equiv 0$, and Φ_1 and Φ_2 are nonzero leads Equation (1) to the convection-diffusion Equation [10,11]. Several mathematical models of fluid mechanics, which simultaneously describe the diffusion and convective [12] mechanisms of energy and matter transfer, are reduced to such an equation. The phonon transport within silicon structures, which is subjected to internal heat generation, can also be explored [13,14]. In [15], the authors proposed the equation considered as a suitable governing equation for the gas flow through a Graphene Oxide membrane. A mathematical model describing the flow of a mixture of ideal gases in a highly porous electrode for fuel cell engineering is proposed in [16]. Its particular case is the well-known Burgers equation [7].

Finally, Equation (1), if $\Phi_2(c)$ is a linear function, which describes the non-stationary thermal conductivity in a medium moving at a constant speed, when the thermal conductivity coefficient and the reaction rate are arbitrary functions of the temperature [17].

Note that the problem is also being studied in the case of several spatial variables, and solutions of different types are constructed. In [18,19], the author considers the anisotropic case and construct weak solutions. In [20], the authors present weak supersolutions for different functional spaces. Analytical travelling waves for the nonlinear convection-diffusion equation are studied in [21], including the use of Lie symmetry [22]. Various models of a similar but more general form are used, for example, in the study of diffusion processes in metallurgy [23], as well as the thermal fields located in the permafrost area [24]. The list could be continued, so the study of Equation (1) is still relevant.

In this paper, we deal with the problem of constructing and studying diffusion-wave-type solutions in the case of power functions Φ_i. The existence and uniqueness theorem is proved. It, unlike the known ones, allows us to set the boundary condition at a moving point. In addition, exact solutions are found and investigated in detail in one particular case. Their construction is reduced to the integration of the Cauchy problem for an ordinary differential equation.

In contrast to similar solutions that we dealt with in [1], this study is more systematic. Firstly, here these Cauchy problems are investigated in a general formulation. Secondly, we do not limit ourselves to considering cases when equations can be integrated explicitly but perform their qualitative analysis and constructed phase portraits, which allowed us to investigate the behavior of solutions. We also construct finite difference schemes and prove their convergence, which, in particular, makes it possible to construct accurate estimates for the solutions obtained.

2. Problem Formulation

If the functions $\Phi_1(T), \Phi_2(T)$ are differentiable, Equation (1) can be written as:

$$T_t = (\Phi_1'(T)T_x)_x + \Phi_2'(T)T_x + \Phi_3(T). \quad (2)$$

We assume that $\Phi_i(T),\ i = 1,2,3$ are power functions:

$$\Phi_1'(T) = \lambda_1 T^{\sigma_1},\ \Phi_2'(T) = \lambda_2 T^{\sigma_2},\ \Phi_3(T) = \lambda_1 T^{\sigma_3},$$

where σ_i, $i = 1,2,3$ are positive constants, $\sigma_1 + \sigma_3 > 1$, λ_i, $i = 1,2,3$ are constants, and $\lambda_1 > 0$.

The substitution $u = \Phi_1'(T) = \lambda_1 T^{\sigma_1}$ and effortless transformations lead Equation (2) to the form:

$$u_t = uu_{xx} + \frac{1}{\sigma}u_x^2 + Au^\theta u_x + Bu^\beta. \tag{3}$$

Here $\sigma = \sigma_1 > 0$, $\theta = \sigma_1/\sigma_2 > 0$, $\beta = \sigma_3/\sigma_1 + 1 - 1/\sigma_1 > 0$, $A = \lambda_2\lambda_1^{-1/\sigma}$, $B = \lambda_3\lambda_1^{1/\sigma - 1/\sigma_3}$. Obviously, Equation (3) has the trivial solution $u \equiv 0$.

Let us set for Equation (3) the boundary conditions:

$$u(t,x)|_{x=a(t)} = f(t),\ f(0) = 0,\ f'(0) \geq 0. \tag{4}$$

Previously, the same conditions for the porous medium Equation [25] were considered. In this paper, we prove the solvability of problem (3) and (4) in the class of analytical functions. Moreover, we show that if there exists a sufficiently smooth solution to problem (3) and (4), then together with the trivial one, it forms a diffusion wave.

3. Main Theorem

Let us formulate and prove the existence and uniqueness theorem. Here and further, an analytical solution means a solution in the class of analytical functions, i.e., it coincides in a neighborhood with its Taylor expansion.

Recall that the diffusion wave-type solution means a piecewise smooth solution to Equation (1), consisting of a trivial $u \equiv 0$ part and a nontrivial $u = u(t,x) \geq 0$ one, continuously joined on some line in the plane of variables t, x. This line is called the wave front.

Theorem 1. *Let the functions $a(t)$ and $f(t)$ be analytical in some neighborhood of $t = 0$; $f'(0) \geq 0$; $[a'(0)]^2 + f'(0) > 0$; and let θ and β be natural (positive integer) numbers. Then problem (3) and (4) has a nonzero analytical solution of diffusion-wave type in some neighborhood of the point $(t = 0, x = 0)$, which is unique if the direction of the diffusion-wave front moving is chosen.*

Proof. Let us give a brief scheme of the further reasoning. First, we construct the solution in the form of a power series. Then we reduce problem (3) and (4), which is not a Cauchy–Kovalevskaya type, to the standard form by the consequence of non-degenerate substitutions. This standard form is subject to the Cauchy–Kovalevskaya theorem.

To simplify the boundary conditions, we make the substitution $t_1 = t, r = x - a(t)$. It is easy to show that the Jacobian of the substitution is nonzero. As a result, we get the problem:

$$u_t - a'(t)u_r = uu_{rr} + \frac{1}{\sigma}u_r^2 + Au^\theta u_r + Bu^\beta, \tag{5}$$

$$u(t,r)|_{r=0} = f(t). \tag{6}$$

Here and to further simplify the notation, the first independent variable retains t without index 1.

We construct the solution to problem (5) and (6) as the series:

$$u(t,r) = \sum_{k,m=0}^{\infty} u_{k,m}\frac{t^k r^m}{k!m!},\ u_{k,m} = \left.\frac{\partial^{k+m}u}{\partial t^k \partial r^m}\right|_{t=r=0}. \tag{7}$$

This method develops the method of special series, which was proposed and widely used in the scientific school of A.F. Sidorov [26,27].

Since the construction essentially coincides [28] (see also [25]), we try to avoid repetitions, focusing on new points in the proof and emerging difficulties.

Since the functions $a(t), f(t)$ are analytical, they allow the expansions:

$$f(t) = \sum_{n=0}^{\infty} f_n \frac{t^n}{n!}, \; a(t) = \sum_{n=0}^{\infty} a_n \frac{t^n}{n!}.$$

Boundary condition (6) implies the equalities $u_{n,0} = f_n$, and $f_0 = 0$. The remaining coefficients of (7) are determined by recursive induction on the total order of differentiation $n = k + m$.

First, we establish the induction base by considering the case $k + m = 1$. As it has been shown, $u_{1,0} = f_1$. Assume, that $t = r = 0$ in (5). Then it is possible to consider the equation obtained as quadratic with respect to $u_{0,1}$ and find its roots:

$$u_{0,1}^{\pm} = \frac{\sigma}{2}\left(-a_1 \pm \sqrt{a_1^2 + 4f_1}\right).$$

Since $f'(0) \geq 0$, $[a'(0)]^2 + f'(0) > 0$, both roots are real.

The direction of the diffusion wave moving depends on the choice of the sign of $u_{0,1}$. The value $u_{0,1}^{-}$ corresponds to a diffusion wave whose front lies to the right of the line $x = a(t)$ in the plane of variables t, x. A diffusion wave whose front is located to the left of the line $x = a(t)$ corresponds to $u_{0,1}^{+}$. These cases can be considered separately, or one can be chosen based on additional reasons. Looking ahead, we note that the procedure for constructing a solution is similar in both cases.

If the sign is chosen, then the series (7) is constructed uniquely.

We differentiate (5) k times with respect to r, $n - k$ times by t, and set $t = r = 0$. After collecting terms, we arrive at the equality:

$$b_{n-k}u_{n-k-1,k+2} + c_k u_{n-k,k+1} + u_{n-k+1,k} = R_{n-k,k}, \tag{8}$$

where:

$$b_{n-k} = -(n-k)f_1, \; c_k = -\left(k + \frac{2}{\sigma}\right)u_{0,1} - a_1.$$

We do not show here the explicit form of $R_{n-k,k}$ since it is cumbersome. Their form for the particular case $A = B = 0$ can be found in [25], where it is presented since the convergence proof technique used there requires direct estimates. Here, we use another technique for constructing the majorant problem based on the hodograph transformation. In this regard, it is enough to point out that $R_{n-k,k}$ depend on the derivatives of the unknown function of order at most n, which are known by the induction hypothesis. The condition $\theta, \beta \in \mathbb{N}$ ensures infinite differentiability of Equation (5).

Changing in (8) k from 0 to n and taking into account that $u_{n+1,0} = f_n$, and $b_0 = 0$, we obtain the following system of linear algebraic equations:

$$\begin{pmatrix} c_0 & b_n & 0 & \dots & 0 & 0 & 0 \\ 1 & c_1 & b_{n-1} & \dots & 0 & 0 & 0 \\ & \dots & & & & \dots & \\ 0 & 0 & 0 & \dots & 1 & c_{n-1} & b_1 \\ 0 & 0 & 0 & \dots & 0 & 1 & c_n \end{pmatrix} \times \begin{pmatrix} u_{n,1} \\ u_{n-1,2} \\ \dots \\ u_{1,n} \\ u_{0,n+1} \end{pmatrix} = \begin{pmatrix} R_{n,0}^{*} \\ R_{n-1,1} \\ \dots \\ R_{1,n-1} \\ R_{0,n} \end{pmatrix}. \tag{9}$$

Here $R_{n,0}^{*} = R_{n,0} - f_{n+1}$. You can see that the matrix A_n of system (9) is tridiagonal of order $n + 1$, and the condition of diagonal dominance is not satisfied. Let us prove that its determinant is nonzero.

Indeed, it is necessary to consider three cases: (1) $f_1 = 0$; (2) $f_1 > 0$, $u_{0,1} = u_{0,1}^{+}$; (3) $f_1 > 0$, $u_{0,1} = u_{0,1}^{-}$.

1. Let $f_1 = 0$. Then $b_k = 0$ for all k, i.e., the matrix A_n is triangular two-diagonal and its determinant is equal to the product of the elements of the main diagonal:

$$\det A_n = \prod_{i=0}^{n} c_i.$$

Two subcases are possible here: (a) $u_{0,1} = u_{0,1}^{+} = 0$, then $c_k = -a_1$ for all k; (b) $u_{0,1} = u_{0,1}^{-} = -a_1\sigma$, then $c_k = (k\sigma + 1)a_1$. For the both subcases $\det A_n \neq 0$ since $a_1 \neq 0$.

2. Let $f_1 > 0$, $u_{0,1} = u_{0,1}^{+}$. Then $b_k < 0$ for $k \geq 1$ and $c_k > 0$ for all k, i.e., all elements on the main diagonal and subdiagonal are positive, and all elements of the superdiagonal are negative. Hence, all the principal minors of the matrix A_n are positive, which means its non-degeneracy.

3. Let $f_1 > 0$, $u_{0,1} = u_{0,1}^{-}$. Then $b_k < 0$ for $k \geq 1$ and $c_k < 0$ for all k. Let us introduce an auxiliary numeric sequence $\Delta_{n,k}$ as follows:

$$\Delta_{n,0} = 1,\ \Delta_{n,1} = c_0 < 0,\ \Delta_{n,k} = c_{k-1}\Delta_{n,k-1} - b_{n-k+2}\Delta_{n,k-2},\ k = 2,3,\ldots,n.$$

It can be shown by induction on k that $\Delta_{n,k}$ consists of two positive terms for even k and two negative ones for odd k. Hence we have that $\Delta_{n,k} \neq 0$ for all admissible n and k. On the other hand, it is easy to show that $\Delta_{n,n} = \det A_n$ by induction on n. Thus, $\det A_n \neq 0$, moreover, $\det A_n > 0$ for even n and $\det A_n < 0$ for odd n.

Thus, we have proved that system (9) is non-degenerate, and the coefficients of series (7) are uniquely determined if one of the two possible values of $u_{0,1}^{\pm}$ is chosen. This finishes the first step of the proof.

We refuse the direct estimates applied in [25] in the proof of convergence. Here we use an alternative methodology, which reduces the problem to a special form previously considered in [1,28].

Since $u_{0,0} = 0$, $u_{0,1}^2 + u_{1,0}^2 \neq 0$, then if series (7) converges, there exists a line $r = g(t)$ in the plane t, r, on which the unknown function vanishes:

$$u|_{r=g(t)} = 0,\ g(0) = 0.$$

In problem (5) and (6), which is equivalent to the original one, let us make the substitution $t_2 = t, s = r - g(t)$. We arrive at the problem that consists of one equation and two boundary conditions:

$$u_t - [a'(t) + g'(t)]u_s = uu_{ss} + \frac{1}{\sigma}u_s^2 + Au^{\theta}u_s + Bu^{\beta}, \tag{10}$$

$$u(t,s)|_{s=-g(t)} = f(t),\ u(t,s)|_{s=0} = 0. \tag{11}$$

To simplify the notation, the first independent variable retains t without index 2.

The function $g(t)$ is still unknown, and it will be determined simultaneously with the construction of the function u. Thus, we obtain one of the problems with a free boundary. The most famous of them for parabolic equations is the Stefan problem [29,30].

The following substitution changes the roles of the unknown function u and the independent variable s, i.e., it is a variant of the hodograph transformation. Equation (10) becomes:

$$us_{uu} = Bu^{\beta}s_u^3 + [s_t + a'(t) + g'(t) + Au^{\theta}]s_u^2 + \frac{1}{\sigma}s_u. \tag{12}$$

Conditions (11) take the form:

$$s(t,u)|_{u=f(t)} = -g(t),\ s(t,u)|_{u=0} = 0. \tag{13}$$

Let us differentiate the first condition of (13) and substitute the resulting expression $[s_t + s_u f'(t)]|_{u=f(t)} = -g'(t)$ into Equation (12). We obtain that:

$$us_{uu} = Bu^{\beta}s_u^3 + \Big\{s_t + a'(t) - [s_t + s_u f'(t)]|_{u=f(t)} + Au^{\theta}\Big\}s_u^2 + \frac{1}{\sigma}s_u. \tag{14}$$

The positive trait of Equation (14) is that it no longer contains the unknown function $g(t)$. The boundary condition for (14) takes the form:

$$s(t,u)|_{u=0} = 0. \tag{15}$$

Having constructed a solution to problem (14) and (15) which, recall, does not contain the function $g(t)$, we can find $g(t)$ from the first condition of (13). Thus, we have decomposed problem (10) and (11), which includes two unknown functions into two separate tasks. They contain one unknown function and can be solved sequentially.

As a result of the substitutions performed, we have obtained the problem with the known diffusion front, which was previously considered in [1]. As already noted, the detailed proof of the similar theorem for the porous medium equation with two spatial variables is given in [28]. In this regard, we will be brief so as to not overload the paper.

Completing the series of substitutions, let us introduce the variable $y = u - f$, which allows us to make the surface $u = f$ as a new coordinate plane. Next, the unknown function is represented as $s(u,y) = us_1(y) + u^2 Z(u,y)$, where s_1 is the known analytical function, and $Z = Z(u,y)$ is a new unknown function. Note that in this case, the second boundary condition of (13) is satisfied automatically, and the problem is reduced to one equation of the form:

$$\begin{aligned}&\Psi_0(y)Z|_{y=0} + \Psi_1(y)u(Z_u|_{y=0}) + \Psi_2(y)u^2(Z_{uu}|_{y=0})\\&+B_0 Z + B_1 u S_u + u^2 Z_{uu} = h_0 + uh_1 + u^2 h_2 + u^3 h_3.\end{aligned} \tag{16}$$

Here $B_0 = 2(1+1/\sigma)$, $B_1 = (4+1/\sigma)$ are constants; Ψ_i, $i = 0,1,2$ and h_j, $j = 0,1,2,3$ are analytical functions of their variables. Moreover, $h_0 = h_0(u,y)$, and the remaining h_j depend on independent variables and derivatives of the function Z with respect to u of order at most $j-1$. The functions $\Psi_i(y)$ are positive for $y = 0$. Thus, Equation (16) obeys Lemma 2 from [28]. Therefore, it is solvable in the class of analytical functions. The construction of the majorant problem and the proof of the existence of its analytical solution are also carried out similarly. □

Remark 1. *We have constructed an analytical solution to problem* (3) *and* (4) *and simultaneously determined the line* $x = a(t) + g(t)$*, which is the diffusion wave front. The non-negative part of the specified solution* $u = u^+$ *and the trivial solution* $u \equiv 0$*, which are joined on the diffusion front, give the required diffusion wave.*

Remark 2. *A particular case of problem* (3) *and* (4)*, when* $f(t) \equiv 0$*, is a problem about the moving of a diffusion wave with a given diffusion front, which obeys the theorem proved in [1].*

4. Exact Solutions

Theorem 1 ensures the existence and uniqueness of the solution to the problem of diffusion wave initiating and, remarkably, gives an algorithm for its construction in the form of a double series. Unfortunately, it is local, and, as attempts to use such constructions for a numerical modeling show [31], the radius of convergence of the series is usually small. Thus, the theorem does not allow us to study the global properties of diffusion waves. Besides, the requirement that the parameters β and θ in Equation (3) are natural numbers significantly limits the generality. In general, these problems are far from being solved, as well as for most other nonlinear degenerate partial differential equations. Therefore, we investigate the properties of diffusion-wave type solutions to Equation (3) for an arbitrary $\beta > 0$ and $\theta > 0$ in the particular case. Exact solutions of parabolic equations are widely used in solving applied problems: From modeling the well clogging process [32] to the description of bubble dynamics [33].

4.1. Reduction to Ordinary Differential Equations (ODEs)

Consider for Equation (3) the boundary condition:

$$u(t,x)|_{x=a(t)} = 0, \tag{17}$$

which, as already noted, is a particular case of (4) when $f \equiv 0$. Problem (3) and (17) is the problem of the diffusion wave moving with a given front.

Note that problem (3) and (17) has the trivial solution $u \equiv 0$. However, in this case, the uniqueness of the solution is violated, and a nonzero solution can also exist. Its existence in the class of analytical functions follows from Theorem 1.

Current and further sections are devoted to finding and studying non-trivial exact solutions to problem (3) and (17), the constructing of which is reduced to the integration of Cauchy problems for ODEs. Previously, we studied this problem in detail for the nonlinear heat equation [34] and for the nonlinear heat equation with a source [35] and found new classes of diffusion-wave type solutions. Those problems corresponded to the case $A = 0$. Here let us consider the case when $A \neq 0$.

Following [1], we look for solutions to Equation (3) having the form:

$$u = \psi(t)v(x - a(t)). \tag{18}$$

Solution (18) is a generalized traveling wave, which becomes a simple traveling wave if $a(t)$ is a linear function. Substituting (18) into Equation (3), we obtain:

$$vv'' + \frac{1}{\sigma}(v')^2 + A\psi^{\theta-1}(t)v^\theta v' + B\psi^{\beta-2}(t)v^\beta + \frac{a'(t)}{\psi(t)}v' - \frac{\psi'(t)}{\psi^2(t)}v = 0. \tag{19}$$

In order for (19) to become an ODE with respect to $v(z)$, $z = x - a(t)$, it is necessary and sufficient to satisfy the conditions:

$$\frac{a'(t)}{\psi(t)} = \text{const}, \ \frac{\psi'(t)}{\psi^2(t)} = \text{const}, \ \psi^{\theta-1}(t) = \text{const}, \ \psi^{\beta-2}(t) = \text{const}. \tag{20}$$

Here the first two conditions form a first-order ODE system. The third and fourth conditions are additional compatibility conditions that can be satisfied, for example, by choosing θ and β.

Let us consider two possible cases.

1. Let $\psi(t) = \psi = \text{const}$. Without losing the generality of consideration, we can set $\psi = 1$. Then $a(t) = \mu t + \eta$, where μ, η are constants, and (19) takes the form of the following ODE:

$$vv'' + \frac{1}{\sigma}(v')^2 + (Av^\theta + \mu)v' + Bv^\beta = 0. \tag{21}$$

We assume that in this case $\eta = 0, \mu > 0$, which also does not reduce the generality of consideration.

2. Let now $\psi(t) \neq \text{const}$. Then from the first two equations of (20) we have that $\psi(t) = \omega/(\mu t + \eta)$, $a(t) = \omega \ln(\mu t + \eta)$, where μ, η, and ω are nonzero constants, $\eta > 0$. You can see that the necessary and sufficient conditions to satisfy the third and fourth equalities of (20) are $\theta = 1, \beta = 2$. Then (19) takes the form of the following ODE:

$$vv'' + \frac{1}{\sigma}(v')^2 + (Av + \mu)v' + Bv^2 + \frac{\mu}{\omega}v = 0. \tag{22}$$

We can bring (21) and (22) to the general form:

$$vv'' + \frac{1}{\sigma}(v')^2 + (Av^\theta + \mu)v' + Bv^\beta + Cv = 0. \tag{23}$$

4.2. Cauchy Conditions for ODEs

It is easy to see that condition (17) becomes $v(0) = 0$ for a solution having the form (18). Then, obviously, Equation (23) has the trivial solution $v \equiv 0$. Assuming $v = 0$ in (23), we obtain the quadratic equation with respect to $v_1 = v'(0)$:

$$\frac{1}{\sigma}v_1^2 + \mu v_1 = 0, \tag{24}$$

which has roots $v_1 = 0$ and $v_1 = -\mu\sigma$. Accordingly, we will consider Equation (23) with the Cauchy conditions of two types:

$$v(0) = 0,\ v'(0) = 0; \tag{25}$$

$$v(0) = 0,\ v'(0) = -\mu\sigma. \tag{26}$$

The trivial solution corresponds to conditions (25). However, as it is shown below, there may also exist a non-trivial solution that is not analytical, i.e., it cannot be represented as a Taylor series.

Theorem 1 implies that problem (23) and (26) for positive integer values of θ and β has the unique analytic solution in the form of a convergent series in powers of z. Unfortunately, the theorem does not hold for non-integer values of these constants.

Note that Equation (23), although it is an ODE, stays still complex to study. First, it is nonlinear. Second, the Cauchy problems inherit singularities from the original statements, which does not allow for using standard methods and theorems of ODE theory. Thus, the general case is quite complex and cannot be explored within the framework of a single article. Therefore, we consider here one of the particular cases. On the one hand, this case is significant and has interesting properties. On the other hand, it gives a clear idea of the difficulties encountered in studying the properties of the obtained classes of exact solutions and what techniques can be applied to overcome them.

5. Traveling Wave. Qualitative Analysis

In this section, we consider the exact solutions having the form of traveling waves, which, as shown above, are described by Equation (21) with Cauchy conditions (25) or (26). We study them using the methods of ODE theory, including qualitative analysis with the construction of a phase portrait and some quantitative estimates.

5.1. Transition to Phase Variables

Using the fact that the equation does not explicitly depend on z, we proceed to the phase plane. Let us introduce a new independent variable w and an unknown function p:

$$w = v^{\theta}, p = v'. \tag{27}$$

The substitution is non-degenerate if $\theta \geq 1$. Equation (21) takes the form:

$$\theta w p \frac{dp}{dw} + \frac{p^2}{\sigma} + Awp + \mu p + Bw^{\beta/\theta} = 0. \tag{28}$$

Let $\theta = \beta \geq 1$, i.e., in the third and fourth terms of Equation (21) v has the same degree. Due to the linear change of variables, we can reduce the number of constants. Let,

$$w = \tilde{w}\tilde{A}, \tilde{A} = \frac{\mu}{A}; \ p = \tilde{p}\tilde{B}, \tilde{B} = \frac{\mu}{\beta}.$$

Then Equation (28) takes the form ($\sim$ is omitted for simplicity):

$$wp\frac{dp}{dw} + \frac{p^2}{\gamma} + wp + p + \alpha w = 0, \tag{29}$$

where $\gamma = \sigma\theta > 0$, $\alpha = B/\mu > 0$. Note that Equation (29) is similar to (39) from [1], however, the appearance of the term pw in (29) significantly complicates the study.

For (29), let us consider solutions corresponding to the initial condition given at $w = 0$. Since $\theta > 0$, nontrivial solutions of this kind generate solutions to the original problem having the diffusion-wave type. Looking ahead, we note that some of them may not have physical meaning.

If we substitute $w = 0$ into Equation (29), we obtain the algebraic relation $p^2(0)/\gamma + p(0) = 0$, which is an analogue of equality (24). You can make sure that it has roots (1) $p(0) = 0$ and (2) $p(0) = -\gamma$, which correspond to conditions (26) and (25), respectively. Now let us consider Equation (29) with Cauchy conditions (1) and (2) in more detail.

5.2. Singular Points

First, we study the singular points of Equation (29). Since it is autonomous, let us turn to the phase plane $(v, v' = w)$. We use the classic technique proposed in [36] (see, also [34]). The following dynamic system corresponds to Equation (29):

$$\frac{dw}{d\zeta} = wp, \quad \frac{dp}{d\zeta} = -\frac{p^2}{\gamma} - p - pw - \alpha w, \tag{30}$$

where $dz = w\,d\zeta$.

Consider now the equilibrium states (singular points) of system (30). There are two equilibrium states $(0, -\gamma)$ and $(0, 0)$.

Let us introduce the following notation:

$$R(w,p) = wp, \quad Q(w,p) = -p^2/\gamma - p - pw - \alpha w,$$

$$M(v,w) = \begin{pmatrix} R_w & R_p \\ Q_w & Q_p \end{pmatrix} = \begin{pmatrix} p & w \\ -p-\alpha & -2p/\gamma - w - 1 \end{pmatrix},$$

$$\Delta(w,p) = \det M(w,p) = -\frac{2p^2}{\gamma} - p + \alpha w,$$

$$\delta(w,p) = \mathrm{Tr}\, M(w,p) = \frac{(\gamma-2)}{\gamma}p - w - 1.$$

Let us define the type of each singular point.

1. Consider the point $(0, -\gamma)$. Since $\Delta(0, -\gamma) = -\gamma \neq 0$, it is a simple equilibrium point. From $\det(M - \lambda E)|_{w=0, p=-\gamma} = (\lambda + \gamma)(\lambda - 1)$, it follows that $\lambda_1 = -\gamma$ and $\lambda_2 = 1$ are the roots of the characteristic equation. Therefore, the point $(0, -\gamma)$ is the topological saddle since $\Delta < 0$, $\lambda_1, \lambda_2 \in \mathbb{R}$ and $\lambda_1\lambda_2 < 0$.

2. Consider the point $(0, 0)$. Since $\Delta(0, 0) = 0$, this is a compound equilibrium point. Here $\delta(0, 0) = -1 \neq 0$, and the equation that is obtained from system (30) by the elimination of the independent variable ζ can be written as:

$$wpdw - [lp - p^2/\gamma - wp - \alpha w]\,dw = 0,$$

where $l = -1$. We represent the solution to the equation:

$$-lp + p^2/\gamma + pw + \alpha w = 0$$

as a series in powers of w, which we substitute into pw. As a result, we have:

$$p = \phi(w) = -\alpha w + \ldots, \quad \xi(w) = (wp)|_{p=\phi(w)} = -\alpha w^2 + \ldots$$

Since the lowest power of w in the expansion $\xi(w)$ equals two, the point $(0, 0)$ is a saddle-node with one nodal and two saddle sectors. The nodal sector is stable because $l < 0$. Moreover, if $\alpha < 0$, then the trajectories of the nodal sector tend to $(0, 0)$ when $\zeta \to -\infty$ on the left of the Op axis. If $\alpha > 0$, as in the considered case, the trajectories of the nodal sector tend to $(0, 0)$ when $\zeta \to +\infty$ on the right of the Op axis.

5.3. Phase Portrait

Let us construct and explore the phase portrait of system (30) for $\gamma, \alpha > 0$. Note that in all the considered cases:

1. The phase trajectories change the direction of motion when passing through the Ow axis, as well as when crossing the quadric $p^2/\gamma + p + pw + \alpha w = 0$, which, in particular, singular points belong;
2. Both singular points have vertical semi-separatrices, since they are located on the Op axis.

Let us first determine the properties of the second-order curve:

$$p^2/\gamma + p + pw + \alpha w = 0.$$

Bringing it to its canonical form, we obtain:

$$\left(p + \frac{\gamma w}{2} + \frac{\gamma}{2}\right)^2 - \frac{\gamma^2}{4}\left(w + 1 - \frac{2\alpha}{\gamma}\right)^2 = \alpha(\gamma - \alpha). \tag{31}$$

It is easy to see that for $\alpha = \gamma$, we have a pair of intersecting straight lines $p_1(w) = -\gamma w - \gamma + \alpha$, $p_2(w) \equiv -\alpha$. If $\alpha \neq \gamma$, then we obtain hyperbolas with the same asymptotes $p = p_1(w)$ and $p = p_2(w)$ and different positions of the branches depending on the sign of the difference $\gamma - \alpha$.

Let us consider all possible cases. Note here that in all cases, there are three semi-separatrices. The first is a monotonically decreasing curve coming to the singular point $(0,0)$ and located in the second quadrant (bold curve S_1 in Figure 1–3). The second and third are vertical semi-separatrices lying on the Op axis.

Case $\gamma = \alpha$. Figure 1 shows the phase portrait of system (30) for this case. As already noted, the quadric (31) degenerates into two intersecting lines (dashed and green lines). Besides the separatrices mentioned above, there is also a separatrix that coincides with the line $p = -\gamma$ (green line). The nodal sector is bounded by the Op axis and the straight line $p = -\gamma$.

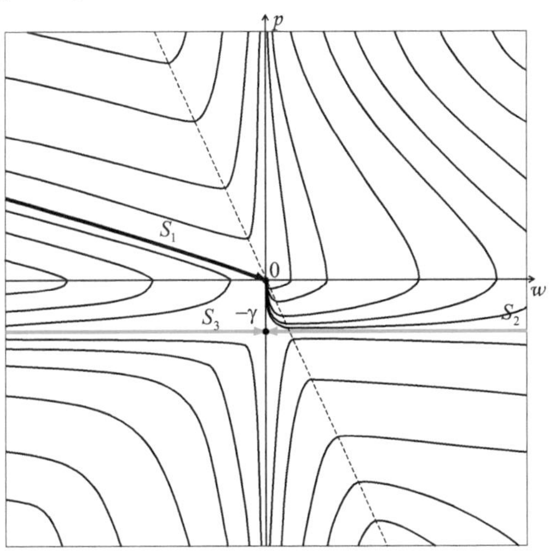

Figure 1. Phase portrait for $\gamma = \alpha$.

Case $\gamma > \alpha$. Figure 2 shows the phase portrait of system (30). You can see that half-hyperbolas (31) are located in the right upper and left lower quarters, into which the lines $p_1(w) = -\gamma w - \gamma + \alpha$, $p_2(w) \equiv -\alpha$ divide the coordinate plane (dashed curves). Here we have two additional separatrices S_2 and S_3 coming into the point $(0, -\gamma)$ (purple curves).

Both S_2 and S_3 are monotonically decreasing functions; S_2 tends to $-\infty$ when $\zeta \to +\infty$; S_3 tends to $p = -\alpha$ when $\zeta \to -\infty$. The nodal sector is bounded by the Op axis and the semi-separatrix S_2 located in the fourth quadrant.

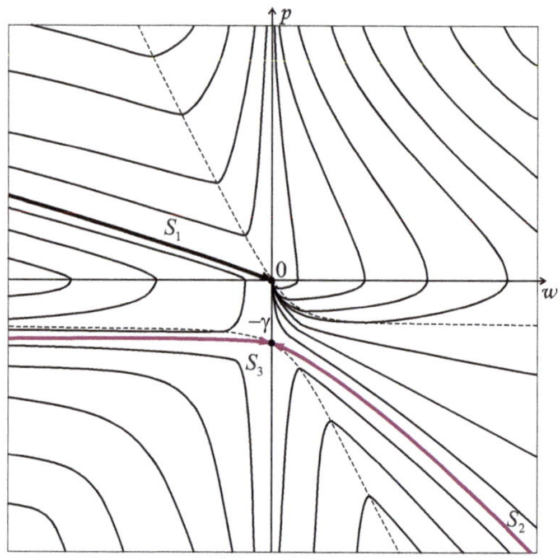

Figure 2. Phase portrait for $\gamma > \alpha$.

Case $\gamma < \alpha$. Here half-hyperbolas (31) are located in the left upper and right lower quarters, into which the lines $p_1(w) = -\gamma w - \gamma + \alpha$, $p_2(w) \equiv -\alpha$ divide the coordinate plane (see Figure 3). Again, in addition to the same separatrices for all cases, we have two semi-separatrices going out the point $(0, -\gamma)$ (blue curves). The separatrix S_2 first increases to the intersection with the Ow axis, then decreases and asymptotically tends to the Op axis when $\zeta \to +\infty$, bounding the nodal sector. The separatrix S_3 is a monotonically increasing function and tends to the line $p = -\alpha$ when $\zeta \to -\infty$.

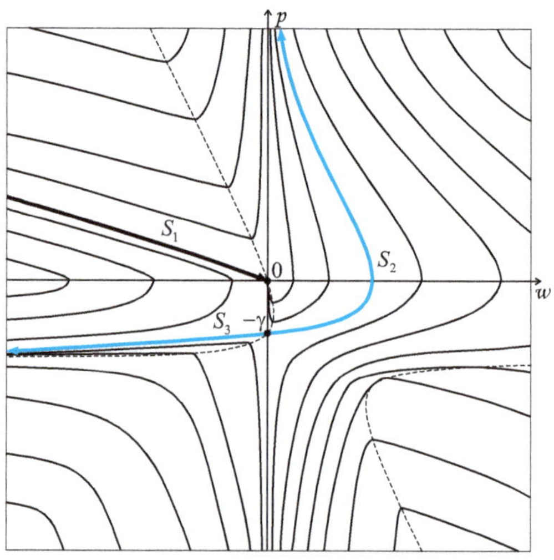

Figure 3. Phase portrait for $\gamma < \alpha$.

The properties of separatrices that do not coincide with the Op axis and the interpretation of the results from the original problem point of view will be discussed below.

6. Zero Initial Condition

Let us first consider the case when the initial condition for Equation (29) has the form:

$$p(0) = 0. \tag{32}$$

Previously, this case has not been considered. The only exception is paper [34], where we showed the existence of a semi-separatrix lying in the second quadrant and passing through the origin for the porous medium equation. However, the properties of the corresponding solution were not studied.

Obviously, in this case, the classical existence theorems are inapplicable due to degeneracy. Therefore, we attempt to eliminate the singularity.

6.1. Solution in the Form of a Series

Following [35], we try to construct an analytical solution to problem (29) and (32) as the series:

$$p(w) = \sum_{k=0}^{\infty} \frac{p_k}{k!} z^k, \; p_k = p^{(k)}(0). \tag{33}$$

Let us construct the coefficients for (33) using the following recurrent procedure. From (32) we have $p_0 = p(0) = 0$. To find p_1, we differentiate Equation (29) with respect to w, set $w = 0, p(0) = 0$, and obtain that $p_1 = p'(0) = -\alpha < 0$. Similarly, we get $p_2 = 2\alpha(1 - \alpha - \alpha/\gamma)$. Thus, the induction base is found.

Assume that $p_0, p_1, \ldots, p_{k-1}, k \geq 3$ are determined. To find p_k, we differentiate Equation (29) k times with respect to w and set $w = 0$. Then we arrive at the equality:

$$k \sum_{i=0}^{k-1} C_{k-1}^i p_i p_{k-i} + \frac{1}{\gamma} \sum_{i=0}^{k} C_k^i p_i p_{k-i} + k p_{k-1} + p_k = 0, \tag{34}$$

where $C_k^i = k!/[i!(k-i)!], k \geq i$. Resolving (34) with respect to p_k and taking into account $p_0 = 0, p_1 = -\alpha$, we have that:

$$p_k = k\left(\alpha k + \frac{2\alpha}{\gamma} - 1\right) p_{k-1} - k \sum_{i=2}^{k-2} C_{k-1}^i p_i p_{k-i} - \frac{1}{\gamma} \sum_{i=2}^{k-2} C_k^i p_i p_{k-i}. \tag{35}$$

You can see that all terms on the right-hand side of (35) are known by the induction hypothesis. Thus, all the coefficients of series (33) are uniquely determined by the formul obtained.

Now we study the properties of the constructed series. To do this, consider:

$$p_2 = 2\alpha\left[1 - \frac{\alpha(1+\gamma)}{\gamma}\right].$$

If $\alpha = \gamma/(1+\gamma)$ we have $p_2 = 0$. Then, it is easy to show by induction on k that $p_3 = p_4 = \ldots = p_k = \ldots = 0$. This means that the series breaks off, and the solution has the form $p = -\alpha w$.

If $\alpha > \gamma/(1+\gamma)$, then $p_2 < 0$, and we can make sure that $p_k < 0, k = 3, \ldots$ Therefore, from (35) we have that:

$$p_k < k\left(\alpha k + \frac{2\alpha}{\gamma} - 1\right) p_{k-1} < 0.$$

Hence we get that:

$$\lim_{k\to\infty} \frac{|p_k|(k-1)!}{|p_{k-1}|k!} > \lim_{k\to\infty} \frac{k(k-1)!}{k!}\left(\alpha k + \frac{2\alpha}{\gamma} - 1\right) = +\infty.$$

Thus, we have proved the divergence of series (33) and the validity of the following proposition.

Proposition 1. *Problem* (29) *and* (32) *has:*

1. *The analytical solution* $p = -\alpha w$, *if* $\alpha = \gamma/(1+\gamma)$;
2. *The solution having the form of a formal power series that converges only at the point* $w = 0$, *if* $\alpha > \gamma/(1+\gamma)$.

Note that for $0 < \alpha < \gamma/(1+\gamma)$ the terms in formula (35) have, generally speaking, different signs, and the question of the convergence or divergence of series (33) is much more challenging. Nevertheless, the results of numerical calculations allow us to make a reasonable assumption that the series diverges.

Remark 3. *There are also simpler examples based on a similar idea. Indeed, consider the Cauchy problem:*

$$xyy' - y + x + 1 = 0, y(0) = 1.$$

We can easily make sure that in this case $y'(0) = 1, y''(0) = 2$, *and* $y^{(k)}(0) \geq k^2 y^{(k-1)}(0) > 0$, $k \geq 2$, *which means the divergence of the Maclaurin series for the function* $y(x)$ *at* $x \neq 0$.

6.2. Euler Polygonal Approximations

As you know, the absence of an analytical solution to the Cauchy problem does not mean that it is impossible to construct a smooth (classical) solution. The simplest example is the problem $y' = \sqrt{x}, y(0) = 0$, which has a unique continuously differentiable solution for $x \geq 0$. In this section, we show that problem (29) and (32) has a similar property for $w \leq 0$, especially since the results of qualitative analysis evidence the existence of such a solution.

We use the classical Euler method. Therefore, it is necessary to construct a finite difference approximation of Equation (29). Calculations have shown that explicit difference schemes, in this case, turn out to be unstable. Therefore, we consider an implicit one, which at an arbitrary point $w_k, k \geq 1$ has the form:

$$w_k p_k \frac{p_k - p_{k-1}}{w_k - w_{k-1}} + \frac{1}{\gamma} p_k^2 + p_k + w_k p_k + \alpha w_k = 0. \tag{36}$$

From Cauchy condition (32) we have that $p_0 = p(0) = 0$. For convenience, we use a finite difference approximation with a constant step h, i.e., $w_k = kh$. Then (36) takes the form

$$\left(k + \frac{1}{\gamma}\right) p_k^2 + (1 + kh - kp_{k-1}) p_k + \alpha kh = 0. \tag{37}$$

The roots of Equation (37) are:

$$p_k = \frac{-1 - kh + kp_{k-1}}{2(k+1/\gamma)} \pm \sqrt{\frac{(-1 - kh + kp_{k-1})^2}{4(k+1/\gamma)^2} - \frac{\alpha kh}{k+1/\gamma}}.$$

We choose the root that corresponds to $+$ sign, otherwise $p_1 \to -\gamma/(1+\gamma)$ if $h \to 0$, i.e., there is a discontinuity of the first kind at zero. So, we have the recurrent sequence:

$$p_0 = 0,\ p_k = \frac{-1 - kh + kp_{k-1}}{2(k+1/\gamma)} + \sqrt{\frac{(-1 - kh + kp_{k-1})^2}{4(k+1/\gamma)^2} - \frac{\alpha kh}{k+1/\gamma}}, k = 1, 2, \ldots \tag{38}$$

For $h > 0$, the radical expression in (38) rapidly becomes negative as k increases, i.e., the scheme is not applicable in this case. This fact goes with the results of the qualitative analysis, which showed that if $\alpha \neq \gamma/(\gamma+1)$ problem (29) and (32) for $w > 0$ does not have a solution.

We investigate the properties of the sequence p_k for $w < 0$, i.e., when $h < 0$. To do this, we formulate the following auxiliary lemma.

Lemma 1. *Let* $x > y$, $A \geq B > 0$. *Then* $x + \sqrt{x^2 + A} - y - \sqrt{y^2 + B} > 0$.

Proof. If $y \geq 0$, then the lemma is obvious. Let $-y > 0$, $y^2 > x^2$. Since $A \geq B$, the inequality holds:

$$x + \sqrt{x^2 + A} - y - \sqrt{y^2 + B} \geq x + \sqrt{x^2 + A} - y - \sqrt{y^2 + A}.$$

To prove the Lemma, it is enough to show that the right-hand side is greater than zero. We use the rule of contraries. Let $x + \sqrt{x^2 + A} - y - \sqrt{y^2 + A} < 0$, then $x - y < \sqrt{y^2 + A} - \sqrt{x^2 + A}$. Since $x > y$, we can square this inequality and collect the terms:

$$xy + A > \sqrt{(y^2 + A)(x^2 + A)}.$$

If we square this inequality one more time and collect the terms, we obtain the inequality:

$$2xy > y^2 + x^2,$$

which is wrong. The contradiction proves the Lemma. □

Now we formulate and prove the lemma about the properties of the sequence p_k.

Lemma 2. *Let $h < 0$, $\alpha > 0$, $\gamma > 0$. Then the sequence p_k is monotonically increasing with respect to k, and the estimate holds:*

$$p_k \geq -kh \min\left\{\alpha, \frac{\gamma}{\gamma+1}\right\}, \; k = 0, 1, 2 \ldots \tag{39}$$

Proof. We carry out the proof by induction on k. Assume for certainty that $\alpha \leq \gamma/(\gamma+1) = 1/(1+1/\gamma)$. Then,

$$p_1 - p_0 = p_1 \geq -\frac{(h+1)\alpha}{2} + \sqrt{\frac{(h+1)^2\alpha^2}{4} - h\alpha^2} = -\frac{(h+1)\alpha}{2} + \sqrt{\frac{(1-h)^2\alpha^2}{4}} = -\alpha h,$$

and the induction base is determined.

Let $0 = p_0 < p_1 < \ldots < p_k$. Consider the difference $p_{k+1} - p_k$. Using (38), we can rewrite it as:

$$p_{k+1} - p_k = \frac{-1 - (k+1)h + (k+1)p_k}{2(k+1+1/\gamma)} - \frac{-1 - kh + kp_{k-1}}{2(k+1/\gamma)} +$$

$$\sqrt{\frac{(-1-(k+1)h+(k+1)p_k)^2}{4(k+1+1/\gamma)^2} - \frac{\alpha(k+1)h}{k+1+1/\gamma}} - \sqrt{\frac{(-1-kh+kp_{k-1})^2}{4(k+1/\gamma)^2} - \frac{\alpha kh}{k+1/\gamma}}.$$

Consider the first difference on the right side:

$$\frac{-1 - (k+1)h + (k+1)p_k}{2(k+1+1/\gamma)} - \frac{-1 - kh + kp_{k-1}}{2(k+1/\gamma)} =$$

$$\frac{(k+1)(k+1/\gamma)(p_k - p_{k-1}) + p_{k-1}/\gamma + 1 - h/\gamma}{2(k+1+1/\gamma)(k+1/\gamma)} > 0.$$

It is valid since all terms and factors are positive both in the numerator and in the denominator by the Lemma condition and the assumption of induction. It is easy to make sure that the inequality holds:

$$-\frac{\alpha(k+1)h}{k+1+1/\gamma} > -\frac{\alpha kh}{k+1/\gamma} > 0.$$

Thus, we can apply Lemma 1 to the difference of the roots, which ensures that it is positive. Therefore, we obtain that $p_{k+1} - p_k > 0$. The monotonic increase of the sequence p_k is proved.

Let us turn to justify estimate (39). We carry out the proof again by induction on k. The induction base was determined earlier. Let $p_i \geq -i\alpha h$, $i = 1, 2 \ldots, k-1$. Then,

$$p_k \geq -\frac{1 + kh + k(k-1)\alpha h}{2(k-1+1/\alpha)} + \sqrt{\frac{(1+kh+k(k-1)\alpha h)^2}{4(k-1+1/\alpha)^2} - \frac{k\alpha h}{k-1+1/\alpha}}$$

$$= -\frac{kh\alpha}{2} - \frac{1}{2(k-1+1/\alpha)} + \sqrt{\left[\frac{1}{2(k-1+1/\alpha)} - \frac{kh\alpha}{2}\right]^2} = -kh\alpha.$$

The case $\alpha \geq \gamma/(\gamma+1)$ is treated similarly. □

Remark 4. *In the case $\alpha = \gamma/(\gamma+1)$, the double unstrict inequality* (39) *becomes an equality, and we get the previously found linear solution $p = -\alpha w$.*

With the help of the lemmas, we now prove the main theorem of this section. Let us introduce the notation:

$$\alpha_m = \min\left\{\alpha, \frac{\gamma}{\gamma+1}\right\}, \alpha_M = \max\left\{\alpha, \frac{\gamma}{\gamma+1}\right\}.$$

Theorem 2. *Problem* (29) *and* (32) *for $w \leq 0$ has a decreasing continuously differentiable solution $p = p(w)$ satisfying the inequality:*

$$\alpha_m w \leq p(w) \leq \alpha_M w \leq 0. \tag{40}$$

Proof. To prove the existence of the solution with the desired properties, we consider and estimate the difference:

$$\Delta p_k = p_k - p_{k-1} = \frac{-1 - kh + kp_{k-1}}{2(k+1/\gamma)} + \sqrt{\frac{(-1-kh+kp_{k-1})^2}{4(k+1/\gamma)^2} - \frac{\alpha kh}{k+1/\gamma}} - p_{k-1} =$$

$$-\frac{1}{2(k+1/\gamma)} + \frac{k(p_{k-1}-h)}{2(k+1/\gamma)} + \sqrt{\left[-\frac{1}{2(k+1/\gamma)} + \frac{k(p_{k-1}-h)}{2(k+1/\gamma)}\right]^2 + \frac{-\alpha kh}{k+1/\gamma}} - p_{k-1}.$$

It follows from Lemma 2 that $\Delta p_k > 0$.

Let first, as in the proof of Lemma 2, $\alpha \leq \gamma/(\gamma+1)$. Then by Lemma 2, the following chain of inequalities holds:

$$\frac{k(p_{k-1}-h)}{k+1/\gamma} \geq \frac{-\alpha kh(k-1+1/\alpha)}{k+1/\gamma} \geq \frac{-\alpha kh(k-1+1+1/\gamma)}{k+1/\gamma} = -\alpha kh > 0.$$

Hence we get that:

$$\Delta p_k \leq -\frac{1}{2(k+1/\gamma)} + \frac{k(p_{k-1}-h)}{2(k+1/\gamma)} + \sqrt{\left[-\frac{1}{2(k+1/\gamma)} + \frac{k(p_{k-1}-h)}{2(k+1/\gamma)}\right]^2 + \frac{k(p_{k-1}-h)}{(k+1/\gamma)^2}}$$

$$-p_{k-1} = -\frac{1}{2(k+1/\gamma)} + \frac{k(p_{k-1}-h)}{2(k+1/\gamma)} + \sqrt{\left[\frac{1}{2(k+1/\gamma)} + \frac{k(p_{k-1}-h)}{2(k+1/\gamma)}\right]^2} - p_{k-1}$$

$$= \frac{k(p_{k-1}-h)}{k+1/\gamma} - p_{k-1} = -\frac{p_{k-1}}{k\gamma+1} - h \leq -\frac{\gamma h}{\gamma+1}.$$

The case $\alpha \geq \gamma/(\gamma+1)$ is treated similarly and gives the estimate $0 < \Delta p_k \leq -\alpha h$. Thus, it has been shown that:

$$0 < \Delta p_k \leq -\max\left\{\alpha, \frac{\gamma}{\gamma+1}\right\}h, \; k = 1, 2, \ldots \tag{41}$$

It follows from (41) that the constructed difference scheme is stable. According to the Lax equivalence theorem, since it also has the approximation property (by construction), it is convergent. This means that the sequence of Euler polygonal lines with vertices at the points (kh, p_k) converges to a continuously differentiable solution to problem (29) and (32) if $k \to \infty, h \to 0$. Moreover, the estimates show that the solution exists for all $w \leq 0$ and decreases.

Inequality (41) also gives the upper estimate from (40). The lower estimate (40) follows from Lemma 2. □

Remark 5. *As the results of the qualitative analysis show, if $w > 0$, Problem* (29) *and* (32) *is solvable only for $\alpha = \gamma/(\gamma+1)$. This explains the divergence of series* (33) *for $\alpha \neq \gamma/(\gamma+1)$.*

7. Nonzero Initial Condition

Let us now consider the second case when the initial condition for Equation (29) is $p(0) = -\gamma$, i.e., the problem:

$$wp\frac{dp}{dw} + \frac{p^2}{\gamma} + wp + p + \alpha w = 0,\ p(0) = -\gamma. \tag{42}$$

Looking ahead, we note that this case leads to results that can be clearly interpreted from the point of view of Problem (3) and (17).

7.1. Solution in the Form of a Series

Let us show that the solution to problem (42) can be found as a power series that converges in a small neighborhood of zero. We construct the series having form (33).

From the boundary condition, we have $p_0 = p(0) = -\gamma$. To find p_1, we differentiate Equation (42) with respect to w, set $w = 0, p(0) = -\gamma$, and obtain that $p_1 = (\alpha - \gamma)/(\gamma+1)$. Similarly, we get:

$$p_2 = \frac{2\alpha(\alpha-\gamma)}{\gamma(\gamma+1)(2\gamma+1)}.$$

Thus, the induction base is found.

Assume that $p_0, p_1, \ldots, p_{k-1}, k \geq 3$ are determined. To find p_k, we differentiate Equation (42) k times with respect to w and set $w = 0$. Then we arrive at the equality (34). Resolving with respect to p_k gives:

$$p_k = \frac{1}{\gamma k+1}\left[\sum_{i=1}^{k-1} C_k^i\left(k-i+\frac{1}{\gamma}\right)p_i p_{k-i} + kp_{k-1}\right]. \tag{43}$$

You can see that all terms on the right-hand side of (43) are known by the induction hypothesis. Thus, all the coefficients of series (33) are uniquely determined by formula (43). If $\gamma = \alpha$, then $p_i = 0,\ i = 1, 2, \ldots$, i.e., the series breaks off and $p \equiv -\gamma = -\alpha$.

The local convergence of series (43) follows from Theorem 1 (see also Theorem 1 in [1]). We have not yet estimated the radius of convergence, but the results of previous studies allow us to make a reasonable assumption that it is small [37]. Thus, we have justified the following proposition.

Proposition 2. *Problem* (42) *has an analytical solution having the form of the locally convergent series* (33), *whose coefficients are determined by formula* (43). *The series breaks off if $\alpha = \gamma$, and the solution is $p = -\gamma$.*

7.2. Euler Polygonal Approximations

The constructed series locally converges in some neighborhood of the point $(0, -\gamma)$. To find the global properties of the solution to problem (42), as above, we use the Euler method. Consider the following finite difference approximation of (42):

$$w_k p_{k-1}\frac{p_k - p_{k-1}}{h} + \frac{1}{\gamma} p_k p_{k-1} + p_{k-1}(1+w_k) + \alpha w_k = 0,$$

where $w_k = kh$. Then we yield the recurrent formula:

$$p_k = \frac{1}{1+1/(k\gamma)}\left(p_{k-1} - h - \frac{1}{k} - \frac{\alpha h}{p_{k-1}}\right). \tag{44}$$

From the Cauchy condition, we have $p_0 = -\gamma$.

Lemma 3. *The following formula is valid:*

$$p_{k+1} = -\gamma - \frac{(k+1)\gamma h}{\gamma+1} - \alpha h \sum_{j=1}^{k+1} \frac{1}{p_{j-1} \prod\limits_{i=j}^{k+1} [1+1/(i\gamma)]}. \tag{45}$$

The lemma is proved by induction on k. The proof is simple and based on direct substitutions.

Lemma 3 is the basis for proving the properties of the difference scheme, which are given below.

Proposition 3. *Sequence* (45) *for $h > 0, \gamma > \alpha$ is decreasing, and $0 < (\gamma-\alpha)h/(\gamma+1) \le p_{k-1} - p_k < (\alpha/\gamma + \gamma)h/(\gamma+1)$, $\lim\limits_{k\to\infty}(p_{k-1} - p_k) = \gamma h/(\gamma+1)$.*

Proof. We carry out the proof by induction on k. Indeed,

$$p_0 - p_1 = -\gamma + \gamma + \frac{(\gamma-\alpha)h}{\gamma+1} = \frac{(\gamma-\alpha)h}{\gamma+1} > 0.$$

Thus, the induction base is found. Let $-\gamma = p_0 > p_1 > \ldots > p_{k-1}$, then

$$p_{k-1} - p_k = \frac{\gamma h}{\gamma+1} + \alpha h \sum_{j=1}^{k} \frac{1}{p_{j-1}\prod\limits_{i=j}^{k}[1+1/(i\gamma)]} - \alpha h \sum_{j=1}^{k-1} \frac{1}{p_{j-1}\prod\limits_{i=j}^{k-1}[1+1/(i\gamma)]} =$$

$$= \frac{\gamma h}{\gamma+1} - \frac{\alpha h}{k\gamma} \sum_{j=1}^{k-1} \frac{1}{p_{j-1}\prod\limits_{i=j}^{k}[1+1/(i\gamma)]} + \frac{\alpha h}{p_{k-1}[1+1/(\gamma k)]} \ge$$

$$\ge \frac{\gamma h}{\gamma+1} - \frac{\alpha h}{k\gamma p_{k-1}} \frac{k\gamma}{\gamma+1} = \frac{\gamma h}{\gamma+1} + \frac{\alpha\gamma h}{(\gamma+1)p_{k-1}} \ge \frac{\gamma h}{\gamma+1} - \frac{\alpha\gamma h}{(\gamma+1)\gamma} = \frac{(\gamma-\alpha)h}{\gamma+1} > 0.$$

On the other hand, the last estimates show that:

$$p_{k-1} - p_k < \frac{\gamma h}{\gamma+1} - \frac{\alpha h}{k\gamma} \sum_{j=1}^{k-1} \frac{1}{p_{j-1}\prod\limits_{i=j}^{k}[1+1/(i\gamma)]}$$

if we cast out the negative term. Hence, we have that:

$$p_{k-1} - p_k < \frac{\gamma h}{\gamma+1} - \frac{\alpha h}{k\gamma p_0} \sum_{j=1}^{k-1} \frac{1}{\prod\limits_{i=j}^{k}[1+1/(i\gamma)]} =$$

$$\frac{\gamma h}{\gamma+1} + \frac{\alpha h}{k\gamma^2}\frac{k}{1+1/\gamma} = \frac{\gamma+\alpha/\gamma}{\gamma+1} h < h.$$

□

Proposition 4. *Sequence* (45) *for* $h > 0, \gamma < \alpha$ *is increasing, and* $0 < (\alpha - \gamma)h/(\gamma + 1) \leq p_k - p_{k-1}$*. Moreover, there exists* $k_* < \gamma(\gamma+1)/[(\alpha - \gamma)h]$ *such that* $p_{k_*} \geq 0$.

Proof. The inequality $p_k - p_{k-1} \geq (\alpha - \gamma)h/(\gamma + 1) > 0$ is proved similarly to Proposition 3. The difference is that due to the change of the sign of $\gamma - \alpha$, the sign of the difference estimate $\Delta p_k = p_k - p_{k-1}$ changes, starting from Δp_1. Hence, in particular, it follows that $p_k \geq -\gamma + (\alpha - \gamma)kh/(\gamma + 1)$. The right side of the last inequality equals to zero for $k_* = \gamma(\gamma+1)/[(\alpha - \gamma)h]$. If the resulting value is not an integer, then it is necessary to round it with excess, and then we obtain $p_{k_*} > 0$. □

Proposition 5. *Sequence* (45) *for* $h < 0, \gamma > \alpha$ *is increasing, and for* $h < 0, \gamma < \alpha$ *is decreasing, and in both cases* $\lim_{k\to\infty} p_k = -\alpha$.

Proof. The increase and decrease of the sequence p_n for $h < 0$ are proved similarly to Propositions 3 and 4, respectively. On the other hand, since the signs in front of p_{k-1} and $1/p_{k-1}$ on the right side of (44) in this case are the same, the limit is not equal to infinity. Obviously, the limiting value p_∞ satisfies the following equation, which is obtained if we tend $k \to \infty$ in (44):

$$p_\infty = p_\infty - h - \frac{\alpha h}{p_\infty}.$$

It is easy to see that the solution is $p_\infty = -\alpha$. □

Theorem 3. *Problem* (44) *has a continuously differentiable solution, which monotonically tends to* $-\alpha$ *when* $w \to -\infty$*. For* $w > 0$*, three cases are possible:*

1. *If* $\gamma > \alpha$*, then the solution is monotonically decreasing, and the estimate holds:*

$$-\gamma - \frac{\alpha/\gamma + \gamma}{\gamma + 1} w < p \leq -\gamma - \frac{\gamma - \alpha}{\gamma + 1} w;$$

2. *If* $\gamma > \alpha$*, then the solution monotonically increases and at some point* $w = w^* < \gamma(\gamma + 1)/(\alpha - \gamma)$ *vanishes;*
3. *If* $\gamma = \alpha$*, then the solution is the constant* $p \equiv -\gamma = -\alpha$.

The theorem statement follows from Propositions 3–5 by the reasoning similar to those we carried out in the proof of Theorem 2.

8. Discussion

This section is devoted to interpreting the results obtained in the previous sections from the point of view of the corresponding diffusion waves properties. Recall that Equation (29) has been obtained from Equation (21) by changing variables. Equation (21), in turn, follows from Equation (3) if the diffusion-wave front $x = a(t)$ is a linear function.

The results for problem (29) and (32) appear to be non-physical. At any rate, we cannot interpret the negative values of w (for which the solution was constructed) from the point of view of applications. The same situation occurs to the «left branches» of the solution to problem (42).

However, the «right branches» of solutions (42), along which $w \geq 0$, allow a clear physical interpretation.

We have the function $p = p^*(w)$, which is the solution to problem 42 for $w \geq 0$. Returning to the space of variables z, w, we obtain that:

$$z = \int_0^w \frac{d\zeta}{p^*(\zeta)}. \tag{46}$$

As shown above, there are three different cases in which the function $p(w)$ behaves differently. Let us consider them separately.

Case $\gamma = \alpha$. Here $p^*(w) \equiv -\gamma$, whence we have that $v = -\gamma z$, i.e., $u = -\sigma\mu(x - \mu t)$ for $\mu\sigma = B/A$ (see Figure 4). Previously, we constructed a similar solution for the porous medium Equation [1]. In this case, the diffusion wave has the form of a plane in the space of variables t, x, and u.

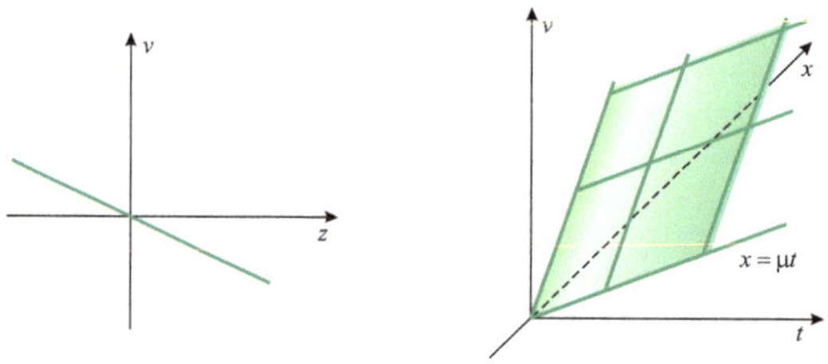

Figure 4. Solution $v(z) = -\gamma z$ and diffusion wave $u = -\sigma\mu(x - \mu t)$.

Case $\gamma > \alpha$. Then $p = p^*(w)$ is infinitely decreasing, and it is bounded upper and below by two straight lines. Hence we have that the function $w(z) = v^\theta(z)$ located between two exponents with negative powers when $z \to \infty$. Returning to the plane of variables v, z, we obtain a monotone infinitely decreasing function, which is defined for all $z \in [0, -\infty)$. The diffusion wave is a cylindrical surface in the space of variables t, x, u, and the unknown function increases with exponential velocity along the generatrix of the cylinder with distance from the wave front (Figure 5).

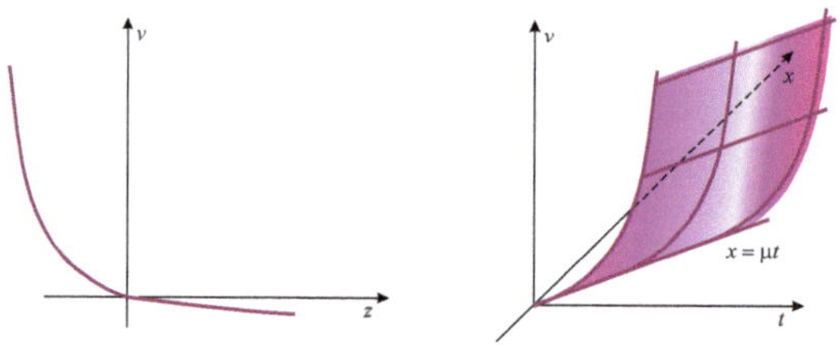

Figure 5. Solution $v(z)$ and the diffusion wave.

Case $\gamma < \alpha$. The study has shown that the function $p(w)$ first increases, and there is a point $w^* > 0$ such that $p(w^*) = 0$, $\lim\limits_{w \to w^* - 0} p'(w) = +\infty$. The point can be determined numerically, since the problem does not have singularities on the interval $[0, w^*)$. Consider the problem:

$$\frac{dw}{dp} = -\frac{wp}{\frac{p^2}{\gamma} + wp + p + \alpha w} = 0, \; w(0) = w^*, \tag{47}$$

where w is an unknown function, and p is an independent variable. It follows from the results of the qualitative analysis that the solution to problem (47) is decreasing on the ray $[0, +\infty)$, and $\lim\limits_{p \to +\infty} w(p) = +0$. Returning to the plane of variables v, z, we get the solution $v = v_*(x)$. From the original problem point of view, there exists a solution $u = v_*(x - \mu t)$, which is a solitary wave (soliton) (see Figure 6).

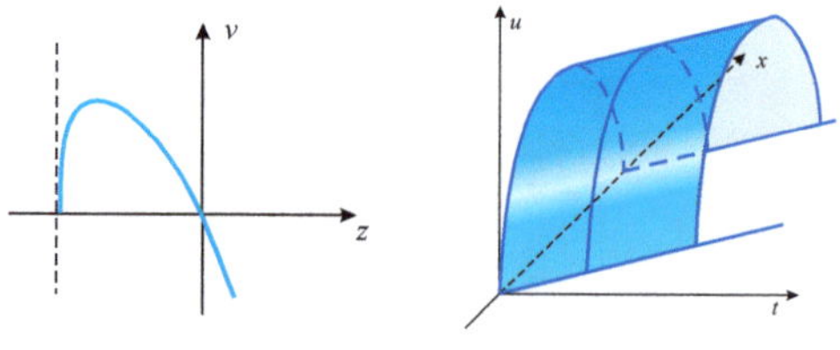

Figure 6. Solution $v(z)$ and the soliton.

Summing up, we note that for different values of the coefficients $\gamma = \sigma\theta > 0$, $\alpha = B/\mu > 0$, we obtained the same basic configurations of diffusion waves that we described earlier (see [1]) for incomplete variants of Equations (3):

- A linear heat wave for the porous medium equation;
- An infinitely increasing wave with a nonzero second derivative with distance from the wave front for the convection–diffusion equation;
- A diffusion wave in the form of a soliton for the generalized porous medium equation (the heat equation with source).

Parameter γ characterizes the diffusion and convection terms, and parameter α characterizes the source and velocity of the wave front. It seems pretty natural that if $\gamma > \alpha$, then the diffusion wave for the complete Equation (3) behaves similarly to the case without a source but with a convection term (the convection-diffusion equation). The case $\gamma < \alpha$ corresponds to the case without convection term but with a source (the generalized porous medium equation). Finally, if the parameters are equal, the diffusion wave behaves similarly to the case when there is neither a source nor a convection term (the porous medium equation).

9. Conclusions

For a second-order one-dimensional singular evolutionary equation with power nonlinearities, we studied diffusion-wave-type solutions propagating along a zero background with a finite velocity. Such properties of solutions usually appear for hyperbolic equations and are atypical for parabolic ones. Apparently, their occurrence is associated with the degeneracy mentioned above, which, in turn, is caused by vanishing the term multiplying the highest (second) derivative. Besides being a fascinating mathematical object, such solutions are also valuable for applications. They allow us to describe nonlinear filtration and diffusion processes with a finite velocity of perturbation propagation by parabolic models. Such models are considered better described physical processes at a distance from the degeneracy line.

This paper is the second step in a large research cycle started in [1]. We have considered an equation with power nonlinearities, a specification of the general equation discussed earlier. The choice of the type of functions was related to the fact that such nonlinearities usually arise in applications. Due to this, we have been able to get more profound results. Thus, in the existence and uniqueness theorem, we have chosen a more complex type of boundary conditions, raising a diffusion wave. As a result, both the technique of constructing the solution and the procedure for proving the convergence of series have become significantly more complicated. Besides, we have studied in detail one of the particular but quite natural cases, where the degree of the convection term coincides with the degree of the source. We have performed both a qualitative analysis of ODEs with the construction of phase portraits and obtained quantitative estimates for the solutions.

The most significant result is that we have shown that all the special cases for incomplete equations take place for the complete equation, and other configurations of diffusion waves do not arise. In addition, a nontrivial solution to the Cauchy problem with zero initial conditions has been found. Although this solution has no physical interpretation since it is negative, its presence is an interesting and non-obvious mathematical fact.

Further research in this direction in the short term, in our opinion, should be associated with the development of a practical computational technique for diffusion waves construction. In this context, the boundary element approach, which we have been developing in recent years in collaboration with colleagues, looks promising. It is also advisable to consider other special cases, for example, to construct and study generalized self-similar solutions to the considered problem.

In the long term, it would be helpful to increase the dimensionality and consider cases where an unknown function depends on two or three spatial variables, as well as consider systems of partial differential equations. In the end, the final stage of the research cycle should be the application of the developed model-algorithmic apparatus for solving applied problems related to modeling diffusion processes occurring in Lake Baikal.

Author Contributions: Conceptualization, A.K.; Formal analysis, A.K.; Investigation, A.K. and A.L.; Methodology, A.K.; Validation, A.L.; Visualization, A.L.; Writing—original draft, A.K.; Writing—review & editing, A.L. All authors contributed equally. All authors have read and agreed to the published version of the manuscript.

Funding: The research was funded by the Ministry of Education and Science of the Russian Federation within the framework of the project "Analytical and numerical methods of mathematical physics in problems of tomography, quantum field theory and fluid mechanics" (no. of state registration: 121041300058-1).

Institutional Review Board Statement: Not applicable.

Informed Consent Statement: Not applicable.

Conflicts of Interest: The authors declare no conflict of interest.

References

1. Kazakov, A. Solutions to Nonlinear Evolutionary Parabolic Equations of the Diffusion Wave Type. *Symmetry* **2021**, *13*, 871. [CrossRef]
2. Friedman, A. *Partial Differential Equations of Parabolic Type*; Prentice-Hall: Englewood Cliffs, NJ, USA, 1964.
3. Ladyzenskaja, O.; Solonnikov, V.; Ural'ceva, N. *Linear and Quasi-Linear Equations of Parabolic Type. Translations of Mathematical Monographs*; American Mathematical Society: Providence, RI, USA, 1988; Volume 23.
4. DiBenedetto, E. *Degenerate Parabolic Equations*; Springer: New York, NY, USA, 1993. [CrossRef]
5. Zeldovich, Y.B.; Raizer, Y.P. *Physics of Shock Waves and High-Temperature Hydrodynamic Phenomena*; Dover Publications: New York, NY, USA, 2002. [CrossRef]
6. Barenblatt, G.; Entov, V.; Ryzhik, V. *Theory of Fluid Flows through Natural Rocks*; Kluwer Academic Publishers: Dordrecht, The Netherlands, 1990.
7. Vazquez, J. *The Porous Medium Equation: Mathematical Theory*; Clarendon Press: Oxford, UK, 2007.
8. Murray, J. *Mathematical Biology: I. An Introduction, Third Edition. Interdisciplinary Applied Mathematics*; Springer: New York, NY, USA, 2002; Volume 17. [CrossRef]
9. Samarskii, A.; Galaktionov, V.; Kurdyumov, S.; Mikhailov, A. *Blow-Up in Quasilinear Parabolic Equations*; Walter de Gruyte: Berlin, Germany, 1995. [CrossRef]
10. Lu, Y.; Klingenbergm, C.; Koley, U.; Lu, X. Decay rate for degenerate convection diffusion equations in both one and several space dimensions. *Acta Math. Sci.* **2015**, *35*, 281–302. [CrossRef]
11. Polyanin, A.D. Functional separable solutions of nonlinear convection-diffusion equations with variable coefficients. *Commun. Nonlinear Sci. Numer. Simul.* **2019**, *73*, 379–390. [CrossRef]
12. Andreev, V.K.; Gaponenko, Y.A.; Goncharova, O.N.; Pukhnachev, V.V. *Mathematical Models of Convection*; Walter de Gruyte: Berlin, Germany, 2012. [CrossRef]
13. Wong, B.; Francoeur, M.; Mengüç, M.P. A Monte Carlo simulation for phonon transport within silicon structures at nanoscales with heat generation. *Int. J. Heat Mass Transf.* **2011**, *54*, 1825–1838. [CrossRef]
14. Valenzuela, C.; del Pino, L.; Curilef, S. Analytical solutions for a nonlinear diffusion equation with convection and reaction. *Phys. A Stat. Mech. Its Appl.* **2014**, *416*, 439–451. [CrossRef]
15. Mrazík, L.; Kříž, P. Porous Medium Equation in Graphene Oxide Membrane: Nonlinear Dependence of Permeability on Pressure Gradient Explained. *Membranes* **2021**, *11*, 665. [CrossRef]
16. Promislow, K.; Stockie, J.M. Adiabatic Relaxation of Convective-Diffusive Gas Transport in a Porous Fuel Cell Electrode. *SIAM J. Appl. Math.* **2001**, *62*, 180–205. [CrossRef]
17. Polyanin, A.D.; Zaitsev, V.F. *Handbook of Nonlinear Partial Differential Equations*, 2nd ed.; Chapman and Hall/CRC: New York, NY, USA, 2012. [CrossRef]
18. Zhan, H. On the Solutions of a Porous Medium Equation with Exponent Variable. *Discret. Dyn. Nat. Soc.* **2019**, *2019*, 1–15. [CrossRef]
19. Zhan, H. On Aanisotropic Parabolic Equations with a Nonlinear Convection Term Depending on the Spatial Variable. *Adv. Differ. Equ.* **2019**, *2019*. [CrossRef]
20. Kinnunen, J.; Lehtelä, P.; Lindqvist, P.; Parviainen, M. Supercaloric Functions for the Porous Medium Equation. *J. Evol. Equ.* **2018**, *19*, 249–270. [CrossRef]
21. Hayek, M. A family of analytical solutions of a nonlinear diffusion–convection equation. *Phys. A Stat. Mech. Its Appl.* **2018**, *490*, 1434–1445. [CrossRef]
22. Pudasaini, S.P.; Hajra, S.G.; Kandel, S.; Khattri, K.B. Analytical solutions to a nonlinear diffusion–advection equation. *Z. Angew. Math. Phys.* **2018**, *69*, 150. [CrossRef]

23. Lekhov, O.S.; Mikhalev, A.V. Calculation of Temperature and Thermoelastic Stresses in the Backups with Unit Collars of Combined Continuous Casting and Deformation during Steel Billet Production. Report 1. *Steel Transl.* **2020**, *50*, 877–881. [CrossRef]
24. Filimonov, M.Y.; Vaganova, N.A. *Simulation of Thermal Fields in the Permafrost With Seasonal Cooling Devices;* Volume 4: Pipelining in Northern and Offshore Environments; Strain-Based Design; Risk and Reliability; Standards and Regulations; American Society of Mechanical Engineers: New York, NY, USA, 2012. [CrossRef]
25. Kazakov, A.; Lempert, A. Existence and Uniqueness of the Solution of the Boundary-Value Problem for a Parabolic Equation of Unsteady Filtration. *J. Appl. Mech. Tech. Phys.* **2013**, *54*, 251–258. [CrossRef]
26. Filimonov, M.Y.; Korzunin, L.G.; Sidorov, A.F. Approximate Methods for Solving Nonlinear Initial Boundary-Value Problems Based on Special Constructions of Series. *Russ. J. Numer. Anal. Math. Model.* **1993**, *8*, 101–126. [CrossRef]
27. Ismaiel, A.; Filimonov, M.Y. *Rotating Range Sensor Approached for Mobile Robot Obstacle Detection and Collision Avoidance Applications;* Thermophysical Basis of Energy Technologies (TBET 2020); AIP Publishing: College Park, MD, USA, 2021. [CrossRef]
28. Kazakov, A.L.; Kuznetsov, P.A. On the Analytic Solutions of a Special Boundary Value Problem for a Nonlinear Heat Equation in Polar Coordinates. *J. Appl. Ind. Math.* **2018**, *812*, 227–235. [CrossRef]
29. Rubinstein, L.I. *The Stefan Problem;* Translations of Mathematical Monographs; American Mathematical Society: Providence, RI, USA, 1971.
30. Gupta, S.C. *The Classical Stefan Problem: Basic Concepts, Modelling and Analysis with Quasi-Analytical Solutions and Methods;* Elsevier: Amsterdam, The Netherlands, 2017.
31. Kazakov, A.; Spevak, L.; Nefedova, O.; Lempert, A. On the Analytical and Numerical Study of a Two-Dimensional Nonlinear Heat Equation with a Source Term. *Symmetry* **2020**, *12*, 921. [CrossRef]
32. Leontiev, N.E.; Roshchin, E.I. Exact Solutions to the Deep Bed Filtration Problem for Low-Concentration Suspension. *Mosc. Univ. Mech. Bull.* **2020**, *75*, 96–101. [CrossRef]
33. Kudryashov, N.A.; Sinelshchikov, D.I. Analytical Solutions for Problems of Bubble Dynamics. *Phys. Lett. A* **2015**, *379*, 798–802. [CrossRef]
34. Kazakov, A.L.; Orlov, S.S.; Orlov, S.S. Construction and study of exact solutions to a nonlinear heat equation. *Sib. Math. J.* **2018**, *59*, 427–441. [CrossRef]
35. Kazakov, A.L. On exact solutions to a heat wave propagation boundary-value problem for a nonlinear heat equation. *Sib. Electron. Math. Rep.* **2019**, *16*, 1057–1068. [CrossRef]
36. Andronov, A.A.; Leontovich, E.A.; Gordon, I.I.; Maier, A.G. *Qualitative Theory of Second-Order Dynamic Systems;* Israel Program for Scientific Translations distributed by Halstead Press, a division of J; Wiley: Jerusalem, NY, USA, 1973.
37. Kazakov, A.L.; Spevak, L.F.; Lee, M.G. On the Construction of Solutions to a Problem with a Free Boundary for the Non-linear Heat Equation. *J. Sib. Fed. Univ. Math. Phys.* **2020**, *13*, 694–707. [CrossRef]

mathematics

Article

Dufour and Soret Effect on Viscous Fluid Flow between Squeezing Plates under the Influence of Variable Magnetic Field

Muhammad Kamran Alam [1,*], Khadija Bibi [1], Aamir Khan [1] and Samad Noeiaghdam [2,3]

1 Department of Pure and Applied Mathematics, The University of Haripur, Haripur 22620, Pakistan; p109954@nu.edu.pk (K.B.); aamir.khan@uoh.edu.pk (A.K.)
2 Industrial Mathematics Laboratory, Baikal School of BRICS, Irkutsk National Research Technical University, 664074 Irkutsk, Russia; snoei@istu.edu or noiagdams@susu.ru
3 Department of Applied Mathematics and Programming, South Ural State University, Lenin Prospect 76, 454080 Chelyabinsk, Russia
* Correspondence: mkalam@uoh.edu.pk

Abstract: The aim of this article is to investigate the effect of mass and heat transfer on unsteady squeeze flow of viscous fluid under the influence of variable magnetic field. The flow is observed in a rotating channel. The unsteady equations of mass and momentum conservation are coupled with the variable magnetic field and energy equations. By using some appropriate similarity transformations, the partial differential equations obtained are then converted into a system of ordinary differential equations and are solved by Homotopy Analysis Method (HAM). The influence of the natural parameters are investigated for the velocity field components, magnetic field components, heat and mass transfer. A direct effect of the squeeze Reynold number is observed on both concentration and temperature. Moreover, increasing the magnetic Reynold number shows an increase in the fluid temperature, but in the case of concentration, an inverse relation is observed. Furthermore, a decreasing effect of the Dufour number is observed on both concentration and temperature distribution. Besides, in case of the Soret number, a direct effect is observed on concentration, but an inverse effect can be seen on temperature distribution. Different effects are shown through graphs in this study and an error analysis is also presented through tables and graphs.

Keywords: unsteady squeeze flow; viscous fluid; heat transfer; mass transfer; Dufour effect; Soret effect; HAM

Citation: Alam, M.K.; Bibi, K.; Khan, A.; Noeiaghdam, S. Dufour and Soret Effect on Viscous Fluid Flow between Squeezing Plates under the Influence of Variable Magnetic Field. *Mathematics* **2021**, *9*, 2404. https://doi.org/10.3390/math9192404

Academic Editors: Vladimir Iosifovich Semenov and Almudena del Pilar Marquez Lozano

Received: 14 August 2021
Accepted: 19 September 2021
Published: 27 September 2021

1. Introduction

The flow of a fluid squeezed between two parallel plates approaching one another is called a squeeze flow. The unsteady squeezing flow between two rotating plates is regarded as one of the most important study areas due to its extensive applications in science and technology. Among these are hydrodynamic lubrication, polymer technology, biomechanics and aerodynamic heating. The interaction of conducting fluids with electromagnetic fields is widely known as Magento-Hydro Dynamics (MHD). The use of an MHD fluid as a lubricant in industrial applications is appealing because it prevents the unanticipated variation of lubricant viscosity with temperature under such high working conditions, and thus, it has gained the interest of many researchers, as the unsteady squeezing flow between parallel plates was considered for viscous MHD fluid by Siddiqui et al. [1]. Further, Erik Sweet [2] investigated the analytical solution for a viscous fluid flow between moving parallel plates in an unstable MHD flow. They used the homotopy analysis method to find the solution, which indicated that the magnetic field's strength has a significant impact on the flow. Later on, Murty et al. [3] observed the electrically conducting fluid in a two-phase MHD convective flow under the action of a constant transverse magnetic field through an inclined channel in a rotating system. Onyango et al. [4] discussed an unsteady MHD flow of viscous fluid between two parallel plates under a constant pressure gradient. Khan et al. [5] observed the flow of a viscous fluid between compressing parallel plates under the influence of a varying magnetic field. They investigated the entropy generation due to magnetic fields, fluid friction and heat transfer in a two-dimensional flow problem. Muhammad et al. [6] discussed the squeezing MHD flow between two parallel plates using Jeffrey fluid. MHD fluid flow between two parallel plates was investigated by Verma et al. [7]. Later on, Hayat et al. [8] analytically treated the squeeze flow of MHD nano fluids between two parallel plates. Moreover, T. Linga Raju [9] discussed the MHD two fluid flow of ionized gases and investigated the effect of hall current on temperature distribution. The effect of magnetohydrodynamics on a fluid film was then observed by E A Hamza [10]; he studied the squeezed flow between two surfaces while rotation was added to the surfaces. Unsteady Couette flow was then studied by Das et al. [11], where the flow was unsteady and the MHD effect was added. The flow was observed in a rotating system.

A viscous fluid flow between rotating parallel plates with varying but constant angular velocities was investigated by Parter et al. [12]. In addition, [13] also made some observations about the flow of viscous fluid between two parallel rotating plates. Further on, [14] K.R. Rajagopal also studied second ordered fluid flowing in a rotating system, and later on, the MHD double diffusive flow of nanofluids was studied by Tripathi et al. [15], where the flow was observed in a rotating channel with viscous dissipation and hall effect.

The MHD flow of viscous fluids in a rotating frame was also studied in cylindrical coordinates as was discussed by Hughes et al. [16]. They examined the lubrication flow of such viscous fluids between rotating parallel disks. In addition, Elshekh et al. [17] observed the flow of fluid film squeezed between rotating parallel disks where an external magnetic field was applied. The influence of a changing magnetic field on the unsteady squeezing flow of viscous fluids between rotating discs was also examined by Shah et al. [18]. The squeezing unsteady flow of MHD fluid between two disks was also discussed by Ganji et al. [19], who observed the flow with suction or injection involved. Between squeezing discs moving at various velocities, the effects of MFD viscosity and magnetic-field-based (MFD) thermosolutal convection of the fluid dynamics are examined by Khan et al. [20].

The unsteady squeeze flow of viscous fluids has been studied in three-dimensional rotating systems by several researchers. Recently, Munawar et al. [21] studied the squeeze flow of viscous fluids in a three-dimensional rotating system. The flow was considered between parallel plates with the lower stretching plate kept porous. Further on, Alzahrani et al. [22] numerically treated the squeezed flow of viscous fluid between rotating parallel plates in a three-dimensional system and examined the effect of Dufour and Soret numbers. Similar work has been done on third grade nanofluids in a three-dimensional rotating system where the thermophoresis effect and Brownian motion was observed by Shah et al. [23]. In addition, the thin film flow of Darcy Forchheimer hydromagnetic nanofuid between rotating parllel disks in a three-dimensional system was discussed by Riasat et al. [24]. They examined the importance of the magnetic Reynold number in such a system. Moreover, Fiza et al [25] examined the flow of Jeffrey fluid in a three-dimensional rotating system.

Srinivas et al. [26] studied the impact of heat transfer on the flow of MHD fluid through a pipe whose walls are expanding or contracting. The simultaneous effect of induced magnetic forces and buoyancy forces on heat transfer was investigated by Sparrow et al. [27]. Furthermore, Khaled et al. [28] discussed the flow of fluid and its heat transfer over a horizontal surface, placed in a free stream, externally squeezed. The effect of Radiation on MHD fluid and its mass transfer was studied by Hayat et al. [29], who observed the flow on a stretching sheet which was kept porous. Mahmood et al. [30] analyzed the flow of viscous fluid over a permeable sensor surface with the squeeze effect and also observed the heat transfer effect for such a flow phenomenon. In addition, the Hiemenz flow over a stretching surface was examined by Tsai et al. [31], where the medium was porous. The heat and mass transfer was studied theoretically for Dufour's and Soret effect. The Heat and Mass transfer of a viscous fluid between squeezing plates was further observed by Mustafa et al. [32]. Furthermore, Yasmin et al. [33] studied non-Newtonian micropolar fluids in a magnetohydrodynamics flow on a curved sheet with stretching effect and discussed the mass and heat transfer for such fluids. Heat and mass transfer in a three-dimensional flow of nanofluids in a rotating system was examined by Sheikholeslami et al. [34]. Noeiaghdam et al. studied the numerical analysis of a natural convection-driven flow of a non-Newtonian power-law fluid in a trapezoidal enclosure with a U-Shaped constructal [35–39].

The above existing literature shows that no study has been conducted so far on the three-dimensional squeeze flow of viscous fluids between two parallel rotating plates under the influence of the variable magnetic field. Additionally, the effect of variable magnetic field on mass transfer in Cartesian coordinates is new work in this area. Hence, the suggested work is the best approach toward such problems and is a way of motivation for researchers bringing a new idea of studying the flow between unsteady rotating parallel plates.

2. Modeling and Formulation of the Physical Problem

We consider an incompressible viscous fluid flow between two squeezing plates, separated by a distance $D(t) = l(1-\beta t)^{1/2}$, where l is the spacing between plates at time $t = 0$ and β is the expansion and contraction parameter whose values lies between $(0, 1)$ for squeezing phenomenon, as shown in Figure 1. With an angular velocity of Ω_u, the upper plate rotates whereas the lower plate is moving with angular velocity Ω_l. The effect of variable magnetic field M is added externally which produces the induced magnetic field B with the following components: B_x, B_y and B_z.

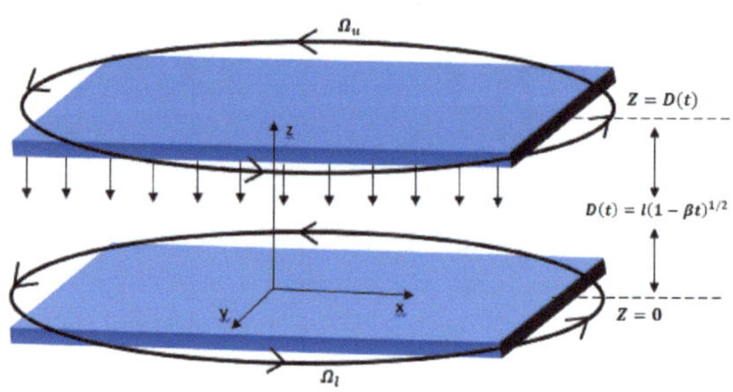

Figure 1. Geometry of the flow problem.

The system of coordinates selected is Cartesian coordinates. The origin is fixed at the center of the lower plate, in which the x-axis is taken along the horizontal axis and the z-axis is perpendicular to the two plates (along the vertical axis). The rotation of plates is along the y-axis. The flow between the plates occurs due to the rotatory motion of plates and the motion of upper plate towards the lower plate, i.e., the squeezing effect. The effect of gravity on fluid is negligible. Now, we will observe the velocity profile of the given fluid, the effect of magnetic field on the velocity of fluid, and the heat and mass transfer for these viscous fluids in a three-dimensional system. Thus, the vector form of continuity, momentum, magnetic field, energy and concentration equations are given by:

$$\frac{\partial u}{\partial x} + \frac{\partial v}{\partial y} + \frac{\partial w}{\partial z} = 0, \tag{1}$$

Navier–Stokes equation x-component:

$$\rho\left[\frac{\partial u}{\partial t}+u\frac{\partial u}{\partial x}+v\frac{\partial u}{\partial y}+w\frac{\partial u}{\partial z}\right]=-\frac{\partial P}{\partial x}+\mu\left[\frac{\partial^2 u}{\partial x^2}+\frac{\partial^2 u}{\partial y^2}+\frac{\partial^2 u}{\partial z^2}\right]+ \frac{1}{\mu_2}\left[B_z\frac{\partial B_x}{\partial z}-B_z\frac{\partial B_z}{\partial x}-B_y\frac{\partial B_y}{\partial x}+B_y\frac{\partial B_x}{\partial y}\right], \tag{2}$$

Navier–Stokes equation y-component:

$$\rho\left[\frac{\partial v}{\partial t}+u\frac{\partial v}{\partial x}+v\frac{\partial v}{\partial y}+w\frac{\partial v}{\partial z}\right]=-\frac{\partial p}{\partial y}+\mu\left[\frac{\partial^2 v}{\partial x^2}+\frac{\partial^2 v}{\partial y^2}+\frac{\partial^2 v}{\partial z^2}\right]+ \frac{1}{\mu_2}\left[B_x\frac{\partial B_y}{\partial x}-B_x\frac{\partial B_x}{\partial y}-B_z\frac{\partial B_z}{\partial y}+B_z\frac{\partial B_y}{\partial z}\right], \tag{3}$$

Navier–Stokes equation z-component:

$$\rho\left[\frac{\partial w}{\partial t}+u\frac{\partial w}{\partial x}+v\frac{\partial w}{\partial y}+w\frac{\partial w}{\partial z}\right]=-\frac{\partial p}{\partial z}+\mu\left[\frac{\partial^2 w}{\partial x^2}+\frac{\partial^2 w}{\partial y^2}+\frac{\partial^2 w}{\partial z^2}\right]+ \frac{1}{\mu_2}\left[B_y\frac{\partial B_z}{\partial y}-B_y\frac{\partial B_y}{\partial z}-B_x\frac{\partial B_x}{\partial z}+B_x\frac{\partial B_z}{\partial x}\right], \tag{4}$$

Magnetic field equation x-component:

$$\frac{\partial B_x}{\partial t}=\left[u\frac{\partial B_y}{\partial y}+B_y\frac{\partial u}{\partial y}-v\frac{\partial B_x}{\partial y}-B_x\frac{\partial v}{\partial y}-w\frac{\partial B_x}{\partial z}-B_x\frac{\partial w}{\partial z}+u\frac{\partial B_z}{\partial z}+B_z\frac{\partial u}{\partial z}\right]+ \frac{1}{\delta\mu_2}\left[\frac{\partial^2 B_x}{\partial x^2}+\frac{\partial^2 B_x}{\partial y^2}+\frac{\partial^2 B_x}{\partial z^2}\right], \tag{5}$$

Magnetic field equation y-component:

$$\frac{\partial B_y}{\partial t}=\left[v\frac{\partial B_z}{\partial z}+B_z\frac{\partial v}{\partial z}-w\frac{\partial B_y}{\partial z}-B_y\frac{\partial w}{\partial z}-u\frac{\partial B_y}{\partial x}-B_y\frac{\partial u}{\partial x}+v\frac{\partial B_x}{\partial x}+B_x\frac{\partial v}{\partial x}\right]+ \frac{1}{\delta\mu_2}\left[\frac{\partial^2 B_y}{\partial x^2}+\frac{\partial^2 B_y}{\partial y^2}+\frac{\partial^2 B_y}{\partial z^2}\right], \tag{6}$$

Magnetic field equation z-component:

$$\frac{\partial B_z}{\partial t}=\left[w\frac{\partial B_x}{\partial x}+B_x\frac{\partial w}{\partial x}-u\frac{\partial B_z}{\partial x}-B_z\frac{\partial u}{\partial x}-v\frac{\partial B_z}{\partial y}-B_z\frac{\partial v}{\partial y}+w\frac{\partial B_y}{\partial y}+B_y\frac{\partial w}{\partial y}\right]+ \frac{1}{\delta\mu_2}\left[\frac{\partial^2 B_z}{\partial x^2}+\frac{\partial^2 B_z}{\partial y^2}+\frac{\partial^2 B_z}{\partial z^2}\right], \tag{7}$$

Heat transfer equation:

$$\rho\left[\frac{\partial T}{\partial t}+u\frac{\partial T}{\partial x}+v\frac{\partial T}{\partial y}+w\frac{\partial T}{\partial z}\right]=\frac{\kappa}{C_p}\left[\frac{\partial^2 T}{\partial x^2}+\frac{\partial^2 T}{\partial y^2}+\frac{\partial^2 T}{\partial z^2}\right]+ \frac{1}{C_p}\nabla q_r+\frac{D\kappa_t}{C_sC_p}\left[\frac{\partial^2 C}{\partial x^2}+\frac{\partial^2 C}{\partial y^2}+\frac{\partial^2 C}{\partial z^2}\right], \tag{8}$$

and similarly, the equation for mass transfer is given by

$$\rho\left[\frac{\partial C}{\partial t}+u\frac{\partial C}{\partial x}+v\frac{\partial C}{\partial y}+w\frac{\partial C}{\partial z}\right]=D\left[\frac{\partial^2 C}{\partial x^2}+\frac{\partial^2 C}{\partial y^2}+\frac{\partial^2 C}{\partial z^2}\right]+ \frac{D\kappa_t}{C_sT_m}\left[\frac{\partial^2 T}{\partial x^2}+\frac{\partial^2 T}{\partial y^2}+\frac{\partial^2 T}{\partial z^2}\right]. \tag{9}$$

where ρ is density of fluid, μ_2 is permittivity of free space, μ is viscosity, P is pressure, V is velocity of fluid, B is magnetic field, T is temperature, C is concentration, κ_t is thermal diffusion ratio, D is diffusion coefficient, C_p is specific heat at constant pressure, κ is conductivity and ∇q_r is the radiative heat flux, given as,

$$\nabla q_r = \frac{16\sigma^s T_u^3}{3\kappa^a}\frac{\partial^2 T}{\partial z^2} \tag{10}$$

Here, κ^a and σ^s are the mean absorption coefficient and Stefan-Boltzmann constant, respectively. The boundary conditions for the above fluid flow are given as follows, at lower plate where $z = 0$, $u = 0$, $v = \frac{\Omega_l x}{1-\beta t}$, $w = 0$, $B_x = B_y = B_z = 0$, $T = T_l$, $C = C_l$.

Moreover, observe the conditions at upper plate where $z = D(t)$, $u = 0$, $v = \frac{\Omega_u x}{1-\beta t}$, $w = \frac{dD(t)}{dt}$, $B_x = 0$ $B_y = \frac{xN_0}{1-\beta t}$, $B_z = \frac{-\beta M_0}{(1-\beta t)^{1/2}}$, $T = T_u$, $C = C_u$. where $D(t) = l(-\beta t)^{1/2}$. Now, using the following transformation to convert the above partial differential equations to ordinary differential equations (ODEs), $u = \frac{\beta x}{(1-\beta t)} f'(\eta)$, $v = \frac{\Omega_l x}{(1-\beta t)} g(\eta)$, $w = \frac{-\beta l}{(1-\beta t)^{1/2}} f(\eta)$, $B_x = \frac{\beta x M_0}{l(1-\beta t)} m'(\eta)$, $B_y = \frac{xN_0}{(1-\beta t)} n(\eta)$, $B_z = \frac{-\beta M_0}{(1-\beta t)^{1/2}} m(\eta)$, $\theta = \frac{T - T_u}{T_l - T_u}$, $\phi = \frac{C - C_u}{C_l - C_u}$, $\eta = \frac{z}{l(1-\beta t)^{1/2}}$.

After non-dimensionlizing the above equations and setting boundary conditions, they will be converted to the following ODEs,

$$\begin{aligned} f'''' &= S_z\left[3f'' + (\eta - 2f)f''' - 2f'f''\right] + \\ &2S_z M_x^2\left[2R_m\left(mm' + \eta mm'' - fmm'' + m^2 f''\right) - m'm''\right], \end{aligned} \tag{11}$$

$$g'' = S_z\left[2g + \eta g' + 2f'g - 2g'f\right] - 2S_z M_x M_y\left[m'n - n'm\right], \tag{12}$$

$$m'' = R_m\left[m + \eta m' - 2m'f + 2f'm\right], \tag{13}$$

$$n'' = R_m\left[2n + \eta n' - 2n'f + 2f'n - 2\left(\frac{M_x}{M_y}\right)\left(g'm - m'g\right)\right], \tag{14}$$

$$\theta''(3 + 4R_d) + 3D_u P_{rn}\phi'' + 3S_z P_{rn}\left(f\theta' - \eta\theta'\right) = 0, \tag{15}$$

$$\phi'' + S_{rt}S_{ch}\theta'' + S_z S_{ch}\left(2f\phi' - \eta\phi'\right) = 0. \tag{16}$$

where $S_z = \frac{\beta l^2}{2\nu}$ denotes the squeezing Reynold number, $M_x = \frac{M_0}{l\sqrt{\rho\mu_2}}$ represents the magnetic field strength along x-axis , $M_y = \frac{N_0}{\Omega_l\sqrt{\rho\mu_2}}$ is the magnetic field strength along x-axis, $R_m = S_z B_t$ is given by $R_m = \left(\frac{\beta l^2}{2\nu}\right)(\nu\sigma\mu_2)$, $R_d = \frac{4\sigma^s T_u^3}{K^a K}$ is the radiation parameter, $D_u = \frac{DK_T K(C_l - C_u)}{C_s \nu C_p (T_l - T_u)\nu}$ is Dufour number and $P_{rn} = \frac{\nu C_p}{K}$ is Prandtl number. The boundary conditions become the following forms, $f(0) = 0$, $f'(0) = 0$, $g(0) = 1$, $m(0) = 0$, $n(0) = 0$, $\theta(0) = 1$, $\phi(0) = 1$. $f(1) = \frac{1}{2}$, $f'(1) = 0$, $g(1) = \frac{\Omega_u}{\Omega_l} = S$, $m(1) = 1$, $n(1) = 1$, $\theta(1) = 0$, $\phi(1) = 0$.

3. Method of Solution

An analytical technique is used to find the solution of the Equations (11)–(16), known as the Homotopy Analysis Method. We will express the functions f, g, m, n, θ and ϕ (where f, g, m, n, θ and ϕ are the functions of η), and $\eta^{q^*}, q^* \geq 0$ are used as a set of base functions,

$$f_n = \sum_{q^*=0}^{\infty} a_{q^*} \eta^{q^*} \tag{17}$$

$$g_n = \sum_{q^*=0}^{\infty} b_{q^*} \eta^{q^*} \tag{18}$$

$$m_n = \sum_{q^*=0}^{\infty} c_{q^*} \eta^{q^*} \tag{19}$$

$$n_n = \sum_{q^*=0}^{\infty} d_{q^*} \eta^{q^*} \tag{20}$$

$$\theta_n = \sum_{q^*=0}^{\infty} f_{q^*} \eta^{q^*} \tag{21}$$

$$\phi_n = \sum_{q^*=0}^{\infty} g_{q^*} \eta^{q^*} \tag{22}$$

where the constant co-efficients $a_{q^*}, b_{q^*}, c_{q^*}, d_{q^*}, f_{q^*}$ and g_{q^*} are to be determined. Initial approximations are chosen as follows:

$$f_0 = 1.5 * \eta^2 - \eta^3; \tag{23}$$

$$g_0 = (S - 1) * \eta + 1; \tag{24}$$

$$m_0 = \eta \tag{25}$$

$$n_0 = \eta \tag{26}$$

$$\theta_0 = 1 - \eta \tag{27}$$

$$\phi_0 = 1 - \eta \tag{28}$$

then, the auxiliary operators are chosen,

$$\begin{aligned} \pounds_f = \frac{\partial^4}{\partial \eta^4}, \pounds_g = \frac{\partial^2}{\partial \eta^2}, \pounds_\theta = \frac{\partial^2}{\partial \eta^2}, \\ \pounds_m = \frac{\partial^2}{\partial \eta^2}, \pounds_n = \frac{\partial^2}{\partial \eta^2}, \pounds_\phi = \frac{\partial^2}{\partial \eta^2} \end{aligned} \tag{29}$$

with the following properties,

$$\pounds_f(q^*_{1^*} \eta^3 + q^*_{2^*} \eta^2 + q^*_{3^*} \eta + q^*_{4^*}) = 0 \tag{30}$$

$$\pounds_g(q^*_{5^*} \eta + q^*_{6^*}) = 0 \tag{31}$$

$$\pounds_m(q^*_{7^*} \eta + q^*_{8^*}) = 0 \tag{32}$$

$$\pounds_n(q^*_{9^*} \eta + q^*_{10^*}) = 0 \tag{33}$$

$$\pounds_\theta(q^*_{11^*} \eta + q^*_{12^*}) = 0 \tag{34}$$

$$\pounds_\theta(q^*_{13^*} \eta + q^*_{14^*}) = 0 \tag{35}$$

where q^*_{1*}, q^*_{2*}, q^*_{3*}, q^*_{4*}, q^*_{5*}, q^*_{6*}, q^*_{7*}, q^*_{8*}, q^*_{9*}, q^*_{10*}, q^*_{11*}, q^*_{12*}, q^*_{13*} and q^*_{14*} are arbitrary constants.

We can obtain the *Zeroth* order deformation as:

$$(1-s)\pounds_g[g(\eta;s)-g_0(\eta)] = s\hbar_g\aleph_g[f(\eta;s),g(\eta;s), m(\eta;s),n(\eta;s)] \tag{36}$$

$$(1-s)\pounds_m[m(\eta;s)-m_0(\eta)] = s\hbar_m\aleph_m[f(\eta;s),m(\eta;s)] \tag{37}$$

$$(1-s)\pounds_n[n(\eta;s)-n_0(\eta)] = s\hbar_n\aleph_n[f(\eta;s),g(\eta;s), m(\eta;s),n(\eta;s)] \tag{38}$$

$$(1-s)\pounds_\theta[n(\eta;s)-\theta_0(\eta)] = s\hbar_\theta\aleph_\theta[f(\eta;s),\theta(\eta;s),\phi(\eta;s)] \tag{39}$$

$$(1-s)\pounds_\phi[n(\eta;s)-\phi_0(\eta)] = s\hbar_\phi\aleph_\phi[f(\eta;s),\theta(\eta;s),\phi(\eta;s)] \tag{40}$$

From Equations (14)–(19), the nonlinear operators are defined as,

$$\aleph_f[f(\eta;s),m(\eta;s)] = \frac{\partial^4 f(\eta;s)}{\partial\eta^4} - S_z\left(3\frac{\partial^2 f(\eta;s)}{\partial\eta^2} + (\eta-2f)\frac{\partial^3 f(\eta;s)}{\partial\eta^3} - 2\frac{\partial f(\eta;s)}{\partial\eta}\frac{\partial^2 f(\eta;s)}{\partial\eta^2}\right) - 2S_zM_z^2\left(M\frac{\partial^3 M(\eta;s)}{\partial\eta^3} - \frac{\partial M(\eta;s)}{\partial\eta}\frac{\partial^2 M(\eta;s)}{\partial\eta^2}\right) \tag{41}$$

$$\aleph_g[f(\eta;s),g(\eta;s),m(\eta;s),n(\eta;s)] = \frac{\partial^2 g(\eta;s)}{\partial\eta^2} - S_z\left(2g+\eta\frac{\partial g(\eta;s)}{\partial\eta} + 2\frac{\partial f(\eta;s)}{\partial\eta}g - 2\frac{\partial g(\eta;s)}{\partial\eta}f\right) - 2S_zM_zM_y\left(\frac{\partial m(\eta;s)}{\partial\eta}n - m\frac{\partial n(\eta;s)}{\partial\eta}\right) \tag{42}$$

$$\aleph_m[f(\eta;s),m(\eta;s)] = \frac{\partial^2 m(\eta;s)}{\partial\eta^2} - R_m\left(m+\eta\frac{\partial m(\eta;s)}{\partial\eta} - 2\left(f\frac{\partial m(\eta;s)}{\partial\eta} - m\frac{\partial f(\eta;s)}{\partial\eta}\right)\right) \tag{43}$$

$$\aleph_n[f(\eta;s),g(\eta;s),m(\eta;s),n(\eta;s)] = \frac{\partial^2 n(\eta;s)}{\partial\eta^2} - R\left(2n+\eta\frac{\partial n(\eta;s)}{\partial\eta} - 2\left(\frac{\partial n(\eta;s)}{\partial\eta}f - n\frac{\partial f(\eta;s)}{\partial\eta}\right) + 2\frac{\partial M_z}{\partial M_y}\left(\frac{\partial g(\eta;s)}{\partial\eta}m - g\frac{\partial m(\eta;s)}{\partial\eta}\right)\right) \tag{44}$$

$$\aleph_\theta[f(\eta;s),\theta(\eta;s),\phi(\eta;s)] = \frac{\partial^2\theta(\eta;s)}{\partial\eta^2}\left(3+4R_d\right) + 3D_uP_{rn}\frac{\partial^2\phi(\eta;s)}{\partial\eta^2} + 3S_zP_{rn}\left(2f\frac{\partial\theta(\eta;s)}{\partial\eta} - \eta\frac{\partial\theta(\eta;s)}{\partial\eta}\right) \tag{45}$$

$$\aleph_\phi[f(\eta;s),\theta(\eta;s),\phi(\eta;s)] = \frac{\partial^2\phi(\eta;s)}{\partial\eta^2} + S_{rt}S_{ch}\frac{\partial^2\theta(\eta;s)}{\partial\eta^2} + S_zS_{ch}\left(2f\frac{\partial\phi(\eta;s)}{\partial\eta} - \eta\frac{\partial\phi(\eta;s)}{\partial\eta}\right) \tag{46}$$

where s is a fixed parameter, nonlinear parameters are $\aleph_f$, $\aleph_g$, $\aleph_m$ and $\aleph_n$, while $\hbar_f$,$\hbar_g$, $\hbar_m$, $\hbar_n$, $\hbar_\theta$ and $\hbar_\phi$ are the nonzero auxiliary parameters.

For $s = 0$ and $s = 1$, we have

$$\begin{aligned} f(\eta,0) &= f_o, \quad f(\eta,1) = f(\eta) \\ g(\eta,0) &= g_o, \quad g(\eta,1) = g(\eta) \\ m(\eta,0) &= m_o, \quad m(\eta,1) = m(\eta) \\ n(\eta,0) &= n_o, \quad n(\eta,1) = n(\eta) \\ \theta(\eta,0) &= \theta_o, \quad \theta(\eta,1) = \theta(\eta) \\ \phi(\eta,0) &= \phi_o, \quad \phi(\eta,1) = \phi(\eta) \end{aligned} \tag{47}$$

so as s varies from 0 to 1, the exact solutions $f(\eta)$, $g(\eta)$, $n(\eta)$, $n(\eta)$, $\theta(\eta)$ and $\phi(\eta)$ can be obtained from initial guesses.

For these functions, the Taylor's series are given by:

$$f(\eta;s) = f_0 + \sum_{n=1}^{\infty} s^n f_n(\eta) \tag{48}$$

$$g(\eta;s) = g_0 + \sum_{n=1}^{\infty} s^n g_n(\eta) \tag{49}$$

$$m(\eta;s) = m_0 + \sum_{n=1}^{\infty} s^n m_n(\eta) \tag{50}$$

$$n(\eta;s) = n_0 + \sum_{n=1}^{\infty} s^n n_n(\eta) \tag{51}$$

$$\theta(\eta;s) = \theta_0 + \sum_{n=1}^{\infty} s^n \theta_n(\eta) \tag{52}$$

$$\phi(\eta;s) = \phi_0 + \sum_{n=1}^{\infty} s^n \phi_n(\eta) \tag{53}$$

$$\begin{aligned} f_n(\eta) &= \frac{1}{n!}\frac{\partial^n f(\eta;s)}{\partial \eta^n}\bigg|_{s=0}, \quad g_n(\eta) = \frac{1}{n!}\frac{\partial^n g(\eta;s)}{\partial \eta^n}\bigg|_{s=0} \\ m_n(\eta) &= \frac{1}{n!}\frac{\partial^n m(\eta;s)}{\partial \eta^n}\bigg|_{s=0}, \quad n_n(\eta) = \frac{1}{n!}\frac{\partial^n n(\eta;s)}{\partial \eta^n}\bigg|_{s=0} \\ \theta_n(\eta) &= \frac{1}{n!}\frac{\partial^n \theta(\eta;s)}{\partial \eta^n}\bigg|_{s=0}, \quad \phi_n(\eta) = \frac{1}{n!}\frac{\partial^n \phi(\eta;s)}{\partial \eta^n}\bigg|_{s=0} \end{aligned} \tag{54}$$

It can be noted that in the above series, convergence strongly depends upon $\hbar_f$, $\hbar_g$, $\hbar_m$, $\hbar_n$, $\hbar_\theta$ and $\hbar_\phi$.

Assuming that these nonzero auxiliary parameters are chosen so that equations converge at $s = 1$, we thus obtain

$$f(\eta) = f_0 + \sum_{n=1}^{\infty} f_n(\eta) \tag{55}$$

$$g(\eta) = g_0 + \sum_{n=1}^{\infty} g_n(\eta) \tag{56}$$

$$m(\eta) = m_0 + \sum_{n=1}^{\infty} m_n(\eta) \tag{57}$$

$$n(\eta) = n_0 + \sum_{n=1}^{\infty} n_n(\eta) \tag{58}$$

$$\theta(\eta) = \theta_0 + \sum_{n=1}^{\infty} \theta_n(\eta) \tag{59}$$

$$\phi(\eta) = \phi_0 + \sum_{n=1}^{\infty} \theta_n(\eta) \tag{60}$$

Differentiating n-times Equations (40)–(45) with respect to s and putting $s = 0$, we have

$$\pounds_f[f_n(\eta) - \chi_n f_{n-1}(\eta)] = \hbar_f R_{f,n}(\eta) \tag{61}$$

$$\pounds_g[g_n(\eta) - \chi_n g_{n-1}(\eta)] = \hbar_g R_{g,n}(\eta) \tag{62}$$

$$\pounds_m[m_n(\eta) - \chi_n m_{n-1}(\eta)] = \hbar_m R_{m,n}(\eta) \tag{63}$$

$$\pounds_n[n_n(\eta) - \chi_n n_{n-1}(\eta)] = \hbar_n R_{n,n}(\eta) \tag{64}$$

$$\pounds_\theta[\theta_n(\eta) - \chi_n \theta_{n-1}(\eta)] = \hbar_\theta R_{\theta,n}(\eta) \tag{65}$$

$$\pounds_\phi[\phi_n(\eta) - \chi_n \phi_{n-1}(\eta)] = \hbar_\phi R_{\phi,n}(\eta) \tag{66}$$

with the given boundary conditions,

$$\begin{aligned} & f_n(0) = 0, \quad f_n'(0) = 0, \quad g_n(0) = 1, \quad m_n(0) = 0, \quad n_n(0) = 0, \\ & \theta_n(0) = 1, \quad \phi_n(0) = 1 \\ & f_n(1) = 0.5, \quad f_n'(1) = 0, \quad g_n(1) = S, \quad m_n(1) = 1, \quad n_n(1) = 1, \\ & \theta_n(1) = 0, \quad \phi_n(1) = 0 \end{aligned} \tag{67}$$

$$\begin{aligned} R_{f,n}(\eta) = {} & f''''_{n-1}(\eta) - S_z\Big(3f''_{n-1}(\eta) + (\eta)f'''_{n-1}(\eta) - \\ & 2f'_{n-1}(\eta)f''_{n-1}(\eta) - 2\textstyle\sum_{j=0}^{n-1} f_j(\eta)f'''_{n-j-1}(\eta)\Big) + 2S_z M_z^2 \\ & \Big(2R_m \textstyle\sum_{j=0}^{n-1} m_j(\eta)\Big[m'_{n-j-1}(\eta) + \eta m''_{n-j-1}(\eta) + \\ & m_{n-j-1}(\eta)f''_{n-j-1}(\eta)\Big] - m'_{n-1}(\eta)m''_{n-1}(\eta)\Big) \end{aligned} \tag{68}$$

$$\begin{aligned} R_{g,n}(\eta) = {} & g''_{n-1}(\eta) - S_z\Big(2g_{n-1}(\eta) + (\eta)g'_{n-1}(\eta) + 2\sum_{j=0}^{n-1} \\ & \Big[g_j(\eta)f'_{n-j-1}(\eta) - f_j(\eta)g'_{n-j-1}(\eta)\Big] + 2S_z M_z M_y \\ & \sum_{j=0}^{n-1}\Big(n_j(\eta)m'_{n-j-1}(\eta) - n_{j(\eta)}m'_{n-j-1}(\eta)\Big) \end{aligned} \tag{69}$$

$$\begin{aligned} R_{m,n}(\eta) = {} & m''_{n-1}(\eta) - R_m\Big[m_{n-1}(\eta) + (\eta)m'_{n-1}(\eta) + \\ & 2\sum_{j=0}^{n-1}\Big(m_j(\eta)f'_{n-j-1}(\eta) - f_j(\eta)m'_{n-j-1}(\eta)\Big)\Big] \end{aligned} \tag{70}$$

$$\begin{aligned} R_{n,n}(\eta) = {} & n''_{n-1}(\eta) - R_m\Big[2n_{n-1}(\eta) + (\eta)n'_{n-1}(\eta) + \\ & 2\sum_{j=0}^{n-1}\Big(n_j(\eta)f'_{n-j-1}(\eta) - f_j(\eta)n'_{n-j-1}(\eta)\Big) \\ & -2\frac{M_z}{M_y}\Big(m_j(\eta)g'_{n-j-1}(\eta) - g_j(\eta)m'_{n-j-1}(\eta)\Big)\Big] \end{aligned} \tag{71}$$

$$\begin{aligned} R_{\theta,n}(\eta) = {} & \theta''_{n-1}(\eta)(3 + 4R_d) + 3D_u P_{rn}\phi''_{n-1}(\eta) + \\ & 3S_z P_{rn}\sum_{j=0}^{n-1}\Big(2f_j(\eta)\theta'_{n-j-1}(\eta) - \eta\theta'_{n-j-1}(\eta)\Big) \end{aligned} \tag{72}$$

$$R_{\phi,n}(\eta) = \phi''_{n-1}(\eta) + S_{rt}S_{ch}\theta''_{n-1}(\eta) + S_z S_{ch} \sum_{j=0}^{n-1} \left(2f_j(\eta)\phi'_{n-j-1}(\eta) - \eta\phi'_{n-j-1}(\eta) \right) \tag{73}$$

and $\chi_n = \begin{cases} 1, \ if \ n > 1, \quad and \ 0, \ if \ n = 1. \end{cases}$

Finally, the general solution can be written as

$$f_n(\eta) = \int_0^\eta \int_0^\eta \int_0^\eta \int_0^\eta \hbar_f R_{f,n}(z) dzdzdzdz + \chi_n f_{n-1} + K_{1*}\eta^3 + K_{2*}\eta^2 + K_{3*}\eta + K_{4*} \tag{74}$$

$$g_n(\eta) = \int_0^\eta \int_0^\eta \hbar_g R_{g,n}(z) dzdz + \chi_n g_{n-1} + q^*_{5*}\eta + q^*_{6*} \tag{75}$$

$$m_n(\eta) = \int_0^\eta \int_0^\eta \hbar_m R_{m,n}(z) dzdz + \chi_n m_{n-1} + q^*_{7*}\eta + q^*_{8*} \tag{76}$$

$$n_n(\eta) = \int_0^\eta \int_0^\eta \hbar_n R_{n,n}(z) dzdz + \chi_n n_{n-1} + q^*_{9*}\eta + q^*_{10*} \tag{77}$$

$$\theta_n(\eta) = \int_0^\eta \int_0^\eta \hbar_\theta R_{\theta,n}(z) dzdz + \chi_n \theta_{n-1} + q^*_{11*}\eta + q^*_{12*} \tag{78}$$

$$\phi_n(\eta) = \int_0^\eta \int_0^\eta \hbar_\phi R_{\phi,n}(z) dzdz + \chi_n \phi_{n-1} + q^*_{13*}\eta + q^*_{14*} \tag{79}$$

thus, for $f(\eta)$, $g(\eta)$, $m(\eta)$, $n(\eta)$, $\theta(\eta)$ and $\phi(\eta)$, the exact solution becomes

$$\begin{aligned} f(\eta) &\approx \sum_{m=0}^{n} f_m(\eta) \\ g(\eta) &\approx \sum_{m=0}^{n} g_m(\eta) \\ m(\eta) &\approx \sum_{m=0}^{n} m_m(\eta) \\ n(\eta) &\approx \sum_{m=0}^{n} n_m(\eta) \\ \theta(\eta) &\approx \sum_{m=0}^{n} \theta_m(\eta) \\ \phi(\eta) &\approx \sum_{m=0}^{n} \phi_m(\eta) \end{aligned} \tag{80}$$

4. Optimal Convergence Control Parameter

It should be noted that the nonzero auxiliary parameters $\hbar_f$, $\hbar_g$, $\hbar_m$, $\hbar_n$, $\hbar_\theta$ and $\hbar_\phi$ are contained in the series solutions (59)–(64), through which the rate of the homotopy series solutions and convergence region can be determined. The average residual error was used to obtain the optimal values of $\hbar_f$, $\hbar_g$, $\hbar_m$, $\hbar_n$, $\hbar_\theta$ and $\hbar_\phi$:

$$\varepsilon_n^f = \frac{1}{L+1} \sum_{j=0}^{L} \left[\aleph_f * \left(\sum_{i=0}^{n} f*(\eta), \sum_{i=0}^{n} m*(\eta) \right)_{m=j\delta m} \right]^2 d\eta \tag{81}$$

$$\varepsilon_n^m = \frac{1}{L+1} \sum_{j=0}^{L} \left[\aleph_m * \left(\sum_{i=0}^{n} f*(\eta), \sum_{i=0}^{n} m*(\eta) \right)_{m=j\delta m} \right]^2 d\eta \tag{82}$$

$$\varepsilon_n^g = \frac{1}{L+1}\sum_{j=0}^{L}\left[\aleph_g * \left(\sum_{i=0}^{n} f*(\eta), \sum_{i=0}^{n} g*(\eta), \sum_{i=0}^{n} m*(\eta), \sum_{i=0}^{n} n*(\eta)\right)_{m=j\delta m}\right]^2 d\eta \quad (83)$$

$$\varepsilon_n^n = \frac{1}{L+1}\sum_{j=0}^{L}\left[\aleph_n * \left(\sum_{i=0}^{n} f*(\eta), \sum_{i=0}^{n} g*(\eta), \sum_{i=0}^{n} m*(\eta), \sum_{i=0}^{n} n*(\eta)\right)_{m=j\delta m}\right]^2 d\eta \quad (84)$$

$$\varepsilon_n^\theta = \frac{1}{L+1}\sum_{j=0}^{L}\left[\aleph_\theta * \left(\sum_{i=0}^{n} f*(\eta), \sum_{i=0}^{n} \theta*(\eta), \sum_{i=0}^{n} \phi*(\eta)\right)_{m=j\delta m}\right]^2 d\eta \quad (85)$$

$$\varepsilon_n^\phi = \frac{1}{L+1}\sum_{j=0}^{L}\left[\aleph_\phi * \left(\sum_{i=0}^{n} *f(\eta), \sum_{i=0}^{n} \theta*(\eta), \sum_{i=0}^{n} \phi*(\eta)\right)_{m=j\delta m}\right]^2 d\eta \quad (86)$$

Additionally,

$$\varepsilon_n^t = \varepsilon_n^f + \varepsilon_n^g + \varepsilon_n^m + \varepsilon_n^n + \varepsilon_n^\theta + \varepsilon_n^\phi \quad (87)$$

where the total squared residual error is ε_n^t. By applying Mathematica package BVPh 2.0 [40], we can minimize total average squared residual error. To acquire the local optimal convergence control parameters, the command Minimize was used.

5. Results and Analysis

Taking 10^{-40} as a maximum residual error, the problem is solved with the HAM BVPh 2.0 package. An investigation is made using 40th-order approximations. The provision of error analysis supports the authentication of results for many relevant physical parameters in Figure 2 and from the results given in Table 1.

Table 2 is provided to determine the equations' inaccuracy from momentum, magnetic field, energy and transportation equations. An increase in the order of approximation can be seen, and the solution obtained from these equations converges to exact analysis.

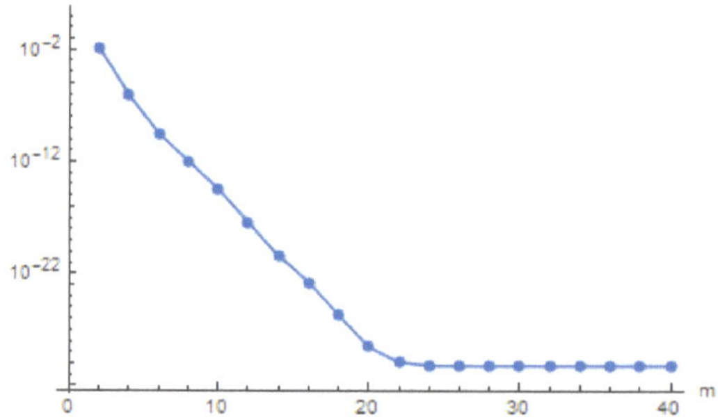

Figure 2. Total residual error with $S_z = -0.25$, $M_x = 1$, $M_y = 3$, $R_m = 0.5$, $S = 1$, $R_d = 1$, $D = 0.5$, $P_{rn} = 0.05$, $S_{rt} = -0.25$ and $S_{ch} = 0.5$.

The total residual error is visually shown in Figure 2 and numerically in Table 1 for different orders of approximation. It is seen that at the 30th order of approximation, the solution converges and the error nearly disappears. Table 2 illustrates the best convergence control parameter values for several approximation orders. From Table 3, we can observe the estimated values of velocity, magnetic field, concentration and temperature for different values of η. It also shows the solution is accurate by verification through the given boundary conditions of the problem (since the boundary conditions can be verified by observing the obtained numerical values at different points). The convergence of HAM solution for skin friction, magnetic flux, heat flux and mass flux for different orders of approximation is given in Table 4. It can be observed that the solution is convergent for fifth order of approximation. The analysis is carried out up to the 40th order of approximation.

Table 1. Estimating the total residual error for different orders of approximation.

m	ε_m^f	ε_m^g	ε_m^m	ε_m^n	ε_m^θ	ε_m^ϕ
1	0.03811	0.00475	0.00066	0.00088	8.4×10^{-9}	1.0×10^{-7}
5	1.5×10^{-9}	1.0×10^{-9}	7.6×10^{-12}	2.1×10^{-8}	1.5×10^{-15}	3.3×10^{-17}
10	1.0×10^{-18}	9.5×10^{-16}	2.8×10^{-21}	2.7×10^{-14}	4.6×10^{-21}	1.2×10^{-23}
15	5.9×10^{-28}	1.0×10^{-21}	1.2×10^{-30}	4.1×10^{-20}	1.0×10^{-26}	3.0×10^{-29}
20	2.8×10^{-30}	1.3×10^{-27}	2.8×10^{-33}	6.5×10^{-26}	2.4×10^{-32}	1.0×10^{-33}
25	2.8×10^{-30}	1.8×10^{-32}	4.5×10^{-33}	1.6×10^{-31}	6.0×10^{-35}	6.8×10^{-34}
30	2.9×10^{-30}	1.2×10^{-32}	3.0×10^{-33}	1.5×10^{-32}	6.2×10^{-35}	6.8×10^{-34}
35	2.9×10^{-30}	1.2×10^{-32}	3.0×10^{-33}	1.5×10^{-32}	6.2×10^{-35}	6.8×10^{-34}
40	2.9×10^{-30}	1.2×10^{-32}	3.0×10^{-33}	1.5×10^{-32}	6.2×10^{-35}	6.8×10^{-34}

Table 2. Optimal values of convergence control parameters in comparison of different orders of approximation.

Order	h_f	h_g	h_m	h_n	h_θ	h_ϕ	ε_m^t
2	−1.0151	−1.0459	−0.8810	−0.7341	−0.1047	−0.1040	0.00203
3	−0.9471	−1.1027	−0.9423	−0.8706	−0.1110	−1.0261	4.90×10^{-6}
4	−0.9678	−1.0804	−0.9360	−0.8889	−0.1053	−0.9786	4.94×10^{-8}
5	−0.9290	−1.0601	−0.9433	−0.9223	−0.1140	−1.0579	2.20×10^{-10}
6	−0.9583	−1.0441	−0.9435	−0.9342	−0.1369	−1.0050	2.87×10^{-12}
7	−1.1097	−1.0693	−0.8652	−0.9240	−0.1172	−1.0967	3.26×10^{-12}

Table 3. Estimated values for velocity, magnetic field components, temperature and concentration in correspondance with different values of η.

η	$f(\eta)$	$g(\eta)$	$m(\eta)$	$n(\eta)$	$\theta(\eta)$	$\phi(\eta)$
0	0	1	0	0	1	1
0.1	0.013945	1.038790	0.083334	0.090262	0.899994	0.899811
0.2	0.051898	1.070930	0.167626	0.179204	0.799991	0.799707
0.3	0.107937	1.095050	0.253977	0.268342	0.699991	0.699719
0.4	0.176060	1.110120	0.343567	0.359189	0.599995	0.599833
0.5	0.250219	1.115470	0.437564	0.453152	0.500000	0.500005
0.6	0.324349	1.110810	0.537016	0.551440	0.400006	0.400177
0.7	0.392389	1.096270	0.642747	0.654974	0.300009	0.300290
0.8	0.448305	1.072300	0.755239	0.764292	0.200010	0.200299
0.9	0.486125	1.039810	0.874512	0.879464	0.100006	0.100192
1	0.5	1	1	1	0	0

Table 4. The convergence of HAM solution for skin friction, magnetic flux, heat flux and mass flux.

m	$f''(\eta)$	$-g'(\eta)$	$-m'(\eta)$	$-n'(\eta)$	$-\theta'(\eta)$	$-\phi'(\eta)$
1	2.9631400	−0.3922459	0.8311292	−0.9204676	1.0000650	1.0020067
5	2.9804441	−0.4157942	−0.8318657	−0.9141364	1.0000668	1.0020675
10	2.9804441	−0.4157877	−0.8318658	−0.9141418	1.0000668	1.002067
15	2.9804441	−0.4157877	−0.8318658	−0.9141418	1.0000668	1.002067
20	2.9804441	−0.4157877	−0.8318658	−0.9141418	1.0000668	1.002067
25	2.9804441	−0.4157877	−0.8318658	−0.9141418	1.0000668	1.002067
30	2.9804441	−0.4157877	−0.8318658	−0.9141418	1.0000668	1.002067
35	2.9804441	−0.4157877	−0.8318658	−0.9141418	1.0000668	1.002067
40	2.9804441	−0.4157877	−0.8318658	−0.9141418	1.0000668	1.002067

In Tables 5–13, the numerical results for skin friction, magnetic flux, heat flux and mass flux are investigated. It is noted that increasing the squeeze parameter S_z decreases $f''(0)$, $-n'(0)$ $-\theta'(0)$ and $-\phi'(0)$, while oppositely $-g'(0)$ and $-m'(0)$ increases with increasing S_z. Increasing the values of M_y shows a certain increasing effect on both $-g'(0)$ and $-n'(0)$ while the effect on $f''(0)$, $-m'(0)$, $-\theta'(0)$ and $-\phi'(0)$ is negligible, as can be seen through Table 6. Similarly, for increasing Mx there is a gradual increase in the values of $f''(0)$, $-g'(0)$ and $-m'(0)$ while the values of $-n'(0)$, $-\theta'(0)$ and $-\phi'(0)$ show gradual decreases, as shown in Table 7.

Table 5. Effect of S_z on skin friction, magnetic flux, heat flux and mass flux.

S_z	$f''(\eta)$	$-g'(\eta)$	$-m'(\eta)$	$-n'(\eta)$	$-\theta'(\eta)$	$-\phi'(\eta)$
−0.01	2.999403	−0.017073	−0.831885	−0.569941	0.999976	0.999247
−0.25	2.985167	−0.467599	−0.831912	−0.553833	0.999397	0.981150
−0.75	2.956173	−1.768705	−0.831980	−0.505905	0.998187	0.943287
−1.25	2.928189	−4.069057	−0.832065	−0.418285	0.996968	0.905179

Table 6. Effect of M_y on skin friction, magnetic flux, heat flux and mass flux.

M_y	$f''(\eta)$	$-g'(\eta)$	$-m'(\eta)$	$-n'(\eta)$	$-\theta'(\eta)$	$-\phi'(\eta)$
1	2.985167	−0.457030	−0.831912	−0.121005	0.999397	0.981150
3	2.985167	−0.467600	−0.831912	−0.553832	0.999397	0.981150
5	2.985167	−0.478166	−0.831912	−0.640398	0.999397	0.981150
7	2.985167	−0.488735	−0.831912	−0.677498	0.999397	0.981150

Table 7. Effect of M_z on skin friction, magnetic flux, heat flux and mass flux.

M_z	$f''(\eta)$	$-g'(\eta)$	$-m'(\eta)$	$-n'(\eta)$	$-\theta'(\eta)$	$-\phi'(\eta)$
1	2.980444	−0.415788	−0.831866	−0.914141	0.999398	0.981167
1.5	2.985167	−0.435894	−0.831912	−0.986660	0.999397	0.981150
2	2.991858	−0.466561	−0.831978	−1.060309	0.999397	0.981125
2.5	3.000600	−0.508235	−0.832064	−1.135544	0.999396	0.981093

Further, Table 8 depicts the impact of magnetic Reynold number on skin friction showing that with increase in the values of R_m, the values of $-g'(0)$ increases gradually, but on the other hand, $f''(0)$, $-m'(0)$, $-n'(0)$ and $-\phi'(0)$ decreases, and the effect on $-\theta'(0)$ is negligible. Similarly, from Table 9, it can be seen that the Dufour number has a direct effect on both $-\theta'(0)$ and $-\phi'(0)$. The effect of Radiation Parameter R_d can be depicted through Table 10, showing that $-\theta'(0)$ and $-\phi'(0)$ increases with increasing R_d, while on the other hand, the effect on $f''(0)$, $-g'(0)$, $-m'(0)$ and $-n'(0)$ is negligible.

Table 11 depicts the impact of Prandtl number on skin friction, showing that with increase in the values of P_{rn}, the values of $-\theta'(0)$ and $-\phi'(0)$ decrease gradually. The effect of S_{rt} number from Table 12 depicts that $-\theta^{'}(0)$ decreases while $-\phi^{'}(0)$ increases by increasing the values of S_{rt}. The effect of Schmidt number S_{ch} can be seen through Table 13. It can be observed that the S_{ch} number has a direct effect $-\theta^{'}(0)$ and inverse effect on $-\phi^{'}(0)$. Furthermore, Figures 3–5 represent the error profile for velocity ($f(\eta)$ and $g(\eta)$) and magnetic field components, -i.e., $m(\eta)$ and $n(\eta)$, as well as temperature and concentration, i.e., $\theta(\eta)$ and $\phi(\eta)$, respectively. A three-dimensional profile for the velocity, magnetic field, concentration and temperature distribution is also represented in Figures 6–8, showing that the flow variables satisfy the given boundary conditions. The impacts of several related flow parameters on velocity and magnetic field components are visually depicted.

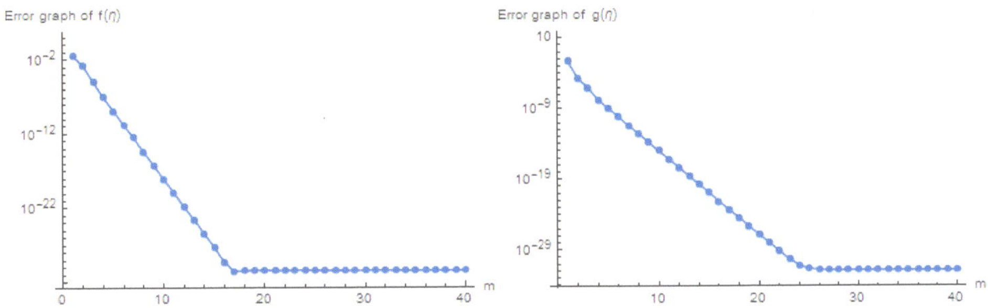

Figure 3. Error profile for $f(\eta)$ and $g(\eta)$ with $S_z = -0.25$, $M_x = 1$, $M_y = 3$, $R_m = 0.5$, $S = 1$, $R_d = 1$, $D = 0.5$, $P_{rn} = 0.05$, $S_{rt} = -0.25$ and $S_{ch} = 0.5$.

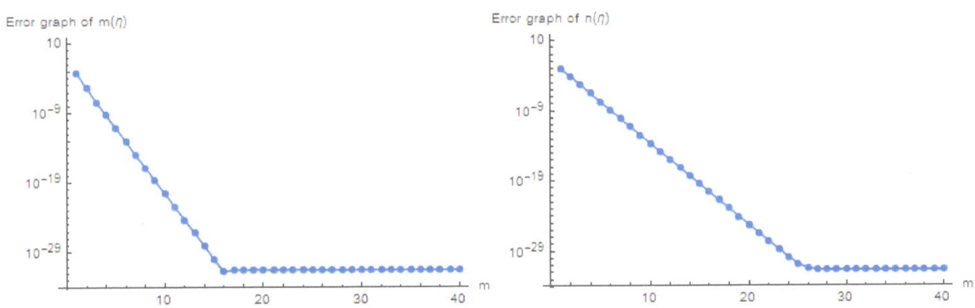

Figure 4. Error profile for $m(\eta)$ and $n(\eta)$ with $S_z = -0.25$, $M_x = 1$, $M_y = 3$, $R_m = 0.5$, $S = 1$, $R_d = 1$, $D = 0.5$, $P_{rn} = 0.05$, $S_{rt} = -0.25$ and $S_{ch} = 0.5$.

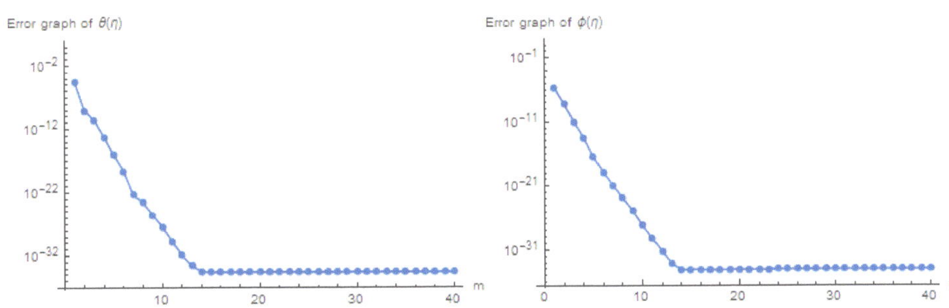

Figure 5. Error profile for $\theta(\eta)$ and $\phi(\eta)$ with $S_z = -0.25$, $M_x = 1$, $M_y = 3$, $R_m = 0.5$, $S = 1$, $R_d = 1$, $D = 0.5$, $P_{rn} = 0.05$, $S_{rt} = -0.25$ and $S_{ch} = 0.5$.

Table 8. Effect of R_m on skin friction, magnetic flux, heat flux and mass flux.

R_m	$f''(\eta)$	$-g'(\eta)$	$-m'(\eta)$	$-n'(\eta)$	$-\theta'(\eta)$	$-\phi'(\eta)$
0.1	2.976998	−0.408251	−0.962689	−0.921462	0.998762	0.981100
0.5	2.977633	−0.414660	−0.831838	−0.663347	0.998762	0.981098
1	2.977560	−0.419649	−0.701154	−0.433268	0.998762	0.981097
1.5	2.976902	−0.422670	−0.597416	−0.271630	0.998762	0.981099

Table 9. Effect of D_u on skin friction, magnetic flux, heat flux and mass flux.

D_u	$f''(\eta)$	$-g'(\eta)$	$-m'(\eta)$	$-n'(\eta)$	$-\theta'(\eta)$	$-\phi'(\eta)$
0.1	2.985167	−0.467599	−0.831912	−0.553833	0.999236	0.981129
0.75	2.985167	−0.467599	−0.831912	−0.553833	0.999498	0.981162
1.25	2.985167	−0.467599	−0.831912	−0.553833	0.999699	0.981188
1.75	2.985167	−0.467599	−0.831912	−0.553833	0.999900	0.981213

Table 10. Effect of R_d on skin friction, magnetic flux, heat flux and mass flux.

R_d	$f''(\eta)$	$-g'(\eta)$	$-m'(\eta)$	$-n'(\eta)$	$-\theta'(\eta)$	$-\phi'(\eta)$
0.1	2.977633	−0.414660	−0.831838	−0.663347	0.998762	0.981098
0.5	2.977633	−0.414660	−0.831838	−0.663347	0.999158	0.981147
1	2.977633	−0.414660	−0.831838	−0.663347	0.999398	0.981177
1.5	2.977633	−0.414660	−0.831838	−0.663347	0.999532	0.981194

Table 11. Effect of P_{rn} on skin friction, magnetic flux, heat flux and mass flux.

P_{rn}	$f''(\eta)$	$-g'(\eta)$	$-m'(\eta)$	$-n'(\eta)$	$-\theta'(\eta)$	$-\phi'(\eta)$
0.05	2.977633	−0.414660	−0.831838	−0.663347	0.999532	0.981194
0.1	2.977633	−0.414660	−0.831838	−0.663347	0.999064	0.981135
0.25	2.977633	−0.414660	−0.831838	−0.663347	0.997668	0.980960
0.5	2.977633	−0.414660	−0.831838	−0.663347	0.995361	0.980671

Table 12. Effect of S_{rt} on skin friction, magnetic flux, heat flux and mass flux.

S_{rt}	$f''(\eta)$	$-g'(\eta)$	$-m'(\eta)$	$-n'(\eta)$	$-\theta'(\eta)$	$-\phi'(\eta)$
0.1	3.000040	−0.222019	−0.831954	−0.622051	0.999758	0.992499
0.5	3.000040	−0.222019	−0.831954	−0.622051	0.999758	0.992548
1	3.000040	−0.222019	−0.831954	−0.622051	0.999758	0.992609
1.5	3.000040	−0.222019	−0.831954	−0.622051	0.999757	0.992670

Table 13. Effect of S_{ch} on skin friction, magnetic flux, heat flux and mass flux.

S_{ch}	$f''(\eta)$	$-g'(\eta)$	$-m'(\eta)$	$-n'(\eta)$	$-\theta'(\eta)$	$-\phi'(\eta)$
0.5	2.985167	−0.478167	−0.831912	−0.640398	0.999397	0.981150
1	2.985167	−0.478167	−0.831912	−0.640398	0.999598	0.962350
1.5	2.985167	−0.478167	−0.831912	−0.640398	0.999799	0.943600
2	2.985167	−0.478167	−0.831912	−0.640398	1.000000	0.924901

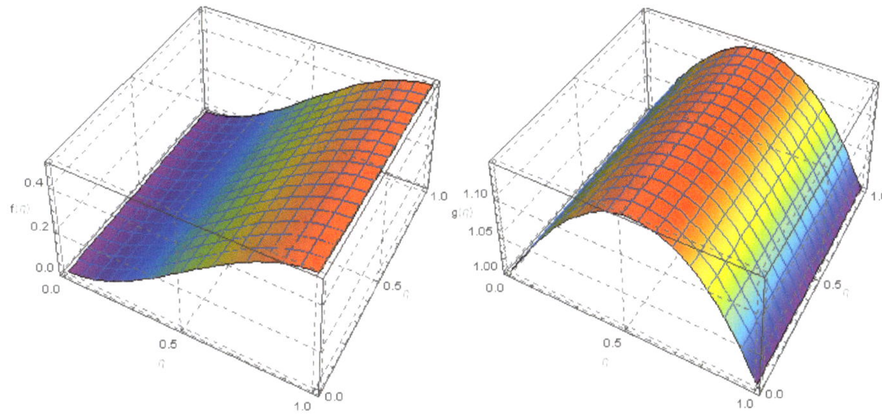

Figure 6. 3D graph for $f(\eta)$ and $g(\eta)$ with $S_z = -0.25$, $M_x = 1$, $M_y = 3$, $R_m = 0.5$, $S = 1$, $R_d = 1$, $D = 0.5$, $P_{rn} = 0.05$, $S_{rt} = -0.25$ and $S_{ch} = 0.5$.

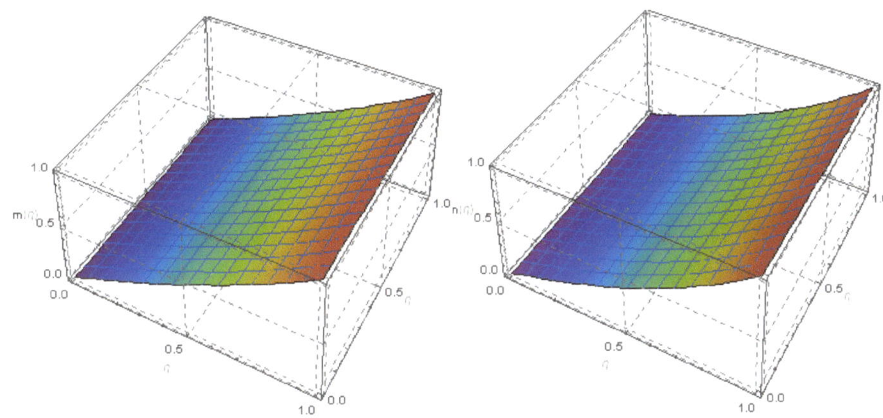

Figure 7. 3D graph for $m(\eta)$ and $n(\eta)$ with $S_z = -0.25$, $M_x = 1$, $M_y = 3$, $R_m = 0.5$, $S = 1$, $R_d = 1$, $D = 0.5$, $P_{rn} = 0.05$, $S_{rt} = -0.25$ and $S_{ch} = 0.5$.

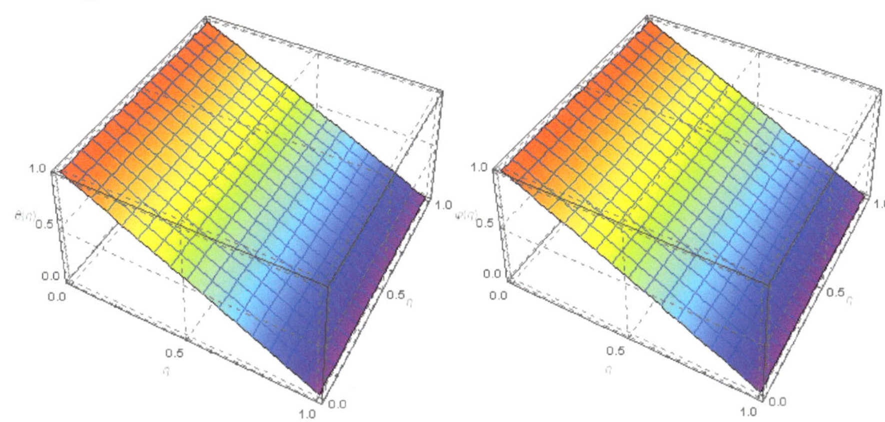

Figure 8. 3D graph for $\theta(\eta)$ and $\phi(\eta)$ with $S_z = -0.25$, $M_x = 1$, $M_y = 3$, $R_m = 0.5$, $S = 1$, $R_d = 1$, $D = 0.5$, $P_{rn} = 0.05$, $S_{rt} = -0.25$ and $S_{ch} = 0.5$.

6. Discussion

In this section, the effects of different involved flow parameters are discussed graphically on velocity and magnetic field components. The effect of squeezing Reynold numbers can be seen in Figures 9–12. It is observed that for fix values of the other parameters, i.e., M_x, M_y, R_m, S, R_d, D, P_{rn}, S_{rt} and S_{ch}, it is clear that increasing the squeeze Reynold number (moving upper disc towards lower disc with increasing order pattern) has a direct effect on the velocity components in both the y- and z-direction. On the other hand, in the x-direction, the velocity increases initially but shows a decreasing effect as $\eta \to 1$, where, as in the case of magnetic field, the increase in squeeze Reynold number results in a decrease in magnetic field component along the z-direction, while a direct relation is observed for the y-component of the magnetic field, i.e., increasing the squeeze number causes an increase in the magnetic field along the y-direction. On both concentration and temperature distribution, the direct effect of the squeeze number is observed.

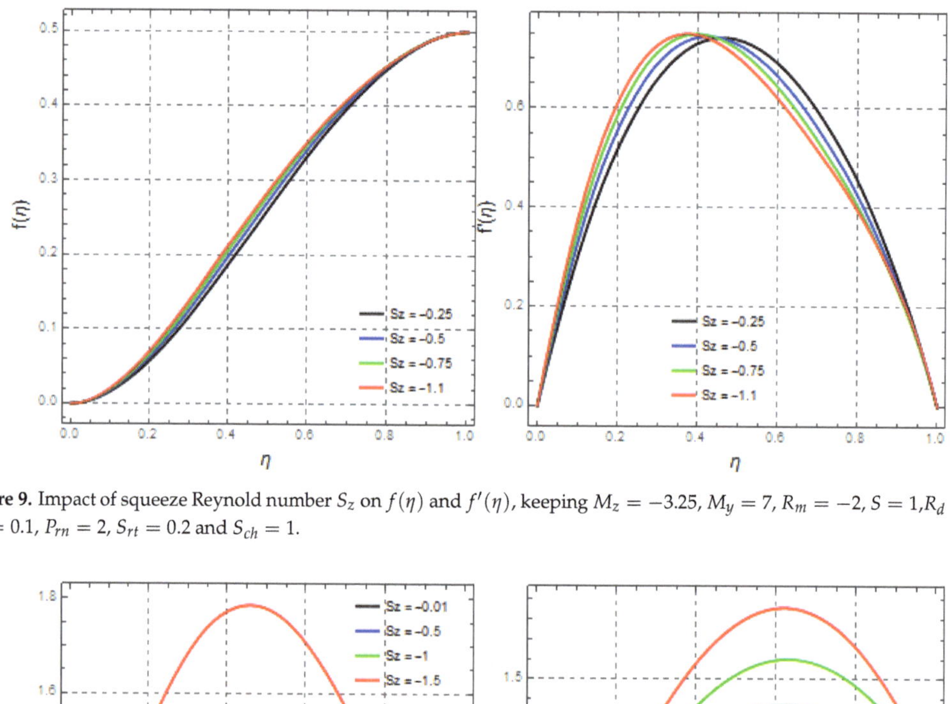

Figure 9. Impact of squeeze Reynold number S_z on $f(\eta)$ and $f'(\eta)$, keeping $M_z = -3.25$, $M_y = 7$, $R_m = -2$, $S = 1$,$R_d = 0.5$, $D_u = 0.1$, $P_{rn} = 2$, $S_{rt} = 0.2$ and $S_{ch} = 1$.

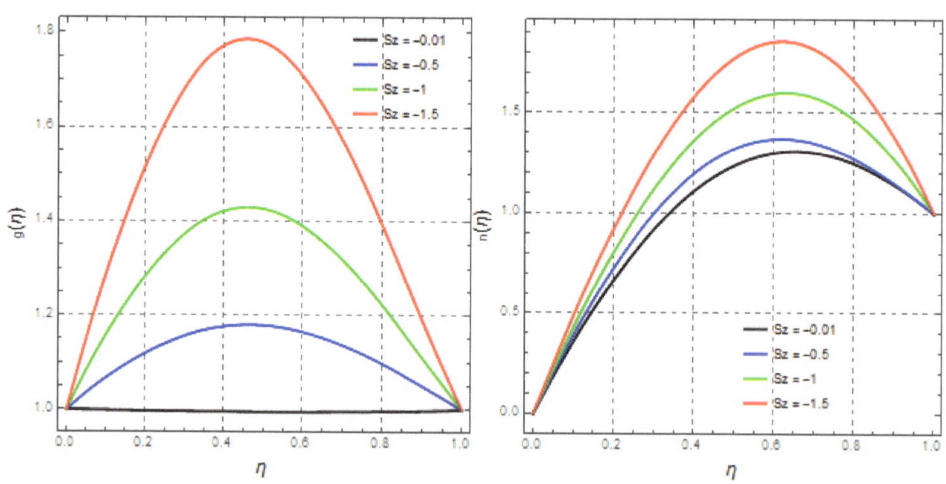

Figure 10. Observing the effect of squeeze Reynold number S_z on $g(\eta)$ and $n(\eta)$ with $M_z = -1$, $M_y = 0.5$, $R_m = -0.75$, $S = 1$, $R_d = 0.5$, $D_u = 0.1$, $P_{rn} = 5$, $S_{rt} = 0.2$ and $S_{ch} = 1$.

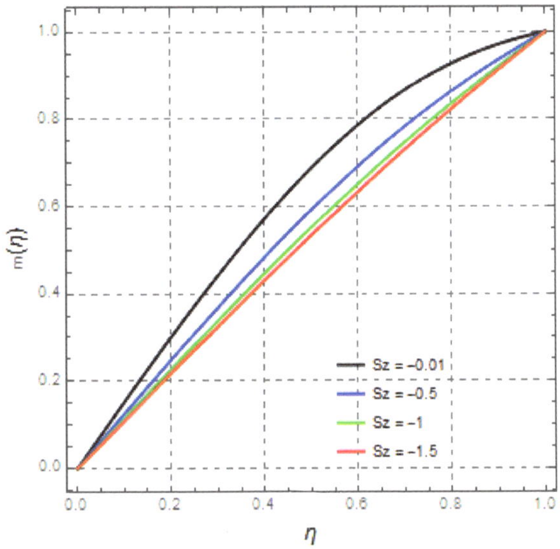

Figure 11. Impact of squeeze Reynold number S_z on $m(\eta)$, keeping $M_z = -1.5$, $M_y = 0.5$, $R_m = -1$, $S = 1$, $R_d = 0.5$, $D_u = 0.1$, $P_{rn} = 2$, $S_{rt} = 0.2$ and $S_{ch} = 1$.

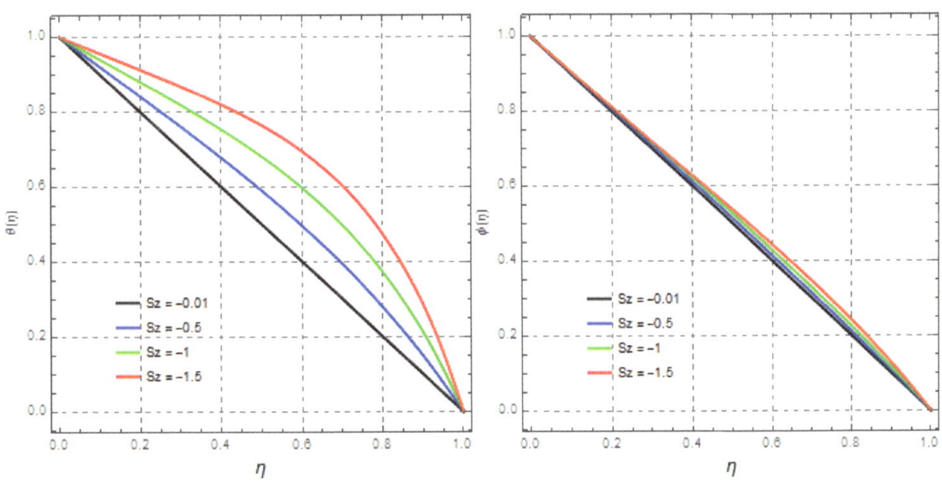

Figure 12. Observing the effect of squeeze Reynold number S_z on $\theta(\eta)$ and $\phi(\eta)$ with $M_z = -1$, $M_y = 0.5$, $R_m = -0.75$, $S = 1$, $R_d = 0.5$, $D_u = 0.1$, $P_{rn} = 5$, $S_{rt} = 0.2$ and $S_{ch} = 1$.

The magnetic field strength M_z is the magnetic field's dimensionless axial strength. For additional parameters with fixed values, from Figures 13–16, it can be seen that increasing M_z generates a decrease in the velocity component $f(\eta)$, whereas decreasing the magnetic field strength along the x-component causes a rise in the velocity $g(\eta)$. The initial velocity along the x-component decreases with increasing magnetic field intensity for the velocity component $f'(\eta)$ but starts to increase as $\eta \to 1$. The figure also shows how M_z has an increasing influence on the magnetic field component $m(\eta)$and $n(\eta)$. When discussing temperature distribution, increasing M_z causes $\theta(\eta)$ to increase as well, whereas decreasing $\phi(\eta)$ causes $phi(\eta)$ to drop. Figures 17 and 18 show the effect of magnetic field intensity M_y, which is the strength of the magnetic field in the y-direction. It is discovered that as the magnetic field intensity increases, both velocity $g(\eta)$ and the magnetic field component $n(\eta)$ drop, indicating an inverse relationship in both cases. To observe the effect of magnetic Reynold number R_m, Figures 19–22 illustrate the relations, showing that an increases in the magnetic Reynold number causes a decrease in the velocity component $f(\eta)$ with fixed values of other parameters, but increases in the velocity component $g(\eta)$ with fixed values of other parameters results in an increase in the value of shear force due to distribution of body force in a non-uniform manner. Body force accelerates near the relative core wall layer because Lorentz force is small near the squeezed plate (because of the current being almost parallel to the magnetic field).

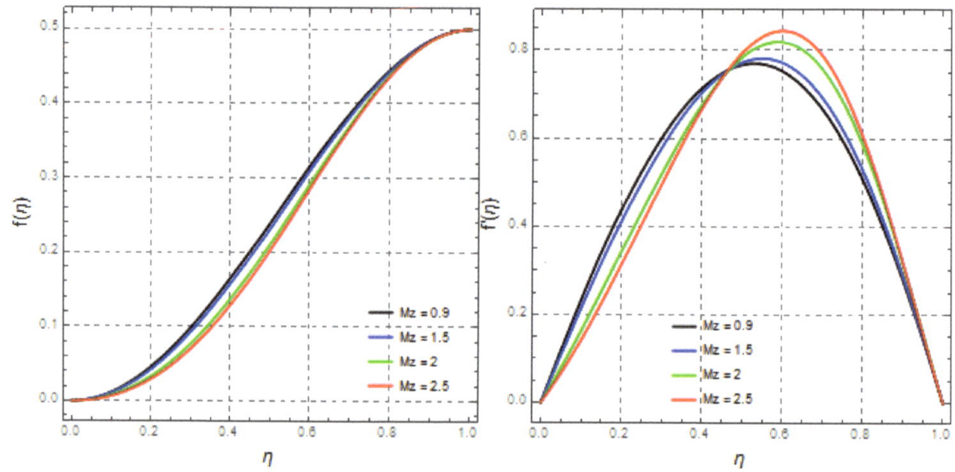

Figure 13. Observing the effect of magnetic strength parameter M_z on $f(\eta)$ and $f'(\eta)$ with $S_z = 2$, $M_y = 3$, $R_m = 1$, $S = 1$, $R_d = 1$, $D_u = 0.1$, $P_{rn} = 1$, $S_{rt} = 0.2$ and $S_{ch} = 1$.

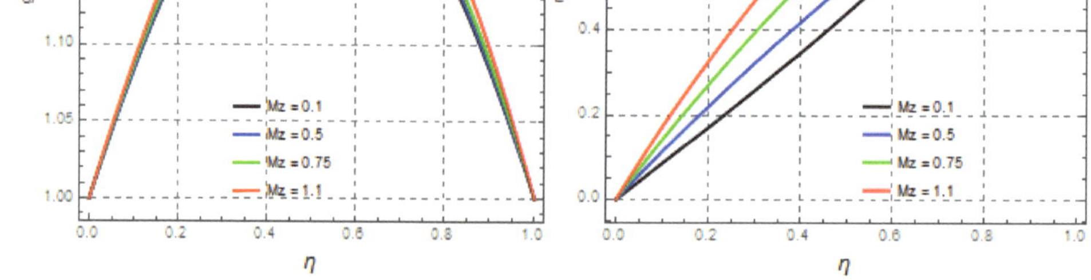

Figure 14. Observing the effect of magnetic strength parameter M_z on $g(\eta)$ and $n(\eta)$ with $S_z = -0.25$, $M_y = 3$, $R_m = 1$, $S = 1$, $R_d = 0.5$, $D_u = 0.1$, $P_{rn} = 1$, $S_{rt} = 2$ and $S_{ch} = 0.5$.

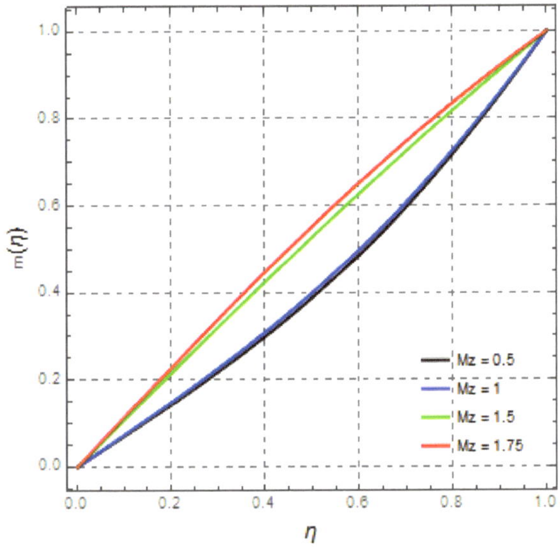

Figure 15. Observing the effect of magnetic strength parameter M_z on $m(\eta)$ with $S_z = -0.25$, $M_y = 3$, $R_m = 1$, $S = 1$, $R_d = 1$, $D_u = 0.1$, $P_{rn} = 1$, $S_{rt} = 2$ and $S_{ch} = 0.5$.

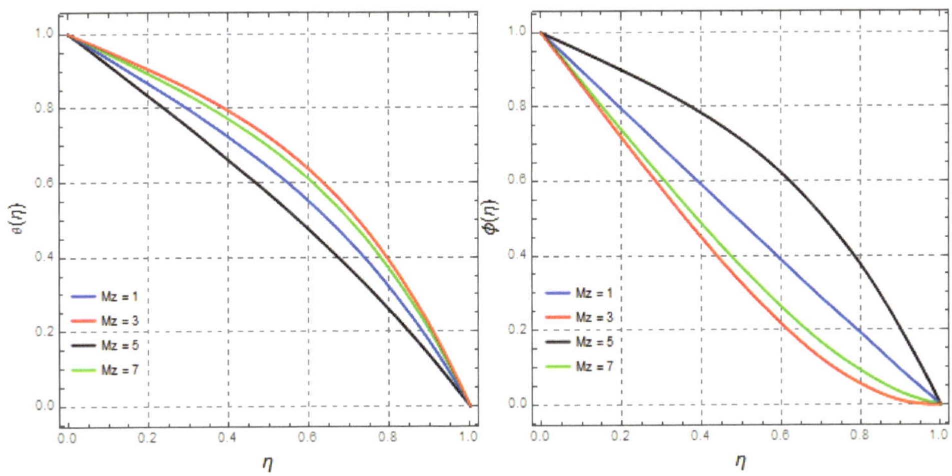

Figure 16. Observing the effect of magnetic strength parameter M_z on $\theta(\eta)$ and $\phi(\eta)$ with $S_z = -1.5$, $M_y = 1$, $R_m = -1$, $S = 1$, $R_d = -1$, $D_u = 0.1$, $P_{rn} = -1$, $S_{rt} = -2$ and $S_{ch} = -0.5$.

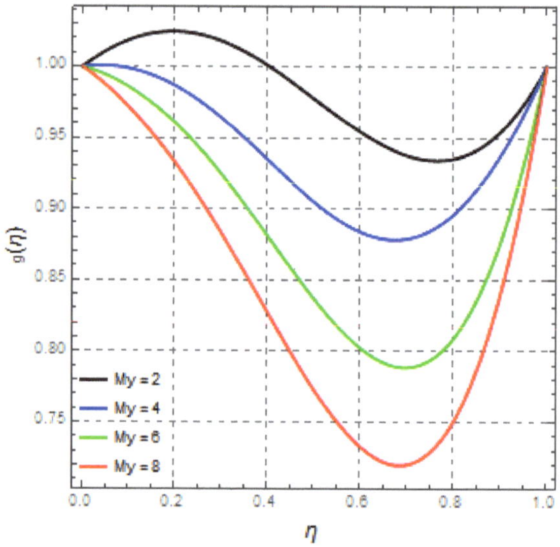

Figure 17. Observing the effect of magnetic strength parameter M_y on $g(\eta)$ with $S_z = -0.5$, $M_z = -0.5$, $R_m = -2$, $S = 1$, $R_d = 0.5$, $D_u = 0.1$, $P_{rn} = 5$, $S_{rt} = 0.2$ and $S_{ch} = 1$.

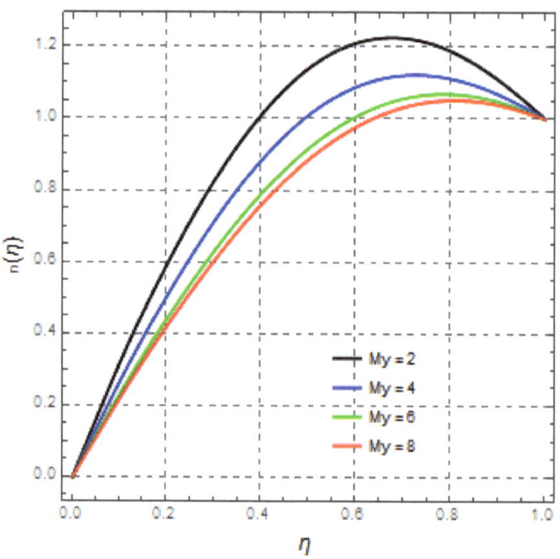

Figure 18. Observing the effect of magnetic strength parameter M_y on $n(\eta)$ with $S_z = -0.5$, $M_z = -1.5$, $R_m = -1$, $S = 1$, $R_d = -1$, $D_u = 2$, $P_{rn} = -0.1$, $S_{rt} = -0.2$ and $S_{ch} = -1$.

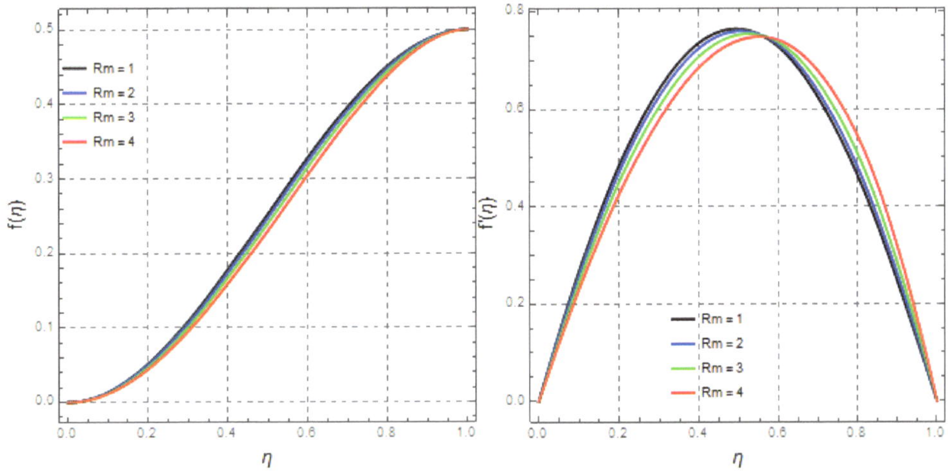

Figure 19. Observing the effect of magnetic Reynold number R_m on $f(\eta)$ and $f'(\eta)$ with $S_z = -1.5$, $M_z = -1.5$, $M_y = 3$, $S = 1$, $R_d = -1$, $D_u = 5$, $P_{rn} = 1$, $S_{rt} = 3$ and $S_{ch} = 1$.

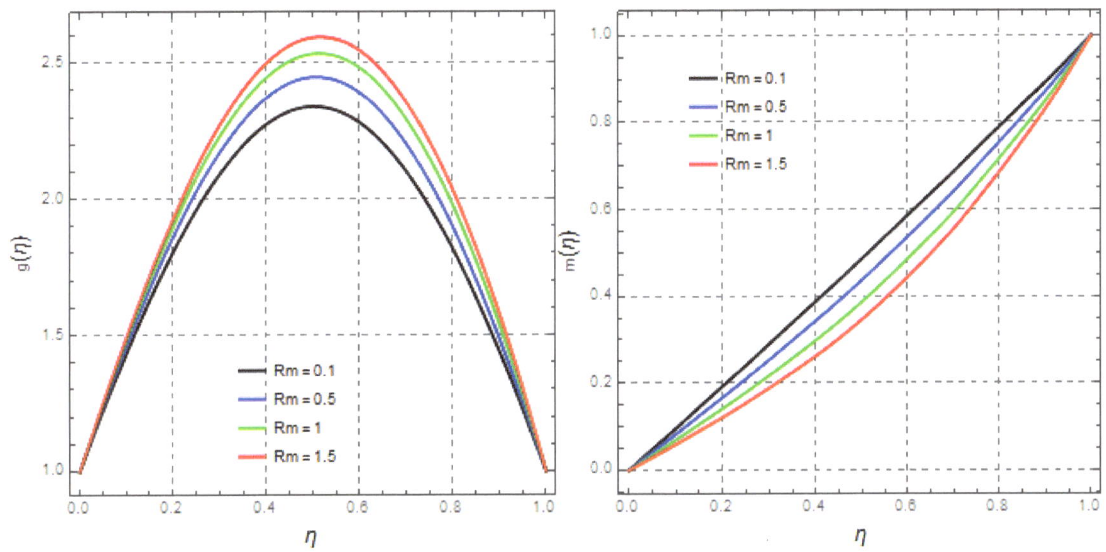

Figure 20. Observing the effect of magnetic Reynold number R_m on $g(\eta)$ and $m(\eta)$ with $S_z = -1.5$, $M_z = -0.75$, $M_y = 1$, $S = 1$, $R_d = -1$, $D_u = 0.1$, $P_{rn} = -1$, $S_{rt} = -2$ and $S_{ch} = -0.5$.

For the x-component of velocity initially, a decrease in the velocity is located but starts increasing as $\eta \to 1$ maximum value of $f'(\eta)$ is observed at the center. In the case of magnetic field component $m(\eta)$, with $R_m = 0.1$, an almost linear profile is observed, whereas when increasing the value of R_m, the profile becomes parabolic. It is also shown that in increasing R_m, the profile for $m(\eta)$ decreases. However, a direct relationship is observed in the case of magnetic field component $n(\eta)$, i.e., when increasing the values of R_m, the profile of $n(\eta)$ also increases. Similarly, for larger values of R_m, the profile becomes more parabolic. Moreover, observing the effect of R_m on $\theta(\eta)$ and $\phi(\eta)$, it can be seen that increasing the magnetic Reynold number has a direct effect on temperature distribution $\theta(\eta)$, while an inverse relation is observed in the case of concentration $\phi(\eta)$.

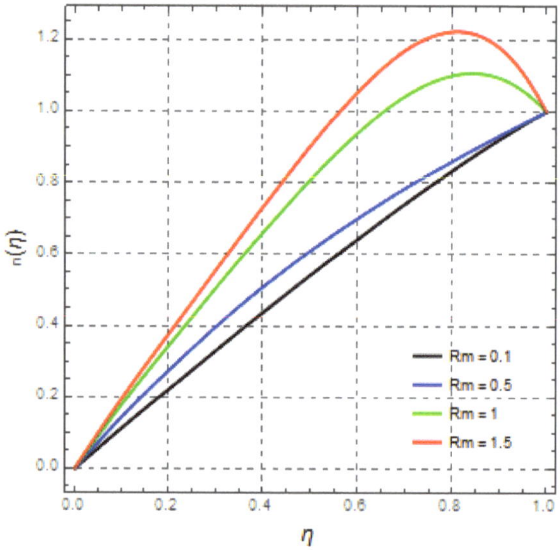

Figure 21. Observing the effect of magnetic Reynold number R_m on $n(\eta)$ with $S_z = -1.5$, $M_z = 0.5$, $M_y = 1$, $S = 1$, $R_d = 1$, $D_u = 0.1$, $P_{rn} = 1$, $S_{rt} = 2$ and $S_{ch} = 0.5$.

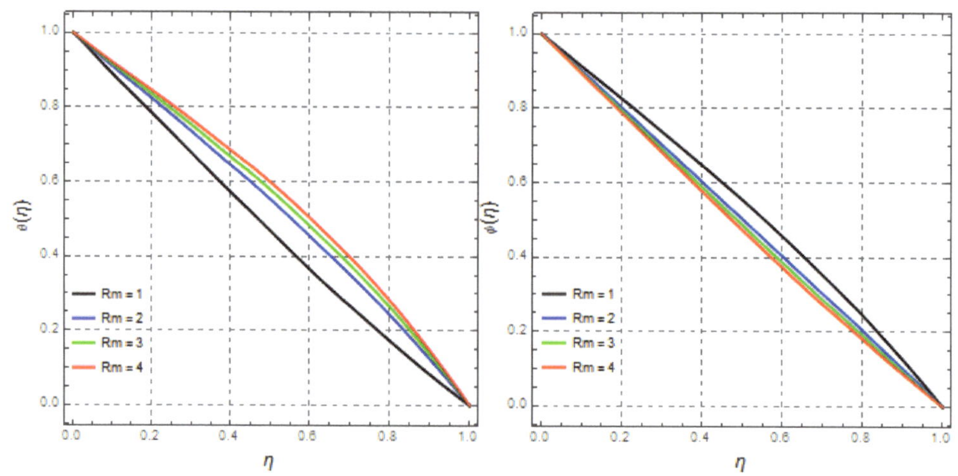

Figure 22. Observing the effect of magnetic Reynold number R_m on $\theta(\eta)$ and $\phi(\eta)$ with $S_z = -1.5$, $M_z = -1.5$, $M_y = 3$, $S = 1$, $R_d = -1$, $D_u = 5$, $P_{rn} = 1$, $S_{rt} = 3$ and $S_{ch} = 1$.

Figure 23 illustrates the effect of the Dufour number on concentration and temperature distributions, showing that in both, cases D_u has a direct effect on $\theta(\eta)$ and $\phi(\eta)$. Similarly, from Figure 24, it can be observed that the Prandtl number P_{rn} has a direct relation with $\theta(\eta)$, as increasing P_{rn} increases the temperature $\theta(\eta)$, whereas an inverse relation is observed in the case of $\phi(\eta)$, i.e., increasing P_{rn}, $\phi(\eta)$ shows a decreasing effect. Figure 25 depicts the Soret effect on temperature and concentration: it can be seen that in increasing the Soret number, heat transfer increases but mass transfer decreases. On the other hand, the opposite case is seen in the case of radiation parameter R_d in Figure 26, i.e., increasing R_d, $\theta(\eta)$ increases while $\phi(\eta)$ decreases. Further, for the Schmidt number, an inverse relation is observed for $\theta(\eta)$, whereas a direct relation can be seen for $\phi(\eta)$, as seen from Figure 27.

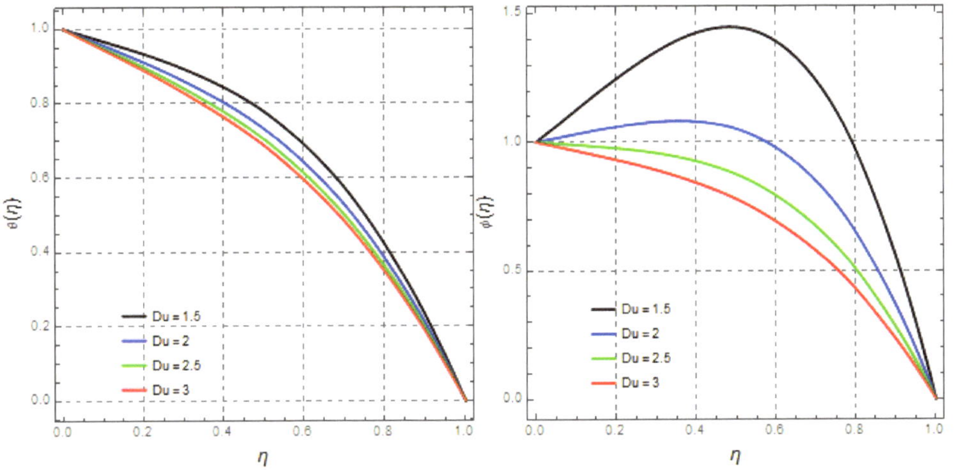

Figure 23. Observing the effect of squeeze Dufour number D_u on $\theta(\eta)$ and $\phi(\eta)$ with $S_z = -1.5$, $M_z = -1.5$, $M_y = 3$, $R_m = -0.5$, $S = 1$, $R_d = -1$, $P_{rn} = -1.5$, $S_{rt} = 0.2$ and $S_{ch} = -1$.

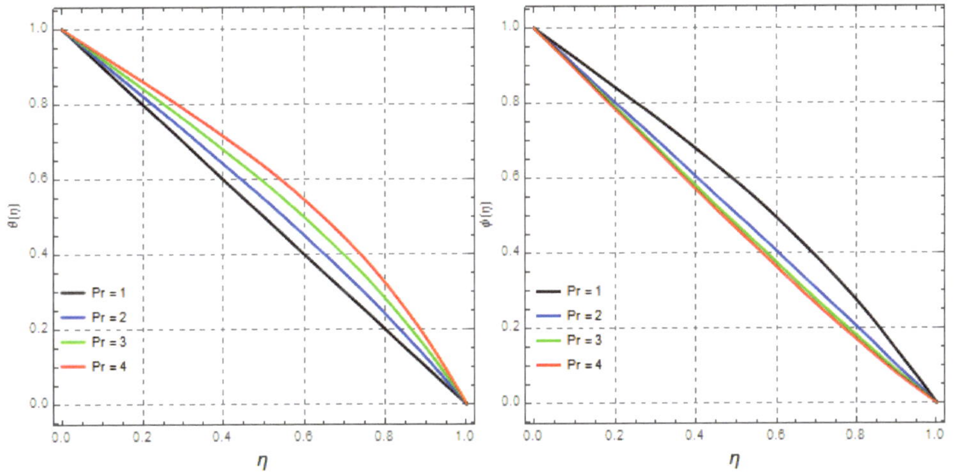

Figure 24. Observing the effect of Prandtl number P_{rn} on $\theta(\eta)$ and $\phi(\eta)$ with $S_z = 1.5$, $M_z = 1.5$, $M_y = 3$, $R_m = 1$, $S = 1$, $R_d = 2$, $D_u = 1$, $S_{rt} = 2$ and $S_{ch} = 1$.

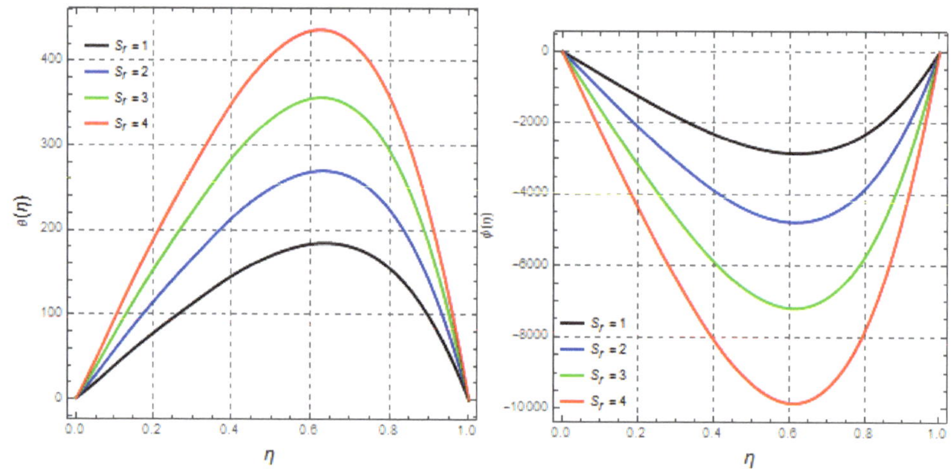

Figure 25. Observing the effect of Soret number S_{rt} on $\theta(\eta)$ and $\phi(\eta)$ with $S_z = -1.5$, $M_z = -1.5$, $M_y = 3$, $R_m = 0.5$, $S = 1$, $R_d = -1$, $D_u = 0.1$, $P_{rn} = 5$ and $S_{ch} = 1$.

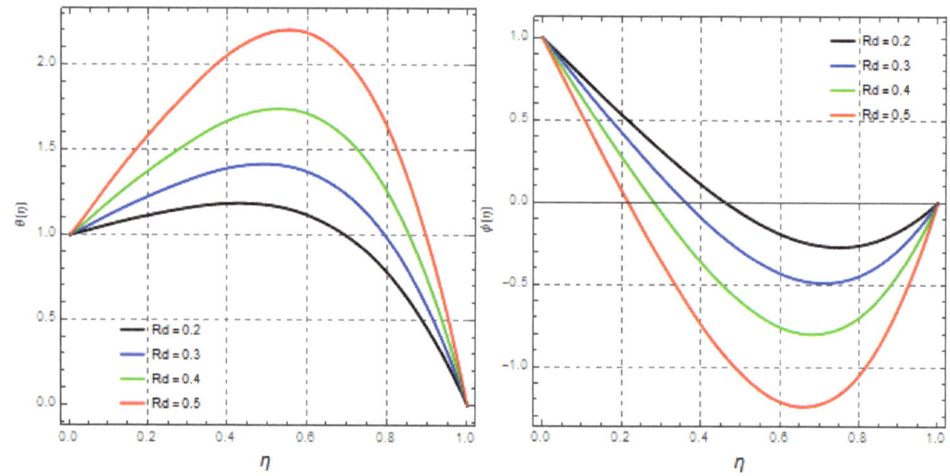

Figure 26. Observing the effect of radiation parameter R_d on $\theta(\eta)$ and $\phi(\eta)$ with $S_z = -1.5$, $M_z = -1.5$, $M_y = 3$, $R_m = -0.5$, $S = 1$, $D_u = 2$, $P_{rn} = 1.5$, $S_{rt} = 2$ and $S_{ch} = -1$.

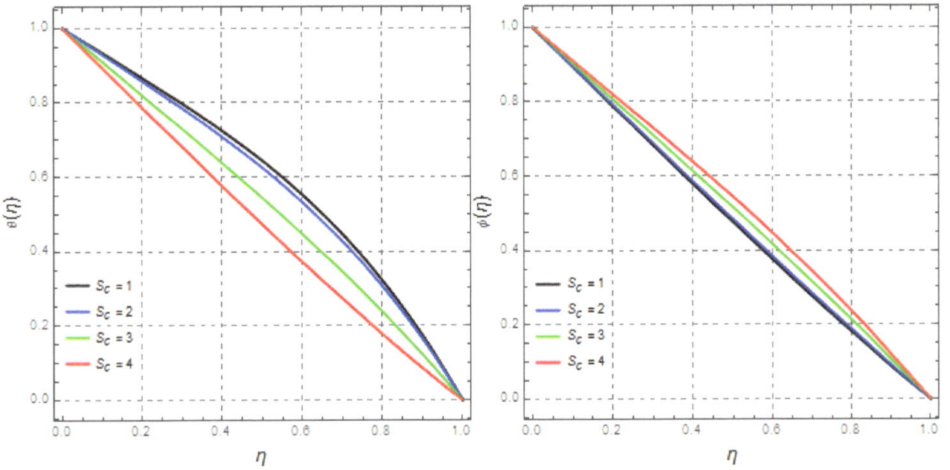

Figure 27. Observing the effect of Schmidt number S_{ch} on $\theta(\eta)$ and $\phi(\eta)$ with $S_z = -1.5$, $M_z = -1.5$, $M_y = 3$, $R_m = -0.5$, $S = 1$, $R_d = -0.5$, $D_u = 5$, $P_{rn} = 1$, and $S_{rt} = 2$.

7. Conclusions

Heat and mass transfer for 3D squeezing MHD flow of viscous fluids is studied in this chapter; flow is considered between two plates rotating at the same angular velocities. The influence of a varying magnetic field is introduced, and the phenomenon is represented using linked governing equations, such as continuity, Navier–Stokes, magnetic field energy, and transport equations. They are then translated to ordinary differential equations using the similarity transformation and solved using the analytical technique (HAM) in Mathematica package BVPh 2.0. The influence of varied parameters on velocity, magnetic field, concentration and temperature distribution was observed by graphs and tables. Below are some conclusions made from the above analysis:

- It is determined that increasing the squeeze effect on the upper plate generates an increase in flow velocity along the y- and z-axis, while the velocity along the x-axis initially increases but then decreases in the upper domain as ($\eta \rightarrow 1$).
- It was also found that increasing the squeeze Reynold number of the magnetic field component reduced the magnetic field's effect along the z-component while increasing the effect along the y-component.
- While squeeze number has a direct effect on both concentration and temperature distribution, an increase in squeeze Reynold number causes an increase in both temperature and mass transfer.
- Moreover, raising the magnetic field strength parameter M_z, which is the strength of the magnetic field along the z-axis, results in a drop in fluid velocity along the z-axis, yet velocity along the y-axis shows a gradual increase by increasing M_z. However, along the x-axis, firstly a decrease in the velocity component is observed, but as $\eta \rightarrow 1$, the velocity begins to increase.
- The magnetic field strength parameter M_z has a direct relationship with the magnetic field component along the z-axis, i.e., increasing the value of M_z has a increasing effect on the value of magnetic field along the z-component. A direct relationship can also be detected along the y-axis.
- Further, for a temperature distribution, increasing M_z shows that $\theta(\eta)$ increases, but ϕ decreases by increasing M_z.
- Further, it is concluded that raising the y-component of the magnetic field strength parameter causes a decrease in the velocity of the fluid along the y-axis and the effect of the magnetic field along the y-axis.

- It is concluded that increasing the magnetic Reynold number R_m causes a decrease in flow velocity along the z-axis. On the other hand, raising R_m has an increasing effect on velocity along y. The flow velocity along the x-axis initially shows a diminishing pattern, but when $\eta \to 1$ increases, the effect increases.
- It is also observed that for the magnetic field, increasing the magnetic Reynold number shows a decrease in the value of the magnetic field along the z-axis, whereas an increasing effect is observed along the y-axis.
- Furthermore, it is also concluded that increasing the magnetic Reynold number shows an increase in the heat transfer, where a decrease in the mass transfer is observed.
- The effect of the Dufour number on concentration and temperature distribution is also observed, and it is concluded that in both cases, D_u has a direct effect on mass and temperature distribution, as increasing the Dufour number increases both mass and heat transfer.
- It is also concluded that an increase in the value of the Prandtl number causes an increase in the value of $\theta(\eta)$, but a decrease in the value of $\phi(\eta)$ is observed.
- For the Soret number, it can be seen that in increasing the Soret number, heat transfer increases but mass transfer decreases. Thus, it has a direct effect on $\theta(\eta)$ and an inverse effect on $\phi(\eta)$.
- It is further concluded that opposite behavior is observed in case of the radiation parameter's effect on temperature distribution, i.e., in increasing its value, $\theta(\eta)$ decreases, while in case of mass transfer, $\phi(\eta)$ increases with an increase in the value of the radiation parameter.
- For the Schmidt number, it is concluded that with an increasing Schmidt number, both heat and mass transfer values increase.

Author Contributions: Conceptualization, M.K.A.; methodology, A.K.; formal analysis, K.B.; funding acquisition, S.N. All authors have read and agreed to the published version of the manuscript.

Funding: This research received no external funding

Institutional Review Board Statement: Not applicable.

Informed Consent Statement: Not applicable.

Data Availability Statement: Not applicable.

Conflicts of Interest: The authors declare no conflict of interest.

Nomenclature

V	Fluid velocity
B	Magnetic field
l	Distance between disks at $t = 0$
μ	Fluid's dynamic viscosity
σ	Electric conductivity
μ_2	Permeability of free space
v	Fluid's kinematic viscosity
ρ	Fluid density
S_z	Squeezing number
P	Fluid pressure
θ	Dimensionless temperature
ϕ	Dimensionless concentration
D	Diffusion coefficient
κ	Thermal conductivity
T	Fluid temperature
C	fluid concentration

R_m	Magnetic Reynold number
S	Squeezing number
M_z	Magnetic strength parameter along z-axis
M_y	Magnetic strength parameter along y-axis
P_r	Prandtl number
S_c	Schmidt number
D_u	Dufour number
S_{rt}	Soret number
R_d	Radiation parameter
C_p	Specific heat at constant pressure

References

1. Siddiqui, A.M.; Irum, S.; Ansari, A.R. Unsteady squeezing flow of a viscous MHD fluid between parallel plates, a solution using the homotopy perturbation method. *Math. Model. Anal.* **2008**, *13*, 565–576. [CrossRef]
2. Sweet, E.; Vajravelu, K.; Van Gorder, R.A.; Pop, I. Analytical solution for the unsteady MHD flow of a viscous fluid between moving parallel plates. *Commun. Nonlinear Sci. Numer. Simul.* **2011**, *16*, 266–273. [CrossRef]
3. Ramachandra Murty, P.S.; Prakash, G.B. MHD Two-Fluid Flow and Heat Transfer between Two Inclined Parallel Plates in a Rotating System. *Hindawi Publ. Corp.* **2014**, *2014*, 256898. [CrossRef] [PubMed]
4. Onyango, E.R.; Kinyanjui, M.; Kimathi, M. Unsteady hydromagnetic flow between parallel plates both moving in the presence of a constant pressure gradient. *Int. J. Eng. Sci. Innov. Technol. (IJESIT)* **2017**, *6*. Available online: http://ir.mksu.ac.ke/handle/123456780/2181 (accessed on 14 August 2021).
5. Khan, M.S.; Shah, R.A.; Khan, A. Effect of variable magnetic field on the flow between two squeezing plates. *Eur. Phys. J. Plus* **2019**, *134*, 219. [CrossRef]
6. Muhammad, T.; Hayat, T.; Alsaedi, A.; Qayyum, A. Hydromagnetic unsteady squeezing flow of Jeffrey fluid between two parallel plates. *Chin. J. Phys.* **2017**, *55*, 1511–1522. [CrossRef]
7. Verma, P.D.; Mathur, A. Magenetohydrodynamic Flow between Two Parallel Plates, One in Uniform Motion and the Other at Rest with Uniform Suction at the Stationary Plate. *Proc. Indian Natl. Sci. Acad. USA* **1969**, *35*, 4.
8. Hayat, T.; Sajjad, R.; Alsaedi, A.; Muhammad, T.; Ellahi, R. On squeezed flow of couple stress nanofluid between two parallel plates. *Results Phys.* **2017**, *7*, 553–561. [CrossRef]
9. Linga Raju, T. MHD heat transfer two-ionized fluids flow between parallel plates with Hall currents. *Results Eng.* **2019**, *4*, 100043. [CrossRef]
10. Hamza, E.A. The magnetohydrodynamic effects on a fluid film squeezed between two rotating surfaces. *J. Phys. Appl. Phys.* **1991**, *24*, 547–554. [CrossRef]
11. Das, S.; Maji, S.L.; Guria, M.; Jana, R.N. Unsteady MHD Couette flow in a rotating system. *Math. Comput. Model.* **2009**, *50*, 1211–1217. [CrossRef]
12. Parter, S.V.; Rajagopal, K.R. Swirling Flow between Rotating Plates. In *The Breadth and Depth of Continuum Mechanics*; Springer: Berlin/Heidelberg, Germany, 1986; ISBN 978-3-642-61634-1.
13. Rajagopal, K.R.; Parter, S.V. *Remarks on the Flow between Two Parallel Rotating Plates;* Technical Summary Report Wisconsin University, Madison; Mathematics Research Center: Madison,WI, USA , 1984.
14. Rajagopal, K.R. The flow of a second order fluid between rotating parallel plates. *J. Non-Newton. Fluid Mech.* **1981**, *9*, 185–190. [CrossRef]
15. Rajat, T.; Gauri, S.; Manoj, M. Double diffusive flow of a hydromagnetic nanofluid in a rotating channel with Hall effect and viscous dissipation: Active and passive control of nanoparticles. *Adv. Powder Technol.* **2017**, *28*, 10.
16. Hughes, W.; Elco, R. Magnetohydrodynamic lubrication flow between parallel rotating disks. *J. Fluid Mech.* **1962**, *13*, 21–32. [CrossRef]
17. Elshekh, S.S.; Elhady, M.K.A.; Ibrahim, F.N. Fluid film squeezed between two rotating disks in the presence of a magnetic field. *Int. J. Eng. Sci.* **1996**, *34*, 1183–1195. [CrossRef]
18. Shah, R.A.; Khan, A.; Shuaib, M. On the study of flow between unsteady squeezing rotating discs with cross diffusion effects under the influence of variable magnetic field. *Heliyon* **2018**, *4*, e00925. [CrossRef] [PubMed]
19. Ganji, D.D.; Abbasi, M.; Rahimi, J.; Gholami, M.; Rahimipetroudi, I. On the MHD squeeze flow between two parallel disks with suction or injection via HAM and HPM. *Front. Mech. Eng.* **2014**, *9*, 270–280. [CrossRef]
20. Khan, A.; Shah, R.A.; Shuaib, M.; Ali, A. Fluid dynamics of the magnetic field dependent thermosolutal convection and viscosity between coaxial contracting discs. *Results Phys.* **2018**, *9*, 923–938. [CrossRef]
21. Munawar, S.; Mehmood, A.; Ali, A. Three-dimensional squeezing flow in a rotating channel of lower stretching porous wall. *Comput. Math. Appl.* **2012**, *64*, 1575–1586. [CrossRef]
22. Alzahrani Abdullah, K.; Zaka, U.M.; Taseer, M. Numerical Treatment for 3D Squeezed Flow in a Rotating Channel With Soret and Dufour Effects. *Front. Phys.* **2020**, *8*, 201. [CrossRef]

23. Shah, Z.; Gul, T.; Islam, S.; Khan, M.A.; Bonyah, E.; Hussain, F.; Mukhtar, S.; Ullah, M. Three dimensional third grade nanofluid flow in a rotating system between parallel plates with Brownian motion and thermophoresis effects. *Results Phys.* **2018**, *10*, 36–45. [CrossRef]
24. Riasat, S.; Ramzan, M.; Kadry, S.; Chu, Y.-M. Significance of Magnetic Reynolds number in a three-dimensional squeezing Darcy-Forchheimer hydromagnetic nanofluid thin-film flow between two rotating disks. *Sci. Rep.* **2020**, *10*, 17208. [CrossRef]
25. Fiza, M.; Alsubie, A.; Ullah, H.; Hamadneh, N.N.; Islam, S.; Khan, I. Three-Dimensional Rotating Flow of MHD Jeffrey Fluid Flow between Two Parallel Plates with Impact of Hall Current. *Math. Probl. Eng.* **2021**, *2021*, 6626411. [CrossRef]
26. Srinivas, S.; Reddy, A.S.; Ramamohan, T.R.; Shukla, A.K. Influence of heat transfer on MHD flow in a pipe with expanding or contracting permeable wall. *Ain Shams Eng. J.* **2014**, *5*, 817–830. [CrossRef]
27. Sparrow, E.M.; Cess, R.D. The effect of a magnetic field on free convection heat transfer. *Int. J. Heat Mass Transf.* **1961**, *3*, 267–274. [CrossRef]
28. Khaled, A.R.A.; Vafai, K.Hydromagnetic squeezed flow and heat transfer over a sensor surface. *Int. J. Eng. Sci.* **2004**, *42*, 509–519. [CrossRef]
29. Tasawar, H.; Muhammad, Q.; Zaheer, A. Radiation and Mass Transfer Effects on the Magnetohydrodynamic Unsteady Flow Induced by a Stretching Sheet. *Zeitschrift für Naturforschung A* **2010**, *65*, 231–239.
30. Mahmood, M.; Asghar, S.; Hossain, M.A. Squeezed flow and heat transfer over a porous surface for viscous fluid. *Heat Mass Transfer* **2007**, *44*, 165–173. [CrossRef]
31. Tsai, R.; Huang, J.S. Heat and mass transfer for Soret and Dufour's effects on Hiemenz flow through porous medium onto a stretching surface. *Int. J. Heat Mass Transf.* **2009**, *52*, 2399–2406. [CrossRef]
32. Mustafa, M.; Hayat, T.; Obaidat, S. On heat and mass transfer in the unsteady squeezing flow between parallel plates. *Meccanica* **2012**, *47*, 1581–1589. [CrossRef]
33. Yasmin, A.; Ali, K.; Ashraf, M. Study of Heat and Mass Transfer in MHD Flow of Micropolar Fluid over a Curved Stretching Sheet. *Sci. Rep.* **2020**, *10*, 4581. [CrossRef] [PubMed]
34. Sheikholeslami, M.; Ganji, D.D. Three dimensional heat and mass transfer in a rotating system using nanofluid. *Powder Technol.* **2014**, *253*, 789–796. [CrossRef]
35. Bilal, S.; Rehman, M.; Noeiaghdam, S.; Ahmad, H.; Akgul, A. Numerical Analysis of Natural Convection Driven Flow of a Non-Newtonian Power-Law Fluid in a Trapezoidal Enclosure with a U-Shaped Constructal. *Energies* **2021**, *14*, 5355. [CrossRef]
36. Noeiaghdam, S.; Sidorov, D. Integral equations: Theories, Approximations and Applications. *Symmetry* **2021**, *13*, 1402. [CrossRef]
37. Ghomanjani, F.; Noeiaghdam, S. Application of Said Ball curve for solving fractional differential-algebraic equations. *Mathematics* **2021**, *9*, 1926. [CrossRef]
38. Zarei, E.; Noeiaghdam, S. Advantages of the Discrete Stochastic Arithmetic to Validate the Results of the Taylor Expansion Method to Solve the Generalized Abel's Integral Equation. *Symmetry* **2021**, *13*, 1370. [CrossRef]
39. Izadi, M.; Yuzbasi, S.; Noeiaghdam, S. Approximating solutions of non-linear Troesch's problem via an efficient quasi-linearization Bessel approach. *Mathematics* **2021**, *9*, 1841. [CrossRef]
40. Zhao, Y.; Liao, S. User Guide to BVPh 2.0. 2002. Available online: http://numericaltank.sjtu.edu.cn/BVPh.htm (accessed on 14 August 2021).

mathematics MDPI

Article

Lie Point Symmetries, Traveling Wave Solutions and Conservation Laws of a Non-linear Viscoelastic Wave Equation

Almudena P. Márquez * and María S. Bruzón

Department of Mathematics, University of Cadiz, Puerto Real, 11510 Cadiz, Spain; m.bruzon@uca.es
* Correspondence: almudena.marquez@uca.es

Abstract: This paper studies a non-linear viscoelastic wave equation, with non-linear damping and source terms, from the point of view of the Lie groups theory. Firstly, we apply Lie's symmetries method to the partial differential equation to classify the Lie point symmetries. Afterwards, we reduce the partial differential equation to some ordinary differential equations, by using the symmetries. Therefore, new analytical solutions are found from the ordinary differential equations. Finally, we derive low-order conservation laws, depending on the form of the damping and source terms, and discuss their physical meaning.

Keywords: viscoelastic wave equation; Lie symmetries; traveling wave solutions; conversation laws

Citation: Márquez, A.P.; Bruzón, M.S. Lie Point Symmetries, Traveling Wave Solutions and Conservation Laws of a Non-linear Viscoelastic Wave Equation. *Mathematics* **2021**, *9*, 2131. https://doi.org/10.3390/math9172131

Academic Editors: Athanasios Tzavaras and Maria C. Mariani

Received: 26 June 2021
Accepted: 27 August 2021
Published: 2 September 2021

1. Introduction

Lately, many viscoelastic wave equations have been considered in the literature. The single viscoelastic wave equation of the form

$$u_{tt} - \Delta u + \int_0^1 h(t-s)\Delta u(x,s)ds + f(u_t) = g(u)$$

in $\Omega \times (0,\infty)$, where Ω is a bounded domain of $\mathbb{R}^N$ $(N \geq 1)$ with initial and boundary conditions, has been extensively studied. Several results concerning non-existence and blow-up solutions in finite time have been proved [1–8].

Furthermore, the non-linear viscoelastic wave equation with damping and source terms

$$u_{tt} - \Delta u + f(u_t) = g(u), \qquad x \in \Omega,\, t > 0, \tag{1}$$

has also been very studied obtaining similar results [9,10]. As in the single viscoelastic wave equation, it is well-known that the damping term $f(u_t)$ assures global existence in the absence of the source term ($g(u) = 0$). The interaction between the damping term and the source term makes the problem more interesting.

Moreover, Messaoudi [11] considered the non-linear viscoelastic wave equation with damping and source terms

$$u_{tt} - \Delta u + a|u_t|^{m-2}u_t = b|u|^{p-2}u, \qquad x \in \Omega,\, t > 0.$$

For this equation, Georgiev and Todorova [12] and Messaoudi [13] analyzed blow-up solutions in different situations.

In general, many authors showed interest in these viscoelastic wave equations. However, in this paper, we focus on studying the viscoelastic wave Equation (1) but from a point of view of the Lie groups theory. In fact, we have published a previous work analyzing this model [14]. Moreover, in this paper we present new results for the model. It is found a complete classification of Lie point symmetries with its associated reductions, new soliton-type solutions, and a complete classification of multipliers and conservation laws with a discussion of their physical meaning.

The resolution of non-linear partial differential equations (PDEs) is a very important field of research in applied mathematics. Symmetry reductions and analytical solutions have many applications in the context of differential equations. For instance, analytical solutions arising from symmetry methods can be used to study properties, such as asymptotic and blow-up behavior. A large amount of literature has been published about the application of the Lie transformation group theory to construct solutions of non-linear PDEs [15–20]. For example, the Fisher equation was studied in [21] to find new analytical solutions. In [22], a $(2+1)$-dimensional Zakharov-Kuznetsov equation was also investigated using Lie symmetry analysis. The authors of [23] analyzed a system of dispersive evolution equations to obtain new exact solutions too. The symmetries leaving the equation invariant can reduce the number of independent variables, transforming the PDEs into ordinary differential equations (ODEs), which one generally easier to solve.

Additionally, there is a similar one-dimensional equation called the "good" Boussinesq equation considering Δ^2. This change would transform the original second-order PDE (1) to a 4-th order PDE, complicating it but of interest. Nevertheless, there have been a few numerical works of this equation in recent years, such as applying a Fourier pseudo-spectral method [24], and a 2-nd order operator splitting numerical scheme for the Equation [25]. The stability and convergence estimates have been presented in these works.

Conservation laws analyze which physical properties of a PDE do not change in the course of time. In particular, local conservation laws are continuity equations yielding conserved quantities of physical importance for all solutions of a given equation. For any PDE, a complete classification of conservation laws can be determined by the multiplier method [26,27]. In [28], the authors obtained the conservation laws and discussed the physical meaning of the corresponding conserved quantities. A classification of conservation laws of a generalized quasilinear KdV equation was provided in [29] too. Moreover, a $(1+1)$-dimensional coupled modified KdV-type system was studied in [30], constructing its conservation laws also using the multiplier method.

To sum up, the aim of this work is to do a complete Lie group classification of Equation (1). Afterwards, we present the reductions obtained from the different symmetries, transforming the PDE into ODEs. Moreover, we obtain traveling wave solutions by comparing Equation (1) and similar equations studied previously [31–33]. Finally, we give a complete classification of the conservation laws admitted by Equation (1).

The structure of the paper is as follows: In Section 2, we study the Lie point symmetries of Equation (1), and in Section 3, we obtain the symmetry reductions, the symmetry variables, and the reduced ODEs. Next, in Section 4, we construct traveling wave solutions using the reduced equations. Then, in Section 5, we present a classification of the conservation laws and the multipliers of Equation (1). Finally, in Section 6, we give some conclusions of the work.

2. Lie Point Symmetries

The idea of the Lie groups theory of symmetry analysis of differential equations relies on the invariance of the equation under a transformation of independent and dependent variables. This transformation sets up a local group of point transformations yielding to a diffeomorphism on the space of independent and dependent variables, so the solutions of the original equation map to other solutions. Any transformation of the independent and dependent variables leads to a transformation of the derivatives [34].

The application of the Lie groups theory to differential equations is completely algorithmic. However, it usually involves many tedious calculations. Nevertheless, we make use of powerful softwares, such as Maple and the needed calculations are done rapidly. Applying the classical Lie method to search for symmetries provides a set of different expressions for the unknown functions $f(u_t)$ and $g(u)$, for which the equation admits symmetries.

In this section, let us briefly describe the classical Lie method and its application to Equation (1), obtaining the symmetry reductions, the symmetry variables and the reduced equations. For details, this method is described in excellent textbooks, such as [17,19] and references therein.

It is consider the one-parameter Lie group of infinitesimal transformations in (x,t,u) given by

$$\begin{aligned} x^* &= x+\epsilon\xi(x,t,u)+\mathcal{O}(\epsilon^2), \\ t^* &= t+\epsilon\tau(x,t,u)+\mathcal{O}(\epsilon^2), \\ u^* &= u+\epsilon\eta(x,t,u)+\mathcal{O}(\epsilon^2), \end{aligned} \tag{2}$$

where ϵ is the group parameter.

The infinitesimal point symmetries constitute the infinitesimal generator

$$\mathbf{v}=\xi(x,t,u)\partial_x+\tau(x,t,u)\partial_t+\eta(x,t,u)\partial_u. \tag{3}$$

Each infinitesimal generator (Equation (3)) is associated with a transformation, determined by solving the system of ODEs

$$\frac{\partial\hat{x}}{\partial\epsilon}=\xi(\hat{x},\hat{t},\hat{u}), \qquad \frac{\partial\hat{t}}{\partial\epsilon}=\tau(\hat{x},\hat{t},\hat{u}), \qquad \frac{\partial\hat{u}}{\partial\epsilon}=\eta(\hat{x},\hat{t},\hat{u}),$$

satisfying the initial conditions

$$\hat{x}|_{\epsilon=0}=x, \qquad \hat{t}|_{\epsilon=0}=t, \qquad \hat{u}|_{\epsilon=0}=u,$$

where ϵ is the group parameter.

We point out that (3) is a point symmetry of Equation (1) if the 2-nd order prolongation of (3) leaves invariant Equation (1). This leads to an overdetermined linear system of determining equations for the infinitesimals $\xi(x,t,u)$, $\tau(x,t,u)$ and $\eta(x,t,u)$, generated by applying the symmetry invariance condition

$$\mathrm{pr}^{(2)}\mathbf{v}(u_{tt}-u_{xx}+f(u_t)-g(u))=0, \quad \text{when} \quad u_{tt}-u_{xx}+f(u_t)-g(u)=0, \tag{4}$$

Here $\mathrm{pr}^{(2)}\mathbf{v}$ represents the second order prolongation of the vector field $\mathbf{v}$, defined by

$$\mathrm{pr}^{(2)}\mathbf{v}=\mathbf{v}+\eta_x\frac{\partial}{\partial u_x}+\eta_t\frac{\partial}{\partial u_t}+\eta_{xx}\frac{\partial}{\partial u_{xx}}+\eta_{xt}\frac{\partial}{\partial u_{xt}}+\eta_{tt}\frac{\partial}{\partial u_{tt}},$$

where the coefficients are given by

$$\begin{aligned} \eta_x &= D_x\eta-u_tD_x\tau-u_xD_x\xi, \\ \eta_t &= D_t\eta-u_tD_t\tau-u_xD_t\xi, \\ \eta_{xx} &= D_x(\eta_x)-u_{xt}D_x\tau-u_{xx}D_x\xi, \\ \eta_{xt} &= D_t(\eta_x)-u_{xt}D_x\tau-u_{xx}D_t\xi, \\ \eta_{tt} &= D_t(\eta_t)-u_{tt}D_t\tau-u_{xt}D_t\xi, \end{aligned}$$

with D_x and D_t the total derivatives of x and t, respectively.

The symmetry determining Equation (4) splits with respect to the t-derivatives and x-derivatives of u, getting an over-determined linear system of equations for the infinitesimals. Here Maple is used for defining the determining equations and then, the commands "rifsimp", "dsolve" and "pdsolve" are used to solve the system. The command "rifsimp" gives a tree containing all solution cases. For each solution case, we use the commands "dsolve" and "pdsolve" to obtain solutions for the over-determined system. Therefore, we proceed to show the classification of all solution cases in Theorem 1.

Theorem 1. *The Lie point symmetries admitted by the non-linear viscoelastic wave Equation (1) for $f(u_t)$ and $g(u)$ arbitrary functions, are generated by the transformations*

$$\begin{aligned} \mathbf{v}_1 &= \partial_x, \\ (\hat{x},\hat{t},\hat{u})_1 &= (x+\epsilon,t,u), \quad \textit{space translation}, \end{aligned}$$

$$\begin{aligned} \mathbf{v}_2 &= \partial_t, \\ (\hat{x},\hat{t},\hat{u})_2 &= (x,t+\epsilon,u), \quad \textit{time translation}. \end{aligned}$$

For some particular functions of $f(u_t)$ and $g(u)$, the non-linear viscoelastic wave Equation (1) admits additional Lie point symmetries, given below.

1. *For $f(u_t) = -\mathrm{e}^{-n}{u_t}^n + f_1$ and $g(u) = ((g_0\,u+g_1)(\frac{2}{n}-1))^{\frac{n}{2-n}} - f_1$, with $n \neq 0,1,2$ and g_0, g_1, f_1 arbitrary constants,*

$$\begin{aligned} \mathbf{v}_3^1 &= g_0\left(\frac{n-1}{n-2}\right)x\partial_x + g_0\left(\frac{n-1}{n-2}\right)t\partial_t + (g_0u+g_1)\partial_u, \\ (\hat{x},\hat{t},\hat{u})_3^1 &= (x\mathrm{e}^{\frac{g_0\,(n-1)\epsilon}{n-2}}, t\mathrm{e}^{\frac{g_0\,(n-1)\epsilon}{n-2}}, \mathrm{e}^{g_0\,\epsilon}u + \mathrm{e}^{g_0\,\epsilon}\int_0^{\epsilon} g_1\,\mathrm{e}^{-g_0\,z_1}\,\mathrm{d}z_1), \quad \textit{scaling and shift}. \end{aligned}$$

2. *For $f(u_t) = au_t^2$ and $g(u) = k\,\mathrm{e}^{cu}$, with $c \neq 0$ and a,k arbitrary constants,*

$$\begin{aligned} \mathbf{v}_3^2 &= -\frac{1}{2}cx\partial_x - \frac{1}{2}ct\partial_t + \partial_u, \\ (\hat{x},\hat{t},\hat{u})_3^2 &= (x\mathrm{e}^{-1/2\,c\epsilon}, t\mathrm{e}^{-1/2\,c\epsilon}, \epsilon+u), \quad \textit{scaling in t and x combined}. \end{aligned}$$

3. *For $f(u_t) = k$ and $g(u) = (au+b)^n - k$, with $n \neq 1$, $c \neq 0$ and a,b,k arbitrary constants,*

$$\begin{aligned} \mathbf{v}_3^3 &= \frac{1}{2}a(n-1)x\partial_x + \frac{1}{2}a(n-1)t\partial_t + (au+b)\partial_u, \\ (\hat{x},\hat{t},\hat{u})_3^3 &= (\mathrm{e}^{1/2\,ak_3\,(n-1)\epsilon}x + \int_0^{\epsilon} k_1\,\mathrm{e}^{-1/2\,ak_3\,(n-1)z_1}\,\mathrm{d}z_1\mathrm{e}^{1/2\,ak_3\,(n-1)\epsilon}, \\ &\quad \mathrm{e}^{1/2\,ak_3\,(n-1)\epsilon}t + \int_0^{\epsilon} k_2\,\mathrm{e}^{-1/2\,ak_3\,(n-1)z_1}\,\mathrm{d}z_1\mathrm{e}^{1/2\,ak_3\,(n-1)\epsilon}, \\ &\quad \mathrm{e}^{-ak_3\,\epsilon}u + \int_0^{\epsilon} -bk_3\,\mathrm{e}^{ak_3\,z_1}\,\mathrm{d}z_1\mathrm{e}^{-ak_3\,\epsilon}), \quad \textit{scaling and shift}. \end{aligned}$$

4. *For $f(u_t) = au_t^{\frac{4}{3}}$ and $g(u) = (cu+k)^2$, with $c \neq 0$ and a,k arbitrary constants,*

$$\begin{aligned} \mathbf{v}_3^4 &= -\frac{1}{2}c\,x\partial_x - \frac{1}{2}c\,t\partial_t + (c\,u+k)\partial_u, \\ (\hat{x},\hat{t},\hat{u})_3^4 &= (x\mathrm{e}^{-1/2\,c\epsilon}, t\mathrm{e}^{-1/2\,c\epsilon}, \mathrm{e}^{c\epsilon}u + \int_0^{\epsilon} k\mathrm{e}^{-cz_1}\,\mathrm{d}z_1\mathrm{e}^{c\epsilon}), \quad \textit{scaling and shift}. \end{aligned}$$

3. Symmetry Reductions

The symmetry variables are found by solving the invariant surface condition

$$\Phi \equiv \xi(x,t,u)u_x + \tau(x,t,u)u_t - \eta(x,t,u) = 0.$$

For Equation (1), a PDE with two independent variables, a single group reduction transforms the PDE into different ODEs.

Reduction 1. For the generator $\lambda\mathbf{v}_1 + \mathbf{v}_2$, we obtain the traveling wave reduction

$$z = x - \lambda t, \qquad u(x,t) = h(z), \tag{5}$$

where $h(z)$ satisfies

$$(\lambda^2-1)h''+f(-\lambda\,h')-g(h)=0. \tag{6}$$

Reduction 2. For the generator $\mathbf{v}_3^1$, we obtain the symmetry reduction

$$z=\tfrac{x}{t},\qquad u=x^{\frac{n-2}{n-1}}h(z)-\tfrac{g_1}{g_0},$$

where $h(z)$ satisfies

$$\begin{aligned}
&z^{\frac{3n^2+2n}{n^2-3n+2}}\left(-z^{\frac{1}{n-1}}\right)^{3n}\Bigg(\Big(g_0{}^{\frac{n}{n-2}}(2-n)^{\frac{n}{n-2}}(n-1)^2 e^{f_2\,n}z^4\\
&-g_0{}^{\frac{n}{n-2}}(2-n)^{\frac{n}{n-2}}(n-1)^2 e^{f_2\,n}z^2\Big)h^{\frac{n}{n-2}+\frac{2}{n-2}}h''\\
&+\Big(2g_0{}^{\frac{n}{n-2}}(2-n)^{\frac{n}{n-2}}(n-1)^2 e^{f_2\,n}z^3\\
&-2g_0{}^{\frac{n}{n-2}}(2-n)^{\frac{n}{n-2}}(n-2)(n-1)e^{f_2\,n}z\Big)h^{\frac{n}{n-2}+\frac{2}{n-2}}h'\\
&+\Big(e^{f_2\,n}\Big(n^{\frac{n}{n-2}}(n^2-2n+1)\Big)h^{\frac{2}{n-2}}\\
&+\Big(g_0{}^{\frac{n}{n-2}}(2-n)^{\frac{n}{n-2}}n-2g_0{}^{\frac{n}{n-2}}(2-n)^{\frac{n}{n-2}}\Big)h^{\frac{2n}{n-2}}\Big)\Bigg)\\
&-\left(g_0{}^{\frac{n}{n-2}}(2-n)^{\frac{n}{n-2}}(n-1)^2 z^{\frac{2n^3}{n^2-3n+2}}\right)h^{\frac{n}{n-2}+\frac{2}{n-2}}(h')^n=0.
\end{aligned}$$

Reduction 3. For the generator $\mathbf{v}_3^2$, the similarity variable and similarity solution are

$$z=\tfrac{x}{t},\qquad u=-\tfrac{2}{c}\ln t+h(z),$$

where $h(z)$ satisfies

$$(c^2z^2-c^2)h''+(a\,z^2c^2)(h')^2+(4\,z\,c\,a+2\,z\,c^2)h'+k\,e^{hc}c^2+4\,a+2\,c=0.$$

Reduction 4. For the generator $\mathbf{v}_3^3$, the invariant solution is

$$z=\frac{x}{t},\qquad u=\frac{1}{c}\Big(t^{-\frac{2}{n-1}}h(z)-b\Big),$$

where $h(z)$ satisfies

$$\begin{aligned}
&(n^2z^2-2nz^2-n^2+z^2+2n)h''-h''h+(2n^2z-2za)h'\\
&+(an^2-2an+a)(h^n)+(2n+2)h=0.
\end{aligned}$$

Reduction 5. For the generator $\mathbf{v}_3^4$, the invariant solution is

$$z=\frac{x}{t},\qquad u=\frac{1}{c}\Big(t^{-2}h(z)-k\Big),$$

where $h(z)$ satisfies

$$(z^2-1)h''+\left(-\sqrt[3]{\tfrac{-h'z-2h}{c}}az+6z\right)h'+c\,h^2+\left(-2\sqrt[3]{\tfrac{-h'z-2h}{c}}a+6\right)h=0.$$

4. Traveling Wave Solutions

In this section we are studying Equation (6) in order to find traveling wave solutions of Equation (1). The other ordinary differential equations obtained are not considered because they are non-autonomous differential equations.

The procedure followed compares Equation (6) with a similar equation, studied before by Kudryashov, whose general solution appears in [31].

The second-order Equation (6) is expressed as

$$h'' = \frac{1}{1-\lambda^2} f(-\lambda h') + \frac{1}{1-\lambda^2} g(h). \tag{7}$$

Kudryashov obtained in [31] the general solution of a second-order ODE, given by

$$h'' = \frac{1}{\lambda}\left(\mu h' + \frac{1}{2}h^2 - \omega h - c_0\right), \tag{8}$$

where c_0 is an arbitrary constant and λ, μ, ω satisfies $\omega = \frac{6\mu^2}{25\lambda}$. The expression of its general solution is given in terms of the Weierstrass elliptic function, considering $g_2 = 0$ and $g_3 = c_1$,

$$h(z) = \omega_k + \frac{6\alpha^2}{25\beta} - \exp\left\{\frac{2z\alpha}{5\beta}\right\}\mathcal{P}\left(c_2 - \frac{5\beta}{\alpha\sqrt{12\beta}}\exp\left\{\frac{z\alpha}{5\beta}\right\}, 0, c_1\right),$$

with c_1, c_2 arbitrary constants.

Equations (7) and (8) are equal if

$$\begin{aligned} f(-\lambda h') &= \tfrac{1-\lambda^2}{\lambda}\mu h', \\ g(h) &= \tfrac{1}{\lambda}\left(\tfrac{1}{2}h^2 - \omega h - c_0\right). \end{aligned}$$

Consequently, the solutions of Equations (7) and (8) are equivalent for the previous expressions of $f(-\lambda h')$ and $g(h)$.

Lastly, the following analytical solution of the original PDE (1) is given by undoing the change of variables (5):

$$u(x,t) = \omega + \frac{6\alpha^2}{25\beta} - \exp\left\{\frac{2(x-\lambda t)\alpha}{5\beta}\right\}\mathcal{P}\left(c_2 - \frac{5\beta}{\alpha\sqrt{12\beta}}\exp\left\{\frac{(x-\lambda t)\alpha}{5\beta}\right\}, 0, c_1\right). \tag{9}$$

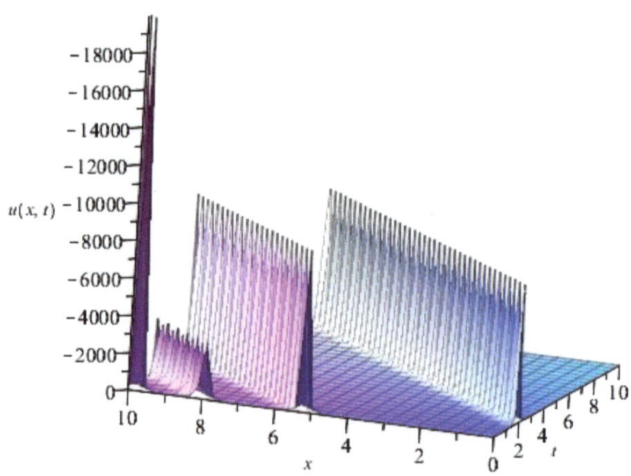

Figure 1. Solution (9), for $\lambda = \alpha = \beta = c_1 = c_2 = 1$.

Solution (9) is a soliton-type traveling wave solution (see Figure 1).

In addition, by the same procedure, the authors of [32] obtained the general solution of a second-order ODE of the form

$$h'' = \frac{2b(\beta q - p) - \alpha}{b(\beta^2+1)} h' + \frac{m_1 n_1}{b(\beta^2+1)} h^2 - \frac{b(p^2+q^2)+a}{b(\beta^2+1)} h.$$

In the same way, we can derive the general solution for this equation, for

$$\begin{aligned} f(-\lambda h') &= \tfrac{2b(\beta q - p) - \alpha}{b(\beta^2+1)}(1-\lambda^2)h', \\ g(h) &= \tfrac{m_1 n_1}{b(\beta^2+1)}(1-\lambda^2)h^2 - \tfrac{b(p^2+q^2)+a}{b(\beta^2+1)}(1-\lambda^2)h. \end{aligned}$$

Furthermore, we can find another solution by using the Jacobi elliptic function method. Let us assume that Equation (6) has a solution of the form

$$h = \alpha\, H^{\beta}(z),$$

where α and β are parameters to be determined. Here, $H(z)$ is a solution of the Jacobi equation

$$(H')^2 = r + p\,H^2 + q\,H^4, \tag{10}$$

with r, p and q constants.

Here H has the expression of an exponential or polynomial function. If H is a solution of Equation (10), then we can distinguish three cases: (i) H is the Jacobi elliptic sine function $sn(z,m)$; (ii) H is the Jacobi elliptic cosine function, $cn(z,m)$; (iii) H is the Jacobi elliptic function of the third kind $dn(z,m)$. However, we focus on the first case.

If $H(z) = sn(z,m)$,

$$h(z) = p\, sn^q(z|m) \tag{11}$$

is a solution of Equation (6). Substituting Equation(11) into Equation (6), we obtain the expressions of $f(-\lambda h')$, $g(h)$, and the parameters that make Equation(11) a solution of Equation (6).

This procedure was applied by Bruzón and Gandarias [33] to a similar equation, obtaining an exact solution. In the same way, an exact solution of Equation (1) is

$$u(x,t) = p\, sn^q(x - \lambda\, t|m). \tag{12}$$

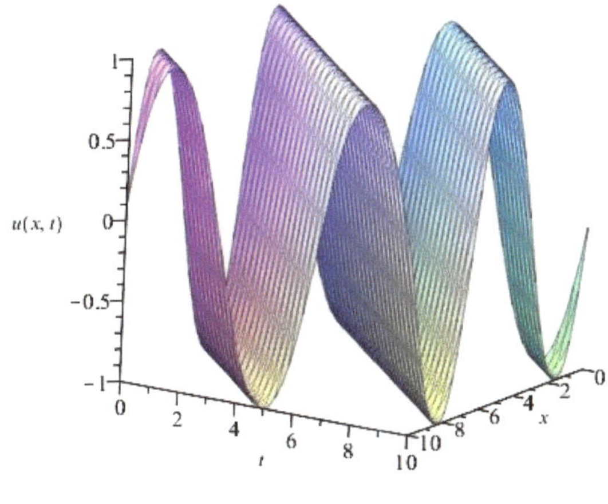

Figure 2. Solution (12) for $\lambda = p = q = 1$ and $m = 0.5$.

Solution (12) is shown in Figure 2. Specifically, the solution is a stable non-linear non-harmonic oscillatory periodic wave.

5. Conservation Laws

The notion of a conservation law is a mathematical formulation of the familiar physical laws of conservation of energy, conservation of momentum and so on. This concept plays an important role in the analysis of basic properties of the solutions. For example, the invariance of a variational principle under a group of time translations implies the conservation of energy for the solutions of the associated Euler-Lagrange equations, and the invariance under a group of spatial translations implies conservation of momentum [19].

Anco and Bluman presented a direct conservation law method for PDEs expressed in normal form. A PDE is in normal form if it can be expressed in a solved form for some leading derivative of u, such that all the other terms in the equation contain neither the leading derivative nor its differential consequences [26].

A local conservation law of the non-linear viscoelastic wave Equation (1) satisfies the space-time divergence expression

$$D_tT + D_xX = (u_{tt} - u_{xx} + f(u_t) - g(u))Q, \tag{13}$$

named the characteristic equation, where T is the conserved density and X the conserved flux. The vector (T, X) is called the conserved current.

The general expression of a low-order multiplier Q, written in terms of u and derivatives of u, depends on those variables that, by derivatives, can lead to a leading derivative of Equation (1). For example, u_{tt} can be derived by the derivative of u_t with respect to t, and u_{xx} by the derivative of u_x with respect to x.

Therefore, it is defined

$$Q(t, x, u, u_t, u_x)$$

as the general expression for a low-order multiplier for the non-linear viscoelastic wave Equation (1).

However, all low-order multipliers are found by solving the determining equation

$$E_u((u_{tt} - u_{xx} + f(u_t) - g(u))Q) = 0, \tag{14}$$

where E_u is the Euler operator with respect to u [19], defined by

$$E_u = \partial_u - D_x\partial_{u_x} - D_t\partial_{u_t} + D_xD_t\partial_{u_{xt}} + D_x^2\partial_{u_{xx}} + \cdots .$$

Hence, splitting the determining Equation (14) with respect to u_{xx}, u_{tt}, u_{tx}, we obtain an overdetermined linear system for $Q, f(u_t), g(u)$.

Thus, a complete classification of multipliers is found by solving the system with the same algorithmic method used for the determining equation for infinitesimal symmetries. The classification of multipliers is shown in Theorem 2. Then, for each multiplier we determine the corresponding conserved density T and flux X, by integrating directly the characteristic Equation (13). For this classification we apply the same Maple commands used for the Lie symmetries classification, "rifsimp", "dsolve", and "pdsolve". Theorem 3 shows the results obtained.

Theorem 2. *All the multipliers admitted by the non-linear viscoelastic wave Equation (1), with $f(u_t) \neq 0$, are given below.*

1. *For $f(u_t) = f_0$ and $g(u)$ arbitrary function, the multipliers are*

$$Q_1 = u_t, \qquad Q_2 = u_x.$$

2. *For* $f(u_t) = f_0$ *and* $g(u) = g_1\, e^{g_0 u} - f_0$*, besides* Q_1 *and* Q_2*, the multiplier is*

$$Q_3 = tu_t + xu_x + \frac{2}{g_0}.$$

3. *For* $f(u_t) = f_0 u_t + f_1$ *and* $g(u)$ *an arbitrary function, the multiplier is*

$$Q_4 = u_x\, e^{f_0 t}.$$

4. *For* $f(u_t) = -g_0 - \frac{1}{u_t f_0 + f_1}$ *and* $g(u) = g_0$*, the multiplier is*

$$Q_5 = f_0 u_t u_x + f_1 u_x.$$

5. *For* $f(u_t) = -\frac{4f_0}{u_t + f_1} + f_2$ *and* $g(u) = \frac{4f_0}{f_1} - f_2$*, the multiplier is*

$$Q_6 = u_t + f_1.$$

Theorem 3. *All the non-trivial low-order conservation laws admitted by the non-linear viscoelastic wave Equation (1), with* $f(u_t) \neq 0$*, are given below.*

1. *For* $f(u_t) = f_0$*,* $g(u)$ *arbitrary function and* $Q_1 = u_t$*, the conservation law is*

$$\begin{aligned} T_1 &= \frac{1}{2} u_x^2 + \frac{1}{2} u_t^2 + \int g(u) + f_0 \, du, \\ X_1 &= -u_t u_x. \end{aligned} \tag{15}$$

2. *For* $f(u_t) = f_0$*,* $g(u)$ *arbitrary function and* $Q_2 = u_x$*, the conservation law is*

$$\begin{aligned} T_2 &= u_t u_x, \\ X_2 &= -\frac{1}{2} u_x^2 - \frac{1}{2} u_t^2 + \int g(u) + f_0 \, du. \end{aligned} \tag{16}$$

3. *For* $f(u_t) = f_0$*,* $g(u) = g_1\, e^{g_0 u} - f_0$ *and* $Q_3 = tu_t + xu_x + \frac{2}{g_0}$*, the conservation law is*

$$\begin{aligned} T_3 &= \frac{1}{2g_0} 2\, t e^{u g_0} g_1 + \left(t u_t^2 + t u_x^2 + 2\, u_x x u_t \right) g_0 + 4\, u_t, \\ X_3 &= \frac{1}{2g_0} 2\, x e^{u g_0} g_1 + \left(-2\, t u_t u_x - x u_t^2 - u_x^2 x \right) g_0 - 4\, u_x. \end{aligned}$$

4. *For* $f(u_t) = f_0 u_t + f_1$*,* $g(u)$ *an arbitrary function and* $Q_4 = u_x\, e^{f_0 t}$*, the conservation law is*

$$\begin{aligned} T_4 &= u_x e^{f_0\, t} u_t, \\ X_4 &= \int e^{f_0\, t} (g(u) + f_1) \, du + \frac{1}{2} \left(-u_t^2 - u_x^2 \right) e^{f_0\, t}. \end{aligned}$$

5. *For* $f(u_t) = -g_0 - \frac{1}{u_t f_0 + f_1}$*,* $g(u) = g_0$ *and* $Q_5 = f_0 u_t u_x + f_1 u_x$*, the conservation law is*

$$\begin{aligned} T_5 &= \frac{1}{6} f_0\, u_x^3 + \frac{1}{2} f_0\, u_t^2 u_x + f_1\, u_x u_t, \\ X_5 &= -\frac{1}{2} f_0\, u_t u_x^2 - \frac{1}{2} u_x^2 f_1 - \frac{1}{6} f_0\, u_t^3 - u - \frac{1}{2} u_t^2 f_1. \end{aligned}$$

6. *For* $f(u_t) = -\frac{4f_0}{u_t + f_1} + f_2$*,* $g(u) = \frac{4f_0}{f_1} - f_2$ *and* $Q_6 = u_t + f_1$*, the conservation law is*

$$\begin{aligned} T_6 &= \frac{1}{2} u_x^2 + \frac{1}{2} u_t^2 + f_1\, u_t + 4\, \frac{f_0\, u}{f_1}, \\ X_6 &= (-u_t - f_1) u_x. \end{aligned}$$

Next, we study the meaning of some of these conservation laws. Every conservation law yields a corresponding conserved integral

$$\mathcal{C}[u] = \int_{\Omega} T\, dx,$$

where Ω is the domain of solutions $u(x,t)$.

For Equation (1), with $f(u_t)$ constant and $g(u)$ non-linear function, conservation law (15) yields conservation of an energy quantity

$$\mathcal{E}[u] = \int_{\Omega} \frac{1}{2}\, {u_x}^2 + \frac{1}{2}\, {u_t}^2 + \int g(u) + f_0 \, \mathrm{d}u\, dx,$$

which represents the total energy for solution $u(x,t)$.

Conservation law (16) yields the conserved quantity

$$\mathcal{M}[u] = \int_{\Omega} u_t u_x\, dx,$$

which is a momentum quantity.

6. Conclusions

In this paper, we have obtained a complete Lie group classification for the viscoelastic wave Equation (1) in the presence of damping and source terms, for different expressions of the functions f and g. Then, we have constructed the corresponding reduced equations. These reductions make easier the resolution of the viscoelastic wave Equation (1) in order to obtain solutions of physical interest, such as solitons. Moreover, we have obtained these traveling wave solutions from the reduced equations by the comparison between Equation (1) and comparable equations studied before by other authors. Furthermore, Lie point symmetries are not the only ones that can be studied. Another symmetries such as contact or potential symmetries can be studied in the future. Finally, we have derived the non-trivial low-order conservation laws by using the multiplier method.

Author Contributions: M.S.B. and A.P.M. contributed equally to this work. Both contributed substantially to obtain the mathematical results and give shape to the manuscript. All authors have read and agreed to the published version of the manuscript.

Funding: This research received no external funding.

Acknowledgments: The support of the *Plan Propio de Investigación de la Universidad de Cádiz* is gratefully acknowledged. The authors also thank the referees for their suggestions to improve the quality of the paper.

Conflicts of Interest: The authors declare no conflict of interest.

References

1. Kafini, M.; Messaoudi, S.A. A blow up result for a viscoelastic system in R^N. *Electron. J. Differ. Equ.* **2006**, *113*, 1–7.
2. Kafini, M.; Messaoudi, S.A. A blow up result in a Cauchy viscoelastic problem. *Appl. Math. Lett.* **2008**, *21*, 549–553. [CrossRef]
3. Wang, Y. A global nonexistence theorem for viscoelastic equations with arbitrary positive initial energy. *Appl. Math. Lett.* **2009**, *22*, 1394–1400. [CrossRef]
4. Kalantarov, V.K.; Ladyzhenskaya, O.A. The occurrence of collapse for quasilinear equations of parabolic and hyperbolic type. *J. Soviet Math.* **1978**, *10*, 53–70. [CrossRef]
5. Haraux, A.; Zuazua, E. Decay estimates for some semilinear damped hyperbolic problems. *Arch. Ration. Mech. Anal.* **1988**, *150*, 191–206. [CrossRef]
6. Ball, J. Remarks on blow up and nonexistence theorems for nonlinear evolutions equations. *Quart. J. Math. Oxford* **1977**, *28*, 473–486. [CrossRef]
7. Messaoudi, S.A. Blow up of solutions with positive initial energy in a nonlinear viscoelastic wave equations. *J. Math. Anal. Appl.* **2006**, *320*, 902–915. [CrossRef]
8. Messaoudi, S.A.; Said-Houari, B. Global nonexistence of positive initial-energy solutions of a system of nonlinear viscoelastic wave equations with damping and source terms. *J. Math. Anal. Appl.* **2010**, *365*, 277–287. [CrossRef]

9. Levine, H.A. Instability and nonexistence of global solutions to nonlinear wave equations of the form. *Trans. Am. Math. Soc.* **1974**, *192*, 1–21.
10. Levine, H.A. Some additional remarks on the nonexistence of global solutions to nonlinear wave equations. *J. Math. Anal.* **1974**, *5*, 138–146. [CrossRef]
11. Messaoudi, S.A. Blow up and global existence in nonlinear viscoelastic wave equations. *Math. Nachr.* **2003**, *260*, 58–66. [CrossRef]
12. Georgiev, V.; Todorova, G. Existence of a solution of the wave equation with nonlinear damping and source term. *J. Differ. Equ.* **1994**, *109*, 295–308. [CrossRef]
13. Messaoudi, S.A. Blow up in a nonlinearly damped wave equation. *Math. Nachr.* **2001**, *231*, 1–7. [CrossRef]
14. Márquez, A.P.; Bruzón, M.S. Symmetry analysis and conservation laws of a family of non-linear viscoelastic wave equations. In Proceedings of the XXVI Congreso de Ecuaciones Diferenciales y Aplicaciones, Asturias, Spain, 14–18 June 2021; Gallego, R., Mateos, M., Eds.; Universidad de Oviedo, Servicio de Publicaciones: Oviedo, Spain, 2021; pp. 284–288.
15. Abramowitz, M.; Stegun, I.A. *Handbook of Mathematical Functions*; Dover: New York, NY, USA, 1972.
16. Bluman, G.W.; Cole, J.D. The general similarity solution of the heat equation. *J. Math. Mech.* **1969**, *18*, 1025–1042.
17. Bluman, G.W.; Kumei, S. *Symmetries and Differential Equations*; Springer: Berlin, Germany, 1989.
18. Champagne, B.; Hereman, W.; Winternitz, P. The computer calculation of Lie point symmetries of large systems of differential equations. *Comput. Phys. Commun.* **1991**, *66*, 319–340. [CrossRef]
19. Olver, P.J. *Applications of Lie Groups to Differential Equations*; Springer: Berlin, Germany, 1986.
20. Ovsiannikov, L.V. *Group Analysis of Differential Equations*; Academic: New York, NY, USA, 1982.
21. Rosa, M.; Camacho, J.C.; Bruzón, M.S.; Gandarias, M.L. Classical and potential symmetries for a generalized Fisher equation. *J. Comput. Appl. Math.* **2017**, *318*, 181–188. [CrossRef]
22. Khalique, C.M.; Adem, K.R. Exact solutions of the (2+1)-dimensional Zakharov–Kuznetsov modified equal width equation using Lie group analysis. *Math. Comput. Model.* **2011**, *54*, 184–189. [CrossRef]
23. Tracinà, R.; Bruzón, M.S.; Gandarias, M.L.; Torrisi, M. Nonlinear self-adjointness, conservation laws, exact solutions of a system of dispersive evolution equations. *Commun. Nonlinear Sci. Numer. Simul.* **2014**, *19*, 3036–3043. [CrossRef]
24. Cheng, K.; Feng, W.; Gottlieb, S.; Wang, C. A Fourier pseudospectral method for the "good" Boussinesq equation with second order temporal accuracy. *Numer. Methods Partial. Differ. Equ.* **2015**, *31*, 202–224. [CrossRef]
25. Zhang, C.; Wang, H.; Huang, J.; Wang, C.; Yue, X. A second order operator splitting numerical scheme for the "good" Boussinesq equation. *Appl. Numer. Math.* **2017**, *119*, 179–193. [CrossRef]
26. Anco, S. Symmetry properties of conservation laws. *Int. J. Mod. Phys. B* **2016**, *30*, 1640003–1640015. [CrossRef]
27. Anco, S. Generalization of Noether's theorem in modern form to non-variational partial differential equations, in Recent progress and Modern Challenges in Applied Mathematics. In *Modeling and Computational Science, Fields Institute Communications*; Springer: New York, NY, USA, 2016; pp. 1–51.
28. Recio, E.; Garrido, T.M.; de la Rosa, R.; Bruzón, M.S. Hamiltonian structure, symmetries and conservation laws for a generalized (2+1)-dimensional double dispersion equation. *Symmetry* **2019**, *11*, 1031. [CrossRef]
29. Bruzón, M.S.; Recio, E.; Garrido, T.M.; de la Rosa, R. Lie symmetries, conservation laws and exact solutions of a generalized quasilinear KdV equation with degenerate dispersion. *Discret. Continous Dyn. Syst. Ser. S* **2020**, *13*, 2691–2701. [CrossRef]
30. Khalique, C.M.; Simbanefayi, I. Conserved quantities, optimal system and explicit solutions of a (1+1)-dimensional generalised coupled mKdV-type system. *J. Adv. Res.* **2021**, *29*, 159–166. [CrossRef]
31. Kudryashov, N. On "new travelling wave solutions" of the KdV and the KdV-Burgers equations. *Commun. Nonlinear Sci. Numer. Simul.* **2009**, *14*, 1891–1900. [CrossRef]
32. Bruzón, M.S.; Garrido, T.M.; Recio, E.; de la Rosa, R. Lie symmetries and travelling wave solutions of the nonlinear waves in the inhomogeneous Fisher-Kolmogorov equation. *Math. Meth. Appl. Sci.* **2019**, *2*, 1–9. [CrossRef]
33. Bruzón, M.S.; Gandarias, M.L. Travelling wave solutions for a generalized double dispersion equation. *Nonlinear Anal.* **2009**, *71*, e2109–e2117. [CrossRef]
34. Oliveri, F. Lie Symmetries of Differential Equations: Classical Results and Recent Contributions. *Symmetry* **2010**, *2*, 658–706. [CrossRef]

mathematics

MDPI

Review

Optical Solitons with Cubic-Quintic-Septic-Nonic Nonlinearities and Quadrupled Power-Law Nonlinearity: An Observation

Islam Samir [1], Ahmed H. Arnous [2], Yakup Yıldırım [3], Anjan Biswas [4,5,6,7,8], Luminita Moraru [9,*] and Simona Moldovanu [10]

1 Department of Physics and Mathematics Engineering, Faculty of Engineering, Ain Shams University, Cairo 11566, Egypt
2 Department of Physics and Engineering Mathematics, Higher Institute of Engineering, El-Shorouk Academy, Cairo 11837, Egypt
3 Department of Computer Engineering, Biruni University, Istanbul 34010, Turkey
4 Department of Mathematics and Physics, Grambling State University, Grambling, LA 71245, USA
5 Mathematical Modeling and Applied Computation (MMAC) Research Group, Department of Mathematics, King Abdulaziz University, Jeddah 21589, Saudi Arabia
6 Department of Applied Mathematics, National Research Nuclear University, 31 Kashirskoe Hwy, 115409 Moscow, Russia
7 Department of Applied Sciences, Cross-Border Faculty, Dunarea de Jos University of Galati, 111 Domneasca Street, 800201 Galati, Romania
8 Department of Mathematics and Applied Mathematics, Sefako Makgatho Health Sciences University, Medunsa 0204, South Africa
9 Department of Chemistry, Physics and Environment, Faculty of Sciences and Environment, Dunarea de Jos University of Galati, 47 Domneasca Street, 800008 Galati, Romania
10 Department of Computer Science and Information Technology, Faculty of Automation, Computers, Electrical Engineering and Electronics, Dunarea de Jos University of Galati, 47 Domneasca Street, 800008 Galati, Romania
* Correspondence: luminita.moraru@ugal.ro

Citation: Samir, I.; Arnous, A.H.; Yıldırım, Y.; Biswas, A.; Moraru, L.; Moldovanu, S. Optical Solitons with Cubic-Quintic-Septic-Nonic Nonlinearities and Quadrupled Power-Law Nonlinearity: An Observation. *Mathematics* **2022**, *10*, 4085. https://doi.org/10.3390/math10214085

Academic Editors: Almudena del Pilar Marquez Lozano, Vladimir Iosifovich Semenov and Ravi P. Agarwal

Received: 17 September 2022
Accepted: 25 October 2022
Published: 2 November 2022

Abstract: The current paper considers the enhanced Kudryashov's technique to retrieve solitons with a governing model having cubic-quintic-septic-nonic and quadrupled structures of self-phase modulation. The results prove that it is redundant to extend the self-phase modulation beyond cubic-quintic nonlinearity or dual-power law of nonlinearity.

Keywords: solitons; dual-power; Kudryashov

MSC: 78A60

1. Introduction

Optical soliton is one of the most important topics of study in nonlinear fiber optics during the present times [1–5]. The dynamics of such solitons is typically described by the nonlinear Schrödinger's equation (NLSE) [6–9] with a singleton form of self-phase modulation (SPM) [10–13] that emerges from the nonlinear refractive index structure of an optical fiber [14–20]. This typically appears with cubic nonlinear structure AKA Kerr law of nonlinearity [21–25] and its generalization to power-law of nonlinear medium [26–31]. The third form of singleton SPM that leads to optical Gaussons, as opposed to optical solitons, is with logarithmic law of nonlinearity [32]. Apart from these three forms with single nonlinear term, the lesser known structures of SPM, sparingly visible, are saturable law and exponential form. The remaining forms of SPM typically contain two or more nonlinear structures that are applicable in various forms of materials such a $LiNbO_3$ crystals. These are cubic-quintic nonlinearity, AKA parabolic form of SPM and its generalization to dual-power form of SPM. Several other forms of refractive index structures have emerged, such as quadratic-cubic (QC) form, generalized QC form, anti-cubic (AC) type, and generalized AC form of nonlinearity. The current paper draws attention to the possible extension of parabolic and dual power-laws of nonlinearity to cubic-quintic-septic-nonic (CQSN) form and its generalization to quadrupled power-law of nonlinearity (QPL) and beyond. Although the case of CQS law along with its generalization to triple power-law has been meaningfully addressed in the past [33,34], this paper carries out the analysis and proves that it is redundant to extend beyond CQS or triple-power law of nonlinear structure. This analysis has been carried out with chromatic dispersion (CD). The detailed analysis follows through with both forms of nonlinear refractive index structures.

Governing Model

$$iq_t + aq_{xx} + F\left(|q|^2\right)q = 0, \tag{1}$$

where the first term stems from temporal evolution, where $i = \sqrt{-1}$, whilst F comes from SPM. x depicts spatial variable, whereas a describes CD. t imply to temporal variable, while $q(x,t)$ denotes the wave profile.

2. The Enhanced Kudryashov's Technique

Consider a governing equation [35–37]

$$F(u, u_x, u_t, u_{xt}, u_{xx}, ...) = 0, \tag{2}$$

where $u = u(x,t)$ is dependent variable, whereas x and t are independent variables.

Step-1: Equation (2) reduces to

$$P(U, -kvU', kU', k^2U'', ...) = 0, \tag{3}$$

by using the restriction

$$\xi = k(x - vt),\ u(x,t) = U(\xi), \tag{4}$$

where v and k are constants.

Step-2: Equation (3) holds the solution structure

$$U(\xi) = \lambda_0 + \sum_{l=1}^{N} \sum_{i+j=l} \lambda_{ij} Q^i(\xi) R^j(\xi), \tag{5}$$

where N stems from the balancing procedure in Equation (3), while $R(\xi)$ and $Q(\xi)$ satisfy the ancillary equations

$$R'(\xi)^2 = R(\xi)^2(1 - \chi R(\xi)^2), \tag{6}$$

and

$$Q'(\xi) = Q(\xi)(\eta Q(\xi) - 1), \tag{7}$$

along with the explicit solutions

$$R(\xi) = \frac{4c}{4c^2 e^{\xi} + \chi e^{-\xi}}, \tag{8}$$

and

$$Q(\xi) = \frac{1}{\eta + be^{\xi}}. \tag{9}$$

Here χ, λ_0, η, $\lambda_{ij}(i,j = 0,1,...,N)$, a and b stand for constants.

Step-3: Putting (5) together with (6) and (7) into (3) leaves us with a system of equations that enables us the much-needed constant parameters in (4)–(9).

3. Optical Solitons

The current section employs the integration tool to retrieve optical solitons to the model having CQSN and QPL nonlinearity structures of SPM.

3.1. CQSN Nonlinearity

In this case, the model shapes up as

$$iq_t + aq_{xx} + \left(b_1|q|^2 + b_2|q|^4 + b_3|q|^6 + b_4|q|^8\right)q = 0. \tag{10}$$

It must be noted that b_j $(1 \leq j \leq 4)$ stem from $\chi^{(j)}$ for $(1 \leq j \leq 4)$ nonlinearities. Although $\chi^{(1)}$ and $\chi^{(2)}$ are substantial for $LiNbO_3$ crystals, $\chi^{(3)}$ and $\chi^{(4)}$ are negligibly small and miniscule. The current paper includes these nonlinearities to study the corresponding NLSE and check on its integrability aspect for the first time. The drawn conclusions will be interesting. It will be observed that these negligible nonlinear contributions must be set to zero for integrability purposes. This would lead to consistency between the Physics and Mathematics of the problem [38]. We consider the solution structure

$$q(x,t) = U(\xi)e^{i\phi(x,t)}, \tag{11}$$

with

$$\xi = k(x - vt), \tag{12}$$

and

$$\phi(x,t) = -\kappa x + \omega t + \theta. \tag{13}$$

Here, $U(\xi)$ comes from the amplitude component, where ξ is the wave variable and v is the velocity. Additionally, $\phi(x,t)$ stems from the phase component, where θ is the phase constant, ω is the angular frequency and κ is the wave number.

Putting (11) into (10) provides us the simplest equations

$$ak^2U'' - U\left(a\kappa^2 + \omega\right) + b_4U^9 + b_3U^7 + b_2U^5 + b_1U^3 = 0, \tag{14}$$

and

$$-kU'(2a\kappa + v) = 0. \tag{15}$$

Equation (15) enables us the soliton velocity

$$v = -2a\kappa. \tag{16}$$

Using the constraint

$$U(\xi) = V(\xi)^{\frac{1}{4}}, \tag{17}$$

Equation (14) stands as

$$4ak^2VV'' - 3ak^2V'^2 - 16V^2\left(a\kappa^2 + \omega\right) + 16b_1V^{\frac{5}{2}} + 16b_3V^{\frac{7}{2}} + 16b_4V^4 + 16b_2V^3 = 0. \tag{18}$$

Setting $b_1 = b_3 = 0$ reduces Equation (18) to

$$4ak^2VV'' - 3ak^2V'^2 - 16V^2\left(a\kappa^2 + \omega\right) + 16b_4V^4 + 16b_2V^3 = 0. \tag{19}$$

It must be noted that in Equation (18), b_1 and b_3 were set to zero simply for Equation (18) to be rendered integrable since these would Free (18) from all terms carrying fractional exponents of V. Thus, only b_2 and b_4 sustain to permit integrability of (18). This is equivalent to studying the governing model with only two non-zero terms, namely b_2 and b_4 terms. This is equivalent to saying that the governing NLSE is integrable with cubic—quintic nonlinear form of refractive index that is present in $LiNbO_3$ crystals. Thus, extending the SPM beyond

$\chi^{(5)}$ nonlinearity is redundant [14,38]. By the implementation of balancing procedure in Equation (19), the solution structure (5) stands as

$$V(\xi) = \lambda_0 + \lambda_{01}R(\xi) + \lambda_{10}Q(\xi). \tag{20}$$

Substituting (20) along with (6) and (7) into (19) gives way to the results:
Result-1:

$$\lambda_0 = \frac{6(a\kappa^2 + w)}{b_2},\ \lambda_{01} = 0,\ \lambda_{10} = -\frac{6\eta(a\kappa^2 + \omega)}{b_2},\ k = \pm 4\sqrt{\frac{a\kappa^2 + \omega}{a}},\ b_4 = -\frac{5b_2^2}{36(a\kappa^2 + \omega)}. \tag{21}$$

Plugging (21) along with (9) into (20) provides us

$$q(x,t) = \left\{ \frac{6(\omega + a\kappa^2)}{b_2} \frac{b \exp\left[\pm 4\sqrt{\frac{\omega + a\kappa^2}{a}}(x - vt) \right]}{\eta + b \exp\left[\pm 4\sqrt{\frac{\omega + a\kappa^2}{a}}(x - vt) \right]} \right\}^{\frac{1}{4}} e^{i(-\kappa x + \omega t + \theta)}. \tag{22}$$

Setting $a(a\kappa^2 + \omega) > 0$ and $\eta = \pm b$ collapses Equation (22) to the dark and singular solitons

$$q(x,t) = \left\{ \frac{3(\omega + a\kappa^2)}{b_2} \left(1 \pm \tanh\left[2\sqrt{\frac{\omega + a\kappa^2}{a}}(x - vt) \right] \right) \right\}^{\frac{1}{4}} e^{i(-\kappa x + \omega t + \theta)}, \tag{23}$$

and

$$q(x,t) = \left\{ \frac{3(\omega + a\kappa^2)}{b_2} \left(1 \pm \coth\left[2\sqrt{\frac{\omega + a\kappa^2}{a}}(x - vt) \right] \right) \right\}^{\frac{1}{4}} e^{i(-\kappa x + \omega t + \theta)}. \tag{24}$$

Result-2:

$$\lambda_0 = \lambda_{01} = 0,\ k = \pm 4\sqrt{\frac{a\kappa^2 + \omega}{a}},\ \lambda_{10} = \frac{6\eta(a\kappa^2 + \omega)}{b_2},\ b_4 = -\frac{5b_2^2}{36(a\kappa^2 + \omega)}. \tag{25}$$

Inserting (25) along with (9) into (20) enables us

$$q(x,t) = \left\{ \frac{6(\omega + a\kappa^2)}{b_2} \frac{\eta}{\eta + b \exp\left[\pm 4\sqrt{\frac{\omega + a\kappa^2}{a}}(x - vt) \right]} \right\}^{\frac{1}{4}} e^{i(-\kappa x + \omega t + \theta)}. \tag{26}$$

Taking $a(a\kappa^2 + \omega) > 0$ and $\eta = \pm b$ turns Equation (26) into the dark and singular solitons

$$q(x,t) = \left\{ \frac{3(\omega + a\kappa^2)}{b_2} \left(1 \mp \tanh\left[2\sqrt{\frac{\omega + a\kappa^2}{a}}(x - vt) \right] \right) \right\}^{\frac{1}{4}} e^{i(-\kappa x + \omega t + \theta)}, \tag{27}$$

and

$$q(x,t) = \left\{ \frac{3(\omega + a\kappa^2)}{b_2} \left(1 \mp \coth\left[2\sqrt{\frac{\omega + a\kappa^2}{a}}(x - vt) \right] \right) \right\}^{\frac{1}{4}} e^{i(-\kappa x + \omega t + \theta)}. \tag{28}$$

Result-3:

$$\lambda_0 = 0,\ \lambda_{10} = 0,\ k = \pm 4\sqrt{\frac{a\kappa^2+\omega}{a}},\ b_2 = 0,\ \lambda_{01} = \sqrt{\frac{5\chi(a\kappa^2+\omega)}{b_4}}. \tag{29}$$

Putting (29) along with (8) into (20) leaves us with

$$q(x,t) = \left\{4\sqrt{5}\sqrt{\frac{a\kappa^2\chi + w\chi}{b_4}}\frac{c}{4c^2\exp\left[\pm 4\sqrt{\frac{\omega+a\kappa^2}{a}}(x-vt)\right] + \chi\exp\left[\mp 4\sqrt{\frac{\omega+a\kappa^2}{a}}(x-vt)\right]}\right\}^{1/4}$$
$$\times e^{i(-\kappa x+\omega t+\theta)}. \tag{30}$$

Setting $a(a\kappa^2+\omega) > 0$ and $\chi = \pm 4c^2$ changes Equation (30) to the bright and singular solitons

$$q(x,t) = \left\{\pm\sqrt{\frac{5(\omega+a\kappa^2)}{b_4}}\operatorname{sech}\left[4\sqrt{\frac{\omega+a\kappa^2}{a}}(x-vt)\right]\right\}^{\frac{1}{4}} e^{i(-\kappa x+\omega t+\theta)}, \tag{31}$$

and

$$q(x,t) = \left\{\pm\sqrt{\frac{5(\omega+a\kappa^2)}{b_4}}\operatorname{csch}\left[4\sqrt{\frac{\omega+a\kappa^2}{a}}(x-vt)\right]\right\}^{\frac{1}{4}} e^{i(-\kappa x+\omega t+\theta)}. \tag{32}$$

3.2. QPL Nonlinearity

In this case, the model sticks out as

$$iq_t + aq_{xx} + \left(b_1|q|^{2n} + b_2|q|^{4n} + b_3|q|^{6n} + b_4|q|^{8n}\right)q = 0, \tag{33}$$

where b_j $(j = 1-4)$ come from QPL nonlinearity. Putting (11) into (33) paves way to the auxiliary equations

$$ak^2U'' - U\left(a\kappa^2+\omega\right) + b_4U^{8n+1} + b_3U^{6n+1} + b_2U^{4n+1} + b_1U^{2n+1} = 0, \tag{34}$$

and

$$-kU'(2a\kappa + v) = 0. \tag{35}$$

Equation (35) leaves us with the soliton velocity

$$v = -2a\kappa. \tag{36}$$

Using the restriction

$$U(\xi) = V(\xi)^{\frac{1}{4n}}, \tag{37}$$

Equation (34) reads as

$$4ak^2nVV'' + ak^2(1-4n)V'^2 - 16n^2V^2\left(a\kappa^2+\omega\right) + 16b_1n^2V^{\frac{5}{2}} + 16b_3n^2V^{\frac{7}{2}} + 16b_4n^2V^4 + 16b_2n^2V^3 = 0. \tag{38}$$

Taking $b_1 = b_3 = 0$ simplifies Equation (38) to

$$4ak^2nVV'' + ak^2(1-4n)V'^2 - 16n^2V^2\left(a\kappa^2+\omega\right) + 16b_4n^2V^4 + 16b_2n^2V^3 = 0. \tag{39}$$

By the implementation of balancing technique in Equation (39), the formal solution (5) turns into

$$V(\xi) = \lambda_0 + \lambda_{01}R(\xi) + \lambda_{10}Q(\xi). \tag{40}$$

Substituting (40) along with (6) and (7) into (39) leaves us with the results:

Result-1:

$$\lambda_0 = \frac{2(2n+1)(a\kappa^2 + w)}{b_2}, \ \lambda_{01} = 0, \ \lambda_{10} = -\frac{2(2n+1)\eta(a\kappa^2+\omega)}{b_2},$$
$$k = \pm 4n\sqrt{\frac{a\kappa^2+\omega}{a}}, \ b_4 = -\frac{(4n+1)b_2^2}{4(2n+10^2)(a\kappa^2+\omega)}. \tag{41}$$

Plugging (41) along with (9) into (40) provides us

$$q(x,t) = \left\{ \frac{2(1+2n)(\omega+a\kappa^2)}{b_2} \frac{b\exp\left[\pm 4n\sqrt{\frac{\omega+a\kappa^2}{a}}(x-vt)\right]}{\eta + b\exp\left[\pm 4\sqrt{\frac{\omega+a\kappa^2}{a}}(x-vt)\right]} \right\}^{\frac{1}{4n}} e^{i(-\kappa x+\omega t+\theta)}. \tag{42}$$

Taking $a(a\kappa^2+\omega) > 0$ and $\eta = \pm b$, the dark and singular solitons stand as

$$q(x,t) = \left\{ \frac{(1+2n)(\omega+a\kappa^2)}{b_2} \left(1 \pm \tanh\left[2n\sqrt{\frac{\omega+a\kappa^2}{a}}(x-vt)\right]\right) \right\}^{\frac{1}{4n}} e^{i(-\kappa x+\omega t+\theta)}, \tag{43}$$

and

$$q(x,t) = \left\{ \frac{(1+2n)(\omega+a\kappa^2)}{b_2} \left(1 \pm \coth\left[2n\sqrt{\frac{\omega+a\kappa^2}{a}}(x-vt)\right]\right) \right\}^{\frac{1}{4n}} e^{i(-\kappa x+\omega t+\theta)}. \tag{44}$$

Result-2:

$$\lambda_0 = \lambda_{01} = 0, \ k = \pm 4n\sqrt{\frac{a\kappa^2+\omega}{a}}, \ \lambda_{10} = \frac{2(2n+1)\eta(a\kappa^2+\omega)}{b_2}, \ b_4 = -\frac{(4n+1)b_2^2}{4(2n+1)^2(a\kappa^2+\omega)}. \tag{45}$$

Inserting (45) along with (9) into (40) enables us

$$q(x,t) = \left\{ \frac{2(1+2n)(\omega+a\kappa^2)}{b_2} \frac{\eta}{\eta + b\exp\left[\pm 4n\sqrt{\frac{\omega+a\kappa^2}{a}}(x-vt)\right]} \right\}^{\frac{1}{4n}} e^{i(-\kappa x+\omega t+\theta)}. \tag{46}$$

Setting $a(a\kappa^2+\omega) > 0$ and $\eta = \pm b$, the dark and singular solitons stick out as

$$q(x,t) = \left\{ \frac{(1+2n)(\omega+a\kappa^2)}{b_2} \left(1 \mp \tanh\left[2n\sqrt{\frac{\omega+a\kappa^2}{a}}(x-vt)\right]\right) \right\}^{\frac{1}{4n}} e^{i(-\kappa x+\omega t+\theta)}, \tag{47}$$

and

$$q(x,t) = \left\{ \frac{(1+2n)(\omega+a\kappa^2)}{b_2} \left(1 \mp \coth\left[2n\sqrt{\frac{\omega+a\kappa^2}{a}}(x-vt)\right]\right) \right\}^{\frac{1}{4n}} e^{i(-\kappa x+\omega t+\theta)}. \tag{48}$$

Result-3:

$$\lambda_0 = 0, \ \lambda_{10} = 0, \ k = \pm 4n\sqrt{\frac{a\kappa^2+\omega}{a}}, \ b_2 = 0, \ \lambda_{01} = \sqrt{\frac{(4n+1)\chi(a\kappa^2+\omega)}{b_4}}. \tag{49}$$

Putting (49) along with (8) into (40) paves way to

$$q(x,t) = \left\{ \sqrt{\frac{a\kappa^2\chi + \omega\chi}{b_4}} \frac{4\sqrt{(4n+1)c}}{4c^2 \exp\left[\pm 4n\sqrt{\frac{\omega+a\kappa^2}{a}}(x-vt)\right] + \chi \exp\left[\mp 4n\sqrt{\frac{\omega+a\kappa^2}{a}}(x-vt)\right]} \right\}^{\frac{1}{4n}}$$
$$\times e^{i(-\kappa x+\omega t+\theta)}. \tag{50}$$

Setting $a(a\kappa^2 + \omega) > 0$ and $\chi = \pm 4c^2$, the bright and singular solitons evolve as

$$q(x,t) = \left\{ \pm \sqrt{\frac{(1+4n)(\omega + a\kappa^2)}{b_4}} \text{sech}\left[4n\sqrt{\frac{\omega + a\kappa^2}{a}}(x-vt)\right] \right\}^{\frac{1}{4n}} e^{i(-\kappa x+\omega t+\theta)}, \tag{51}$$

and

$$q(x,t) = \left\{ \pm \sqrt{\frac{(1+4n)(\omega + a\kappa^2)}{b_4}} \text{csch}\left[4n\sqrt{\frac{\omega + a\kappa^2}{a}}(x-vt)\right] \right\}^{\frac{1}{4n}} e^{i(-\kappa x+\omega t+\theta)}. \tag{52}$$

4. An Observation

This paper simply shows that the NLSE with CD for CQSN or QPL nonlinearity, it is redundant to extend the nonlinear structure of SPM beyond the quintic form or its corresponding generalization in the QPL nonlinear structure. The results fall back to the case of QC or dual-power law of nonlinearity structure, respectively. In both forms of SPM structures, one is compelled to choose $b_1 = b_3 = 0$ thus collapsing the NLSE given by (10) or (33) to the form of parabolic law of nonlinearity or dual-power law of nonlinearity respectively. The respective exponents of the coefficients of b_2 and b_4 can be renamed from $(4,8)$ and $(4n,8n)$ to $(2,4)$ and $(2n,4n)$, respectively, so that the results for the soliton structure collapse and conform to the pre-existing results known earlier [39]. The extension to CQS and triple-power forms of SPM is also studied in [40].

5. Conclusions

The current paper derives 1-soliton solutions to the model with CD having CQSN and QPL nonlinearity structures of SPM. In both cases it was established that the extension beyond septic form of nonlinearity and its generalized form is redundant. It is only with dual-power and parabolic forms of nonlinear refractive index structure the model would make sense. Any extension that is beyond septic or its generalized form would collapse to parabolic dual-power laws. This true with CD being the source of dispersion terms. Additional form(s) of dispersion sources have not been examined yet. This is, thus, an open problem and will be later investigated. The results are yet to be released and are currently awaited. This would subsequently lead to a very interesting structure of the results.

Author Contributions: Conceptualization, I.S.; methodology, A.H.A.; software, Y.Y.; writing—original draft preparation, A.B.; writing—review and editing, L.M.; project administration, S.M. All authors have read and agreed to the published version of the manuscript.

Funding: This work was supported by the project "DINAMIC", Contract no. 12PFE/2021.

Institutional Review Board Statement: Not applicable.

Informed Consent Statement: Not applicable.

Data Availability Statement: Not applicable.

Conflicts of Interest: The authors declare no conflicts of interest.

References

1. Arnous, A.H.; Biswas, A.; Ekici, M.; Alzahrani, A.K.; Belic, M.R. Optical solitons and conservation laws of Kudryashov's equation with improved modified extended tanh-function. *Optik* **2021**, *225*, 165406. [CrossRef]
2. Manafian, J.; Lakestani, M. Optical solitons with Biswas-Milovic equation for Kerr law nonlinearity. *Eur. Phys. J. Plus* **2015**, *130*, 61. [CrossRef]
3. Arnous, A.H. Optical solitons to the cubic quartic Bragg gratings with anti-cubic nonlinearity using new approach. *Optik* **2022**, *251*, 168356. [CrossRef]
4. Arnous, A.H.; Zhou, Q.; Biswas, A.; Guggilla, P.; Khan, S.; Yıldırım, Y.; Alshomrani, A.S.; Alshehri, H.M. Optical solitons in fiber Bragg gratings with cubic–quartic dispersive reflectivity by enhanced Kudryashov's approach. *Phys. Lett. A* **2022**, *422*, 127797. [CrossRef]
5. Arnous, A.H.; Mirzazadeh, M.; Zhou, Q.; Moshokoa, S.P.; Biswas, A.; Belic, M. Soliton solutions to resonant nonlinear schrodinger's equation with time-dependent coefficients by modified simple equation method. *Optik* **2016**, *127*, 11450–11459. [CrossRef]
6. Kudryashov, N.A. Method for finding highly dispersive optical solitons of nonlinear differential equations. *Optik* **2020**, *206*, 163550. [CrossRef]
7. Kudryashov, N.A. Highly dispersive optical solitons of equation with various polynomial nonlinearity law. *Chaos Solitons Fractals* **2020**, *140*, 110202. [CrossRef]
8. Kudryashov, N.A. Optical Solitons of the Generalized Nonlinear Schrödinger Equation with Kerr Nonlinearity and Dispersion of Unrestricted Order. *Mathematics* **2022**, *10*, 3409. [CrossRef]
9. Kudryashov, N.A. Bright and dark solitons in a nonlinear saturable medium. *Phys. Lett. A* **2022**, *427*, 127913. [CrossRef]
10. Kudryashov, N.A. Stationary solitons of the model with nonlinear chromatic dispersion and arbitrary refractive index. *Optik* **2022**, *259*, 168888. [CrossRef]
11. Kudryashov, N.A. Highly dispersive optical solitons of an equation with arbitrary refractive index. *Regul. Chaotic Dyn.* **2020**, *25*, 537–543. [CrossRef]
12. Ozisik, M.; Cinar, M.; Secer, A.; Bayram, M. Optical solitons with Kudryashov's sextic power-law nonlinearity. *Optik* **2022**, *261*, 169202. [CrossRef]
13. Ozisik, M.; Secer, A.; Bayram, M. On the examination of optical soliton pulses of Manakov system with auxiliary equation technique. *Optik* **2022**, *220*, 169800. [CrossRef]
14. Biswas, A.; Konar, S. *Introduction to Non-Kerr Law Optical Solitons*; Chapman and Hall/CRC: New York, NY, USA, 2006.
15. Biswas, A.; Milovic, D.; Edwards, M. *Mathematical Theory of Dispersion-Managed Optical Solitons*; Springer Science & Business Media: Berlin/Heidelberg, Germany, 2010.
16. Tariq, K.U.; Wazwaz, A.M.; Ahmed, A. On some optical soliton structures to the Lakshmanan-Porsezian-Daniel model with a set of nonlinearities. *Opt. Quantum Electron.* **2022**, *54*, 1–34. [CrossRef]
17. Yu, W.; Zhang, H.; Wazwaz, A.M.; Liu, W. The collision dynamics between double-hump solitons in two mode optical fibers. *Results Phys.* **2021**, *28*, 104618. [CrossRef]
18. Darvishi, M.T.; Najafi, M.; Wazwaz, A.M. Some optical soliton solutions of space-time conformable fractional Schrödinger-type models. *Phys. Scr.* **2021**, *96*, 065213. [CrossRef]
19. Ozisik, M.; Secer, A.; Bayram, M.; Aydin, H. An encyclopedia of Kudryashov's integrability approaches applicable to optoelectronic devices. *Optik* **2022**, *265*, 169499. [CrossRef]
20. Secer, A. Stochastic optical solitons with multiplicative white noise via Itô calculus. *Optik* **2022**, *268*, 169831. [CrossRef]
21. Wang, M.Y. Optical solitons of the perturbed nonlinear Schrödinger equation in Kerr media. *Optik* **2021**, *243*, 167382. [CrossRef]
22. Wang, M.Y. Highly dispersive optical solitons of perturbed nonlinear Schrödinger equation with Kudryashov's sextic-power law nonlinear. *Optik* **2022**, *267*, 169631. [CrossRef]
23. Wazwaz, A.M. Bright and dark optical solitons for (3+1)-dimensional Schrödinger equation with cubic-quintic-septic nonlinearities. *Optik* **2021**, *225*, 165752. [CrossRef]
24. Darvishi, M.; Najafi, M.; Wazwaz, A.M. Conformable space-time fractional nonlinear (1+1)-dimensional Schrödinger-type models and their traveling wave solutions. *Chaos Solitons Fractals* **2021**, *150*, 111187. [CrossRef]
25. Liu, W.; Zhang, Y.; Wazwaz, A.M.; Zhou, Q. Analytic study on triple-S, triple-triangle structure interactions for solitons in inhomogeneous multi-mode fiber. *Appl. Math. Comput.* **2019**, *361*, 325–331. [CrossRef]
26. Ozisik, M.; Bayram, M.; Secer, A.; Cinar, M. Optical soliton solutions of the Chen-Lee-Liu equation in the presence of perturbation and the effect of the inter-modal dispersion, self-steepening and nonlinear dispersion. *Opt. Quantum Electron.* **2022**, *54*, 792. [CrossRef]
27. Esen, H.; Secer, A.; Ozisik, M.; Bayram, M. Analytical soliton solutions of the higher order cubic-quintic nonlinear Schrödinger equation and the influence of the model's parameters. *J. Appl. Phys.* **2022**, *132*, 053103. [CrossRef]
28. Ozdemir, N.; Esen, H.; Secer, A.; Bayram, M.; Yusuf, A.; Sulaiman, T.A. Optical solitons and other solutions to the Hirota-Maccari system with conformable, M-truncated and beta derivatives. *Mod. Phys. Lett. B* **2022**, *36*, 2150625. [CrossRef]
29. Esen, H.; Ozdemir, N.; Secer, A.; Bayram, M.; Sulaiman, T.A.; Yusuf, A. Solitary wave solutions of chiral nonlinear Schrödinger equations. *Mod. Phys. Lett. B* **2021**, *35*, 2150472. [CrossRef]

30. Sulaiman, T.A.; Yusuf, A.; Abdel-Khalek, S.; Bayram, M.; Ahmad, H. Nonautonomous complex wave solutions to the (2+1)-dimensional variable-coefficients nonlinear Chiral Schrödinger equation. *Results Phys.* **2020**, *19*, 103604. [CrossRef]
31. Altun, S.; Ozisik, M.; Secer, A.; Bayram, M. Optical solitons for Biswas-Milovic equation using the new Kudryashov's scheme. *Optik* **2022**, *270*, 170045. [CrossRef]
32. Bialynicki-Birula, I.; Mycielski, J. Gaussons: solitons of the logarithmic Schrödinger equation. *Phys. Scr.* **1979**, *20*, 539. [CrossRef]
33. Kohl, R.W.; Biswas, A.; Ekici, M.; Zhou, Q.; Khan, S.; Alshomrani, A.S.; Belic, M.R. Highly dispersive optical soliton perturbation with cubic-quintic-septic refractive index by semi-inverse variational principle. *Optik* **2019**, *199*, 163322. [CrossRef]
34. Kohl, R.W.; Biswas, A.; Zhou, Q.; Ekici, M.; Alzahrani, A.K.; Belic, M.R. Optical soliton perturbation with polynomial and triple-power laws of refractive index by semi-inverse variational principle. *Chaos Solitons Fractals* **2020**, *135*, 109765. [CrossRef]
35. Arnous, A.H. Optical solitons with Biswas-Milovic equation in magneto-optic waveguide having Kudryashov's law of refractive index. *Optik* **2021**, *247*, 167987. [CrossRef]
36. Arnous, A.H.; Biswas, A.; Yıldırım, Y.; Zhou, Q.; Liu, W.; Alshomrani, A.S.; Alshehri, H.M. Cubic-quartic optical soliton perturbation with complex Ginzburg-Landau equation by the enhanced Kudryashov's method. *Chaos Solitons Fractals* **2022**, *155*, 111748. [CrossRef]
37. Arnous, A.H.; Ullah, M.Z.; Moshokoa, S.P.; Zhou, Q.; Triki, H.; Mirzazadeh, M.; Biswas, A. Optical solitons in nonlinear directional couplers with trial function scheme. *Nonlinear Dyn.* **2017**, *88*, 1891–1915. [CrossRef]
38. Akhmediev, N.N.; Ankiewicz, A. *Nonlinear Pulses and Beams*; Springer: Berlin/Heidelberg, Germany, 1997.
39. Kudryashov, N.A. Method for finding optical solitons of generalized nonlinear Schrödinger equations. *Optik* **2022**, *261*, 169163. [CrossRef]
40. Kudryashov, N.A. Highly dispersive optical solitons of the generalized nonlinear eighth-order Schrödinger equation. *Optik* **2020**, *206*, 164335. [CrossRef]

mathematics

MDPI

Review

Operators and Boundary Problems in Finance, Economics and Insurance: Peculiarities, Efficient Methods and Outstanding Problems

Sergei Levendorskiĭ

Calico Science Consulting, Austin, TX 78748, USA; levendorskii@gmail.com

Abstract: The price V of a contingent claim in finance, insurance and economics is defined as an expectation of a stochastic expression. If the underlying uncertainty is modeled as a strong Markov process X, the Feynman–Kac theorem suggests that V is the unique solution of a boundary problem for a parabolic equation. In the case of PDO with constant symbols, simple probabilistic tools explained in this paper can be used to explicitly calculate expectations under very weak conditions on the process and study the regularity of the solution. Assuming that the Feynman–Kac theorem holds, and a more general boundary problem can be localized, the local results can be used to study the existence and regularity of solutions, and derive efficient numerical methods. In the paper, difficulties for the realization of this program are analyzed, several outstanding problems are listed, and several closely efficient methods are outlined.

Keywords: Feynman–Kac theorem; Lévy processes; affine processes; quadratic term structure models; European options; barrier options; American options; fractional differential equations; method of lines; smooth pasting principle; sinh-acceleration

MSC: 28-08; 30-08; 32-08; 35A02; 35A20; 35A21; 35A22; 35B33; 35B40; 35B44; 35C15; 35C20; 35E15; 3E20; 35H10; 35J70; 35K65; 35P10; 35R10; 35R11; 35R35; 35S05; 35S16; 42-08; 42A85; 42B10; 42B37; 60-08; 60E10; 60G35; 60G51; 65M20; 65M70; 65N45; 65N75; 65R10; 65T20; 90-08; 91-10; 91G60

Citation: Levendorskiĭ, S. Operators and Boundary Problems in Finance, Economics and Insurance: Peculiarities, Efficient Methods and Outstanding Problems. *Mathematics* **2022**, *10*, 1028. https://doi.org/10.3390/math10071028

Academic Editors: Almudena del Pilar Marquez Lozano and Vladimir Iosifovich Semenov

Received: 28 January 2022
Accepted: 17 March 2022
Published: 23 March 2022

Publisher's Note: MDPI stays neutral with regard to jurisdictional claims in published maps and institutional affiliations.

1. Introduction

1.1. Expectations and Boundary Problems

The prices (values) of contingent claims in finance, insurance and economics are defined as expectations of certain stochastic expressions. In many cases, the underlying uncertainty is modeled as a strong Markov process X (with killing) on $\mathbb{R}^n$ or its subset $\mathcal{D}$, with the infinitesimal generator L. Let $V(t,x)$ denote the price of an option or other contingent claim in the market at time t and $X_t = x$. The Feynman–Kac theorem suggests that V is the unique solution of a boundary problem for the parabolic equation:

$$(\partial_t + L)V(t,x) + g(t,x) = 0. \tag{1}$$

In the case of diffusion models, L is a differential operator of order 2; in jump-diffusion models, L is an integro-differential operator, and hence, a pseudo-differential operator (PDO). We use the representation of L as a PDO because this facilitates the study of the regularity of solutions, and naturally leads to the construction of efficient numerical methods for option pricing. The function g represents the stream of payoffs $\{g(t,X_t)\}_{t\ge 0}$ that the owner of the contingent claim is entitled to. The value function V satisfies (1) in the open region U in the time-state space, where the derivative security remains alive. As the process (t,X_t) leaves U, the owner of the contingent claim is entitled to an instantaneous payoff $G(\tau_{U^c}, X_{\tau_{U^c}})$, where τ_{U^c} is the hitting time of $U^c =: \mathcal{D}\setminus U$ by $\{(t,X_t)\}$.

The solution of the boundary problem can be used to calculate the expectation, and the properties of the expectation can be used to formulate appropriate boundary and regularity conditions. Unfortunately, this scheme is very difficult to realize in almost all cases, including many popular diffusion models, and the majority of the results in the literature are obtained assuming (without proof) that the boundary problem for the expectation has the unique solution. Furthermore, it is assumed (also without proof) that the expectation defines a sufficiently regular function so that Ito's formula can be applied; then, the solution of the boundary problem is the expectation. The necessity of proving these crucial assumptions is brushed under the carpet. Even in many diffusion models, the proof of the Feynman–Kac theorem is lacking. See [1,2] for details; one of the main reason is a complicated degeneration of L at the boundary of the state space, which makes it difficult to apply standard tools for the analysis of degenerate elliptic operators. If L degenerates at the boundary of a half-space, then general tools [3–5] can be applied, with some modifications, but in the case of degeneration at two and more transversal hyperplanes, the study of the regularity of solutions is more difficult. Furthermore, even if a weak regularity of the solution of the boundary problem is established, the proof of the Feynman–Kac theorem remains quite non-trivial. In the case of jump-diffusion models, additional subtleties emerge, even in the simplest case of Lévy models (the infinitesimal generator is a PDO with the constant symbol). In the majority of empirical studies, when Lévy models are calibrated to the real data, L is of the form $\partial_t + \langle \mu, \partial_x \rangle - a(D_x)$, where $a(D)$ is an elliptic PDO of order $\nu \in (0,1)$, with the symbol analytic in a tube domain. Hence, the operator of the boundary problem (1) becomes a parabolic operator of the standard form, with an elliptic stationary part, only after an appropriate change of variables in the (t,x)-space. This peculiarity leads to several non-standard properties of solutions, which we analyze in this paper. If one has in view applications of the general theory of fractional differential equations to finance, the class of operators of this kind deserves to be regarded as a model class rather than standard fractional differential operators $\partial_t - |\Delta|^\nu$, where $|\Delta|^\nu$ is the fractional Laplacian. (Note that $|\Delta|^\nu$, $\nu \neq 2$, cannot be used in the standard popular models in finance, economics and insurance.) An additional important feature of the infinitesimal generators of Lévy models used in finance is the existence of analytic continuation not only to tube domains but to conesas well. This observation is the basis of efficient numerical methods that evaluate integrals in pricing formulas, which we outline in this paper. These methods can be used in other situations, for instance, to evaluate special functions and stable distributions [6–9].

1.2. Black–Scholes Model and Diffusion Models

In the Black–Scholes model, the only source of uncertainty is the stock price $S_t = e^{X_t}$, where X_t is the Brownian motion (BM) on $\mathbb{R}$, with drift. The infinitesimal generator L is the second-order elliptic differential operator. If the market with several stocks and/or several sources of uncertainty is considered, e.g., volatility and/or stochastic interest rate, then the underlying process X is of a more complicated nature.

Typically, the region U where (1) holds is of the form $\{(t,x) \mid t \in (0,T), x \in U(t)\}$, where $U(t)$ are open subsets of $\mathcal{D}$. If X is a diffusion, $X_{\tau_{U^c}}$ is at the boundary of U, and then V satisfies the terminal condition:

$$V(T,x) = G(T,x),\ x \in U(T), \tag{2}$$

and the boundary condition:

$$V(t,x) = G(t,x),\ x \in \partial U(t). \tag{3}$$

If U is unbounded, appropriate restrictions on the rate of growth of V at infinity are imposed. Boundary problems of this sort arise for the majority of contingent claims in finance and insurance, and for numerous value functions in economics, e.g., real options and stochastic games. In some important cases such as lookback or Asian options, the underlying process is not Markovian; hence, pricing is more involved. In this paper, we consider boundary problems of the form (1), (2), (3), including free boundary problems. In applications, the approximate pricing of contingent claims of a complicated structure is reduced to a sequence of embedded options, equivalently, to a sequence of boundary problems of the form (1), (2), (3). Hence, this study of the regularity of the solutions of the latter is important for studying other types of contingent claims as well. Efficient numerical procedures for the solutions of standard boundary problems can be (and are) used as building blocks to solve more complicated problems.

In the Black–Scholes model, a simple affine change of variables reduces (1) to the backward parabolic equation $V_t + V_{xx} - rV = 0$; hence, the host of standard well-known methods can be used to price numerous contingent claims. Black and Scholes informally derived (1), constructing the perfect continuously changing hedging portfolio; later, Merton gave essentially equivalent proof constructing the perfect continuously changing replicating portfolio. Both proofs contain fundamental mathematical errors but can be made accurate for wide classes of diffusion models. Unfortunately, both proofs fail for jump-diffusion models—the statements and proofs of several popular imperfect substitutes for perfect hedging and replication are also incorrect (see [10]), and the reason for the failure is fundamental rather than technical.

1.3. The Case of Jump-Diffusions

In the Merton–Black–Scholes theory, the pricing equation is derived from the absence of frictions and no-arbitrage assumption. The no-arbitrage assumption means that it is impossible to construct a portfolio of securities traded in the market such that, at the expiration date, the value of the portfolio is non-negative with probability 1, and positive with positive probability. Leaving aside an important theoretical background from economics and finance, and the technical conditions necessary to make the statements mathematically accurate, the absence of frictions and the no-arbitrage assumption imply that the discounted prices of all securities traded in the market must be local martingales under a probability measure $\mathbb{Q}$ on the filtered probability space where the process X lives. The crucial point is that $\mathbb{Q} \neq \mathbb{P}$, where $\mathbb{P}$ is the historic or physical probability measure estimated using the time series for the prices of securities already traded on the market. The pricing measure $\mathbb{Q}$ may not assign non-zero probabilities to events of zero measure under $\mathbb{P}$ and vice versa; hence, $\mathbb{Q}$ must be equivalent to $\mathbb{P}$. This explains the name for $\mathbb{Q}$: an equivalent martingale measure (EMM; another name is risk-neutral measure). The discounting can be taken into account as the killing of the Markov process, and $L = L^{\mathbb{Q}}$ in (1) is the infinitesimal generator of the process with killing, under $\mathbb{Q}$. Then, $\{V(t, X_t)\}$, the discounted price process, is a local martingale if for any $t < \tau_{U^c}$ and stopping time $\tau \geq t$ s.t. $\tau \leq \tau_{U^c}$, $\mathbb{E}^{\mathbb{Q}}[V(\tau, X_\tau) \mid X_t = x] = V(t, x)$.

The boundary problem (1), (2), (3) is used to calculate the prices of new securities. In (sufficiently regular) diffusion models, there exists a unique EMM; hence, one can calculate the price of any contingent claim. In financial markets and insurance, a great variety of contingent claims are constructed and sold, and the implicit assumption is that one can theoretically calculate the price of new securities.

The three conditions are as follows: (1) a perfect hedge is possible; (2) perfect replication is possible; (3) there exists a unique EMM are equivalent (naturally, under subtle technical conditions)—and markets satisfying these conditions are called complete markets. The Feynman–Kac theorem gives the representation of the price as the expectation of the payoff stream under $\mathbb{Q}$:

$$V(t,x) = \mathbb{E}^{\mathbb{Q}}\left[\int_t^{\tau_{U^c}} g(s, X_s)ds + G(\tau_{U^c}, V_{\tau_{U^c}})\right]. \tag{4}$$

If X is a jump process or jump-diffusion process, then the hedging/replicating argument is not applicable. Fortunately, according to the general economic theory, if the market does not admit arbitrage, there exists an EMM such that the price of each contingent claim is given by (4) (an accurate mathematical statement imposes additional subtle conditions on X); an EMM is not unique, and one may choose an EMM one believes in. In real financial markets and numerous empirical studies, the measures calibrated to prices of the underlying financial instruments and options are not only different; quite often, the measures are not equivalent. Several examples can be found in the well-known empirical study in [11]: prices of certain stocks are calibrated to processes of infinite variation, whereas the prices of options—to processes of finite variation, which contradicts the well-known conditions on equivalent probability measures in [12] (the authors of [11] did not notice this discrepancy of their calibration results). One of the important reasons for discrepancies of this kind is the inaccuracy of popular numerical methods used for pricing and calibration. Inaccurate methods are constructed and used because crucial qualitative properties of jump-diffusion models and prices $V(t,x)$ in these models are not taken into account. Examples of errors of popular methods, including the analysis of implications for risk management, can be found in [2,13–20]. The aims of this paper are to list and explain the irregularity of $V(t,x)$ (and the free boundary, in the case of options of American type) in several standard situations, and explain how to design efficient numerical methods that work well in wide classes of jump-diffusion models.

In incomplete markets and the jump-diffusion models which are models of incomplete markets, one cannot use the hedging or replication argument to derive the equation for contingent claims. However, one can start by choosing an EMM and defining the price by (4). In many cases of interest, the representation (4) and the Fourier/Laplace transform technique suffice to express the price in the form of an integral, and an efficient numerical procedure for the evaluation of the designed integral. In some popular classes of models, the derivation of pricing formulas is based on (1), although the rigorous justification is lacking. Informally, one can use Dynkin's formula:

$$V(t,x) = \mathbb{E}^{\mathbb{Q}}\left[\int_t^{\tau_{U^c}} (-\partial_s - L)V(s,X_s)ds + G(\tau_{U^c}, V_{\tau_{U^c}})\right] \tag{5}$$

to conclude that (1) must hold. Thus, (1) is the backward Kolmogorov equation. In [21,22], the formal derivation of (1) is given for Lévy processes satisfying the (ACP)-condition (absolute continuity of potential measures) in the infinite time horizon case (stationary problem) and the (ACT)-condition (absolute continuity of transition measures) in the non-stationary case—in ([12], Section 41), one can find equivalent conditions for the (ACP)- and (ACT)-conditions; the equation is understood in the sense of generalized functions. A similar proof can be given for wider classes of strong Markov processes satisfying the same conditions. However, the author is unaware of a published proof in the general setting. See [1,2] for a review of partial results, and [23–25] for proofs in several special cases.

The next important complication in the case of jump-diffusion processes is the form of the boundary condition which becomes non-local:

$$V(t,x) = G(t,x),\ x \in U(t)^c. \tag{6}$$

Note that the standard boundary problems for fractional differential equations are local although the form of the conditions is non-standard. The non-standard boundary conditions are formally invented in order for the Cauchy problem to be well defined. In the literature, one can find discussions about a proper choice of the boundary conditions. On the contrary, the non-local boundary condition (6) is a *part of the definition* of the value function, and cannot be replaced for convenience with a local condition.

One can use the boundary problem (1), (2), (6) as follows:

(1) To prove the existence and uniqueness of the solution in the class of sufficiently regular functions; sufficiently regular means that Dynkin's formula (5) is valid;
(2) Applying (5) to conclude that V equals the RHS of (4).

Unfortunately, this scheme is very difficult to realize in many models, including many popular diffusion models, and the majority of the results in the literature are obtained assuming without proof that the boundary problem has the unique solution, and the solution is sufficiently regular so that Ito's formula (or, more generally, Dynkin's formula) can be applied; then, the solution satisfies (4) and (5), hence, the problem of calculation of the expectation (4) is solved. The necessity of proving this crucial assumption is brushed under the carpet.

The main difficulty for the proof stems from the irregularity of the value function V, which makes general theorems of Feynman–Kac type available in the literature not applicable; the irregularity of V is the artifact of the properties of the infinitesimal generators. However, for several classes of contingent claims and types of models, V given by (4) can be explicitly calculated using the Fourier/Laplace technique. In the case of Lévy processes (in terms of PDO, the case of operators with constant symbols), and problems of several basic types (boundary problems with flat boundaries), either no or very weak conditions on the process are needed. The result is an integral depending on (t, x) as a parameter, in one or more dimensions. An explicit analytical expression can be used to study the regularity of solutions; in a number of important case, the integral can be efficiently and quickly calculated. Since the study of the regularity of solutions of boundary problems can be localized (see [26]), the author hopes that the results for operators with constant symbols and boundary problems with flat boundaries can be used to prove the correspondence between the stochastic expressions and boundary problems with curved boundaries, and operators with state- and time-dependent symbols.

1.4. Structure of This Paper

Lévy models are considered in Section 2, and the pricing of European options using the Fourier transform technique (solution of the Cauchy problem) and efficient numerical realizations are in Section 3. To price barrier options (boundary problems for parabolic operators), the Laplace transform, Wiener–Hopf factorization and maturity randomization (method of lines) are needed (Section 4). Explicit formulas involve multi-dimensional integrals, and efficient numerical calculations based on contour deformations become more involved. In Section 5, we analyze the peculiarities of the early exercise boundary and the prices of American options (solutions of free boundary problems), and present a general stable scheme based on the method of lines. The proof of convergence is based on the probabilistic interpretation of the operator form of the Wiener–Hopf factorization. In Section 6, we explain how the methods of Sections 4 and 5 are modified to price options in regime-switching models (solve systems of boundary problems), and how to approximate more complicated stochastic volatility models and models with stochastic interest rates with regime-switching models. In Section 7, we outline the structure of several exactly solvable models with non-constant symbols and list several outstanding problems for these classes of models, which seem to be non-trivial and interesting from the point of view of the theory of boundary problems for PDE and PDO. Finally, in Section 8, we summarize the results presented in this paper, review several other groups of methods and, wherever possible, outline the relative advantages of different methods.

2. Lévy Models or PDO with Constant Symbols

2.1. Lévy Processes

Let X be the Lévy process on the filtered probability space $(\Omega; \mathcal{F}; \{\mathcal{F}_t\}_{t\geq 0}; \mathbb{Q})$. $\mathbb{E} = \mathbb{E}^{\mathbb{Q}}$ denotes the expectation operator under $\mathbb{Q}$. We use the definition in [21,27] of the characteristic exponent $\psi(\xi) = \psi^{\mathbb{Q}}(\xi)$ of a Lévy process X on $\mathbb{R}^n$, under $\mathbb{Q}$, which is marginally different from the definition in [12]. Namely, ψ is definable from:

$$\mathbb{E}[e^{i\langle \xi, X_t \rangle}] = e^{-t\psi(\xi)}, \quad \xi \in \mathbb{R}^n,\ t \geq 0, \tag{7}$$

where $\langle a, b \rangle = \sum_{j=1}^n a_j b_j$. This definition of the characteristic exponent implies that the infinitesimal generator L_X of X is the pseudo-differential operator (pdo) $-\psi(D)$; equivalently, L_X acts of oscillating exponents are as follows: $L_X e^{i\langle x,\xi\rangle} = -\psi(\xi) e^{ix\xi}$.

Denote $D = \{x \mid |x| \leq 1\}$. The following theorem (Lévy–Khintchine theorem) can be found in many monographs, e.g., ([12], Thm. 8.1).

Theorem 1. *(i) Let X be a Lévy process on* $\mathbb{R}^n$*. Then, its characteristic exponent admits the representation:*

$$\psi(\xi) = \frac{1}{2}\langle A\xi, \xi\rangle - i\langle\gamma, \xi\rangle + \int_{\mathbb{R}^n}(1 - e^{i\langle x,\xi\rangle} + i\langle x, \xi\rangle \mathbf{1}_D(x))F(dx), \tag{8}$$

where A is a symmetric non-negative-definite $n \times n$ *matrix,* $\gamma \in \mathbb{R}^n$*, and* $F(dx)$ *is a measure on* $\mathbb{R}^n$ *satisfying:*

$$F(\{0\}) = 0, \quad \int_{\mathbb{R}^n}(|x|^2 \wedge 1)F(dx) < \infty. \tag{9}$$

(ii) The representation (8) is unique;

(iii) Conversely, if A is a symmetric non-negative-definite $n \times n$ *matrix,* $\gamma \in \mathbb{R}^n$*, and F is a measure on* $\mathbb{R}^n$ *satisfying (9), then there exists a Lévy process X with the characteristic exponent (7).*

The triple (A, F, γ) is called the *generating triplet* of X. The A and F are called the *Gaussian covariance matrix* and *Lévy measure* of X. When $F = 0$, X is Gaussian, and if $A = 0$, X is called purely non-Gaussian.

Essentially, the term $-i\langle x, \xi\rangle \mathbf{1}_D(x)$ in (8) is needed to ensure the convergence of the integral, and hence different functions can be (and are) used instead of $c(x) := \mathbf{1}_D(x)$, for instance, $c(x) = 1/(1 + |x|^2)$; the A and F are independent of the choice of c. If F satisfies the condition:

$$F(\{0\}) = 0, \quad \int_{\mathbb{R}^n}(|x| \wedge 1)F(dx) < \infty, \tag{10}$$

which is stronger than (9), then (8) can be simplified:

$$\psi(\xi) = \frac{1}{2}\langle A\xi, \xi\rangle - i\langle\gamma_0, \xi\rangle + \int_{\mathbb{R}^n}(1 - e^{i\langle x,\xi\rangle})F(dx), \tag{11}$$

where $\gamma_0 = \gamma - \int_{\mathbb{R}^n} x\mathbf{1}_D(x)F(dx)$.

If the sample paths of a Lévy process have bounded variation on every compact time interval a. s., we say that the Lévy process has *bounded variation*. A Lévy process has bounded variation if and only if $A = 0$ and (10) holds (see, e.g., [28], p. 15).

2.2. Examples of Lévy Processes on $\mathbb{R}$

Example 1. *The Lévy density of a (pure jump) one-dimensional stable Lévy process X of index* $\alpha \in (0, 2)$ *is of the form*

$$F(dy) \quad = \quad |y|^{-\alpha-1}(c_+\mathbf{1}_{(0,+\infty)}(y) + c_-\mathbf{1}_{(-\infty,0)}(y))dy, \tag{12}$$

where $c_\pm \geq 0$ *and* $c_+ + c_- > 0$*. If* $\alpha \neq 1$*, then, substituting (12) into the Lévy–Khintchine formula with* $\sigma^2 = 0$*, one easily derives* $\psi_{st}(\xi) = -i\mu\xi + \psi^0_{st}(\alpha, C_+, \xi)$*, where* μ *can be expressed in terms of* α*,* $c_\pm$ *and b:*

$$\psi^0_{st}(\alpha, C_+, \xi) = C_+\xi^\alpha \mathbf{1}_{(0,+\infty)}(\xi) + C_-(-\xi)^\alpha \mathbf{1}_{(-\infty,0)}(\xi), \tag{13}$$

$C_- = \overline{C_+}$*, and* $C_+ = C_+(\alpha, c_+, c_-) = -c_+\Gamma(-\alpha)e^{-i\pi\alpha/2} - c_-\Gamma(-\alpha)e^{i\pi\alpha/2}$*. See [9]. For somewhat different (naturally, equivalent) formulas, see [29] and ([12], Thm.14.15). In the case* $\alpha = 1$*, the formula for* ψ^0 *is different (see [9]):*

$$\psi^0(\xi) = \sigma|\xi|(1 + i(2\beta/\pi)\,\mathrm{sign}\,\xi \ln|\xi|), \tag{14}$$

where $\sigma = (c_+ + c_-)\pi/2$*,* $\beta = (c_+ - c_-)/(c_+ + c_-)$*. This is a version of Zolotarev's parametrizations [30] for stable processes of index 1. See [9,12] for further references. Note that the symbol of the infinitesimal generator L is non-smooth at zero.*

The tails of the probability distributions of stable Lévy processes exhibit slow polynomial decay. Hence, the second moment of the distribution of X_t is infinite. In real financial markets, the distributions have finite second moments, which makes stable processes unsuitable. The simplest way to ensure that the second moment is finite is to consider processes with exponentially decaying tails; the symbols of L are analytic in a strip. These equivalent properties were used as the basis for the definition of a general class of *Regular Lévy processes of exponential type* (RLPE) [21,22]. Below, we list several popular sub-classes of RLPE. The reader can easily observe that, in all these examples, the characteristic exponent $\psi(\xi) = -i\mu\xi + \psi^0(\xi)$ admits analytic continuation to the union of a strip and cone $\mathcal{C}$ around $\mathbb{R}$, and $\mathrm{Re}\,\psi^0(\xi) \to +\infty$ as $\xi \to \infty$ in a sub-cone $\mathcal{C}_+ \subset \mathcal{C}$. In [31], the definition of the general class of processes enjoying these properties was given. The suggested name *SINH-regular* processes reflects the fact that the integrals in pricing formulas can be efficiently calculated using changes of variables of the form:

$$\xi = \chi_{\omega_1,b,\omega}(y) = i\omega_1 + b\sinh(i\omega + y) \tag{15}$$

(*sinh-acceleration*). Lévy processes used in quantitative finance are SINH-regular processes.

Example 2. *KoBoL processes. Modifying the Lévy density (12):*

$$F(dy) = (c_+ y^{-\nu_+-1} e^{\lambda_- y} \mathbf{1}_{(0,+\infty)}(y) + c_-(-y)^{-\nu_--1} e^{\lambda_+ y} \mathbf{1}_{(-\infty,0)}(y))dy, \tag{16}$$

where $\nu_\pm \in (0,2)$, $c_\pm \geq 0$, $c_+ + c_- > 0$ and $\lambda_- < 0 < \lambda_+$, we obtain a class of Lévy processes with exponentially decaying tails. In the case $\nu_- \neq 1, \nu_+ \neq 1$, the characteristic exponent of the KoBoL process is of the form $\psi(\xi) = -i\mu\xi + \psi^0(\xi)$, where:

$$\psi^0(\xi) = c_-\Gamma(-\nu_-)[\lambda_+^{\nu_-} - (\lambda_+ + i\xi)^{\nu_-}] + c_+\Gamma(-\nu_+)[(-\lambda_-)^{\nu_+} - (-\lambda_- - i\xi)^{\nu_+}]; \tag{17}$$

if either $\nu_+ = 1$ or $\nu_- = 1$, then the formula for ψ^0 is different. See [21,27]. In particular, if $\nu_+ = \nu_- = 1$, then:

$$\begin{aligned}\psi^0(\xi) &= c_+((-\lambda_-)\ln(-\lambda_-) - (-\lambda_- - i\xi)\ln(-\lambda_- - i\xi)) \\ &\quad + c_-(\lambda_+ \ln\lambda_+ - (\lambda_+ + i\xi)\ln(\lambda_+ + i\xi)).\end{aligned} \tag{18}$$

In the symmetric case $\nu_- = \nu_+ \in (0,2) \setminus \{1\}, -\lambda_- = \lambda_+, c_+ = c_-$, the class of processes with the Lévy density (16) was introduced by Koponen [32], who derived a rather inconvenient formula for the characteristic exponent (different from (17)) and suggested a somewhat misleading name truncated Lévy processes. The generalization (16) was introduced in [27], where the characteristic exponent was calculated and several option pricing problems were solved; the name Koponen's family of truncated Lévy processes was used. Starting with [21], the name KoBoL processes is used. In [11], a subclass of KoBoL with $\nu_- = \nu_+ \neq 1$ and $c_+ = c_-$ was called CGMY model and labels for the parameters of the KoBoL model were changed.

Rosinski [33] suggested the name exponentially tilted stable Lévy processes and gave a general definition of a class which is a subclass of the class RLPE.

Example 3. *Normal inverse Gaussian (NIG) processes, and the generalization: normal tempered stable (NTS) processes are constructed in [34,35], respectively. The characteristic exponent is given by*

$$\psi^0(\xi) = \delta[(\alpha^2 + (\xi + i\beta)^2)^{\nu/2} - (\alpha^2 - \beta^2)^{\nu/2}], \tag{19}$$

where $\nu \in (0,2)$, $\delta > 0$, $|\beta| < \alpha$; NIG obtains with $\nu = 1$.

Example 4. *Variance Gamma processes (VGPs) were introduced to finance in [36]. The characteristic exponent can be written in the form:*

$$\psi^0(\xi) = c[\ln(\alpha^2 - (\beta + i\xi)^2) - \ln(\alpha^2 - \beta^2)], \tag{20}$$

where $\alpha > |\beta| \geq 0, c > 0$.

Example 5. *In the Merton model [37], the characteristic exponent is given by*

$$\psi^0(\xi) = \frac{\sigma^2}{2}\xi^2 + \lambda \cdot (1 - e^{im\xi - \frac{s^2}{2}\xi^2}), \tag{21}$$

where $\sigma, s, \lambda > 0$ and $\mu, m \in \mathbb{R}$.

Example 6. *The hyper-exponential jump-diffusion processes (HEJD model) were introduced in [38,39] and studied in detail in [38,40]). The characteristic exponent is of the form:*

$$\psi^0(\xi) = \frac{\sigma^2}{2}\xi^2 + \lambda^+ \cdot \sum_{j=1}^{n^+} \frac{ip_j^+\xi}{i\xi - \alpha_j^+} + \lambda^- \cdot \sum_{k=1}^{n^-} \frac{ip_k^-\xi}{i\xi + \alpha_k^-}, \tag{22}$$

where $n^\pm$ are positive integers and $\alpha_j^\pm, \lambda^\pm, p_j^\pm > 0$ satisfy $\sum_{j=1}^{n^\pm} p_j^\pm = 1$. A double-exponential jump diffusion model introduced to finance in [41] (and well known for decades) can be obtained as a special case of hyper-exponential jump-diffusion models by taking $n^+ = n^- = 1$.

Example 7. *Other classes of Lévy processes with rational characteristic exponents and non-trivial BM components [21,38,40,42,43].*

Example 8. *The characteristic exponents of the processes of the β-class constructed in [44] are of the form:*

$$\begin{aligned}\psi^0(\xi) &= \frac{\sigma^2}{2}\xi^2 + \frac{c_1}{\beta_1}\left\{B(\alpha_1, 1-\gamma_1) - B(\alpha_1 - \frac{i\xi}{\beta_1}, 1-\gamma_1)\right\} \\ &+ \frac{c_2}{\beta_2}\left\{B(\alpha_2, 1-\gamma_2) - B(\alpha_2 + \frac{i\xi}{\beta_2}, 1-\gamma_2)\right\},\end{aligned} \tag{23}$$

where $c_j \geq 0, \alpha_j, \beta_j > 0$ and $\gamma_j \in (0,3) \setminus \{1,2\}$, and $B(x,y)$ is the Beta-function. Evidently, all poles of ψ^0 are on $i\mathbb{R}_+ \setminus 0$.

Example 9. *The Lévy density and characteristic exponent of the meromorphic Lévy processes introduced in [45] are defined by almost the same formulas as in the HEJD model. The difference is that the sums are infinite. A natural condition $\alpha_j^\pm \to +\infty$ as $j \to +\infty$ is imposed, and the requirement that an infinite sum defines a Lévy density is equivalent to $\sum_{j\geq 0} p_j^\pm (\alpha_j^\pm)^{-2} < +\infty$. The formula for ψ^0 is (22) with the infinite numbers of terms; the poles of $\psi(\xi)$ are on $i\mathbb{R} \setminus 0$.*

Remark 1. *If one-factor Lévy models are calibrated to prices of stocks and indices in financial markets, then, in the majority of cases, KoBoL processes of order $\nu \in (0,1)$, without the diffusion component (or very small diffusion component) give the best fit. Typically, processes with a jump part of finite activity such as the Merton model, HEJD model and other models with rational characteristic exponents do not calibrate well to the real data but one may try to use simple models to approximate involved ones. There are publications where HEJD are used to approximate KoBoL dynamics. However, it is evident that the approximations of operators of order $\nu \in (0,1)$ by operators of order 2 cannot work well near the boundary. The reader can find examples in [46] which illustrate this point. A very flexible multi-parameter β-family and meromorphic processes can be used to approximate KoBoL processes, but these approximations are much less efficient than direct calculations in the KoBoL model.*

For applications to qualitative problems in economics, processes with rational characteristic exponents are more suitable due to the triviality of the Wiener–Hopf factorization. See, e.g., [47–52].

2.3. Examples of Lévy Processes on $\mathbb{R}^n$

Example 10. *In the notation used in this paper, the characterization of non-trivial stable Lévy processes on $\mathbb{R}^n$ of index $\alpha \in (0,2)$ ([12], Thm. 14.10) is as follows. There exist $\mu \in \mathbb{R}^n$ and a finite non-zero measure $G_{st}(d\phi)$ on the unit sphere $S_{n-1} := \{\xi \in \mathbb{R}^n \mid |\xi| = 1\}$ such that if $\alpha \neq 1$, then:*

$$\psi(\xi) = -i\langle\mu,\xi\rangle + \int_{S_{n-1}} |\langle\xi,\phi\rangle|^\alpha \Big(1 + i\tan\frac{\alpha\pi}{2}\mathrm{sign}\,\langle\xi,\phi\rangle\Big) G_{st}(d\phi) \tag{24}$$

is the characteristic exponent of a stable Lévy process of index α, and if $\alpha = 1$, then:

$$\psi(\xi) = -i\langle\mu,\xi\rangle + \int_{S_{n-1}} |\langle\xi,\phi\rangle| \Big(1 - i\frac{2}{\pi}\tan\frac{\alpha\pi}{2}\langle\xi,\phi\rangle\ln|\langle\xi,\phi\rangle|\Big) G_{st}(d\phi) \tag{25}$$

Conversely, for any $\mu \in \mathbb{R}^n$ and a finite non-zero measure $G_{st}(d\phi)$ on S_{n-1}, (24) and (25) define the characteristic exponents of a stable Lévy process of index $\alpha \neq 1$ and $\alpha = 1$, respectively.

A straightforward multi-dimensional analog of RLPE class was introduced in ([21], Sect. 9.1.4). The characteristic exponent of an RLPE admits analytic continuation to a tube domain U containing $\mathbb{R}^n$, and stabilizes to a positively homogeneous function as $\xi \to \infty$ remaining in U. As in the 1D-case, we impose conditions on ψ^0 in the representation:

$$\psi(\xi) = -i\langle\mu,\xi\rangle + \psi^0(\xi), \tag{26}$$

where $\mu \in \mathbb{R}^n$.

Example 11. *Consider the multi-factor KoBoL family of pure jump Lévy processes constructed in ([21], Section 9.1.1). Let $\alpha \in (0,2)$, and let $G(dx)$ be a finite non-zero measure and λ a positive continuous function on the unit sphere S_{n-1}. Then:*

$$F(dx) = \rho^{-\alpha-1}\exp(-\lambda(\phi)\rho)d\rho\, G(d\phi) \tag{27}$$

is a Lévy measure. If $\alpha \in (0,2), \alpha \neq 1$, the characteristic exponent is:

$$\psi^0(\xi) = \Gamma(-\alpha)\int_{S_{n-1}} [\lambda(\phi)^\alpha - (\lambda(\phi) - i\langle\xi,\phi\rangle)^\alpha] G(d\phi) \tag{28}$$

and if $\alpha = 1$:

$$\psi^0(\xi) = \int_{S_{n-1}} [\lambda(\phi)\ln\lambda(\phi) - (\lambda(\phi) - i\langle\xi,\phi\rangle)\ln(\lambda(\phi) - i\langle\xi,\phi\rangle)] G(d\phi). \tag{29}$$

Passing to the limit $\lambda \downarrow 0$ in (28) and (29), one can easily derive alternative representations of the characteristic exponents of stable Lévy processes.

Example 12. *The class of multi-factor* normal tempered stable *(NTS) Lévy processes constructed in ([21], Section 9.1.2) can be defined by*

$$\psi^0(\xi) = \delta\Big\{[(\alpha^2 + \langle A(\xi - i\beta),(\xi - i\beta)\rangle]^{\nu/2} - [\alpha^2 - \langle A\beta,\beta\rangle]^{\nu/2}\Big\}, \tag{30}$$

where $\delta, \alpha > 0$, A is a positive-definite matrix, $\beta \in \mathbb{R}^n$, $\alpha^2 - \langle A\beta,\beta\rangle > 0$.

The following is a straightforward definition of the class of SINH-regular processes in [31], in the simplest form.

Definition 1. *We say that X is a SINH-regular process of order (ν, ν') if the characteristic exponent is of the form ψ^0 admits analytic continuation to a domain of the form $\mathcal{U} = iD + \mathcal{C}$, where D is an open set containing $\{0\}$, and $\mathcal{C}$ is an open cone $\mathcal{C}$ around $\mathbb{R}^n$ s.t.:*

$$|\psi^0(\xi)| \le C(1+|\xi|)^\nu,\ \xi \in \mathcal{U}, \tag{31}$$

where C is independent of ξ, and there exists $\mathcal{U}_+ \subset \mathcal{U}$, of the same form: $\mathcal{U}_+ = iD_+ + \mathcal{C}_+$, s.t.:

$$\operatorname{Re}\psi^0(\xi) \ge c|\xi|^{\nu'} - C,\ \xi \in \mathcal{U}_+, \tag{32}$$

where $C, c > 0$ are independent of ξ.

2.4. General Remarks

1. Stable Lévy processes can be characterized as the limiting case of SINH-regular processes when the tube domain $iD + \mathbb{R}^n$ of analyticity shrinks to $\{0\}$; the conditions are valid in $\mathcal{C}$ only. The calculation of expectations in stable Lévy models can be efficiently performed by either modifying the sinh-acceleration technique or approximating stable Lévy processes with SINH-regular ones [8,9].
2. The definition in [31] allows for $\mathcal{U}_+$ to be adjacent to $\mathbb{R}$, and Definition 1 can be generalized in a similar fashion.
3. In [31], the bounds (31) and (32) are formulated for ψ rather than ψ^0. One can easily derive the bounds for ψ using the bounds for ψ^0.
4. It is easy to see that NTS processes are SINH-regular but a multi-factor KoBoL X is SINH-regular only if X is a mixture of independent KoBoL in 1D.
5. VGP and their multi-factor generalizations are SINH-regular if we replace the weight $|\xi|^\nu$ with $\ln(2+|\xi|)$.
6. As in [31], more general weight functions can be used. It can be shown that, in the case of pure jump processes, $\psi^0(\xi) = o(|\xi|^2)$ as $(\mathcal{C} \ni)\xi \to \infty$; hence, one can use the upper bound with $\nu = 2$ in all cases.
7. If $\nu = \nu'$, then ψ^0 is an elliptic symbol. If $\nu' < \nu$, then, typically, $\nu - \nu' < 1$, hence, ψ^0 is hypo-elliptic.
8. If in the representation (26), $\mu \ne 0$, and $\nu < 1$, then the principal symbol of $\psi(\xi)$ is $-i\langle \mu, \xi \rangle$, which leads to the irregularity of the Wiener–Hopf factors and the solutions of the boundary problems.
9. In the case of Cauchy problems in the whole space, the drift can be eliminated by the change of variables $x = x' - t\mu$, and the same change of variables is implicit in efficient numerical methods for the Fourier inversion methods based on the conformal deformation of lines and hyperplanes of integration. Formally, the same change of variables can be applied in the case of more general boundary problems but the change makes a flat boundary (typical in pricing problems for standard barrier options) non-flat. When the boundary is flat, explicit pricing formulas can be derived and the study of the (ir)regularity of the solutions simplified.

3. Pricing European Options in Lévy Models and Cauchy Problems in $\mathbb{R}^n$ for Operators with Constant Symbols

3.1. Exact Formulas

Let $r \ge 0$ be the constant riskless rate, X a Lévy process in $\mathbb{R}^n$ under an EMM $\mathbb{Q}$ chosen for pricing, and $G(X_T)$ the payoff of the contingent claim at maturity date T. The price $V(t,x)$ of the claim at time t and $X_t = x$ is given by

$$V(t,x) = \mathbb{E}^{\mathbb{Q}}\left[e^{-r(T-t)}G(X_T) \mid X_t = x\right]. \tag{33}$$

Assume that the characteristic exponent ψ of X is analytic in a tube domain $iD + \mathbb{R}^n$, where D is an open set containing $\{0\}$, and the Fourier transform:

$$\hat{G}(\xi) = (2\pi)^{-n} \int_{\mathbb{R}^n} e^{-i\langle x,\xi\rangle} G(x)dx \tag{34}$$

is analytic in a tube domain $iD' + \mathbb{R}^n$, where $D' \cap D$ is an non-empty open set. Let the product $e^{-(T-t)\psi(\xi)}\hat{G}(\xi)$ be in $L_1(i\omega + \mathbb{R}^n)$ for $\omega \in D' \cap D$. Decomposing G in the Fourier integral, substituting into (33), and using Fubini's theorem to justify the change of the order of integration and taking expectation, we obtain, for any $\omega \in D \cap D'$:

$$V(x,t) = (2\pi)^{-n} \int_{\mathrm{Im}\,\xi=\omega} e^{i\langle x,\xi\rangle-(T-t)(r+\psi(\xi))}\hat{G}(\xi)d\xi. \tag{35}$$

The evident representation (35) was derived in [21,53,54]. Using the PDO notation, $V(t,x) = e^{-(T-t)(r+\psi(D_x))}G(x)$, where the operator on the RHS acts in appropriate spaces with exponential weights. It is easy to verify that V is a solution of the Cauchy problem

$$(\partial_t - \psi(D_x) - r)V(t,x) = 0,\ t < T, x \in \mathbb{R}^n, \tag{36}$$

subject to $V(T,x) = G(x), x \in \mathbb{R}^n$, and appropriate bounds on V as $x \to \pm\infty$. Equation (36) is understood in the sense of generalized functions, and under certain regularity conditions, the solution is unique. See [53,54] and ([21], Chapt. 15) for details. Thus, in this case, the equivalence of the pricing problem and the Cauchy problem is established.

3.2. Efficient Numerical Realizations in 1D Case

In [21,53,54], the integral on the RHS is numerically realized using the trapezoid rule or Simpson rule, and the standard real-analytical error bounds are used to give prescriptions for the choice of the numerical scheme to satisfy a given error tolerance $\epsilon > 0$. Since the integrand is analytic in a strip around the line of integration, it is significantly more efficient to use the simplified trapezoid rule, the reason being that the error of the infinite trapezoid rule decays as $\exp[-2\pi d/\zeta]$, where d is the half-width of the strip of analyticity around the line of integration, and ζ is the step. See, e.g., Thm. 3.2.1 in [55]. Thus, if the strip of analyticity is not too narrow, it is relatively easy to satisfy a very small error tolerance for the discretization error. Note that popular variations of this straightforward approach such as the Carr–Madan method [56] and COS method [57–59] introduce additional errors which are difficult to control, and lead to systematic errors in practically important situations. See [2,13,14,16–20,31] for the analysis of errors of the Carr–Madan method and COS.

In many cases of interest, the integrand slowly decays at infinity, and a very large number of terms of the truncated sum (simplified trapezoid rule) are needed to satisfy even a moderate error tolerance. However, in the case of standard European options, and in the case of piece-wise polynomial approximations of complicated payoffs [16,46,60], $\hat{G}$ is meromorphic with a finite number of simple poles; in [20], approximations with an infinite number of poles appear. If X is SINH-regular of order (ν,ν') with $\nu' > 0$, one can use an appropriate conformal deformation and the corresponding change of variables to reduce calculations to the case of an integrand which is analytic in a strip around the line of integration and decays at infinity faster than exponentially. The complexity of the numerical scheme based on the sinh-acceleration (15) is of the order of $E \ln E$, where $E = \ln(1/\epsilon)$; in the case of VGP with $\mu = 0$, of the order of $O(E^2)$. See [31] for details. Note that the parameter ω in (15) is chosen so that the oscillating factor $e^{i(x+(T-t)\mu)\xi}$ becomes a quickly decaying one. The idea is similar to the idea of the saddle point method. However, simpler universal families of conformal deformations are easier to use, especially when the deformations of several lines of integration are needed, and the deformations must be in a certain agreement [10,61].

3.3. Stable Lévy Processes and Fractional Differential Operators

In the case of stable Lévy processes, the expectation can be finite only if G is bounded or increases at a sufficiently small polynomial rate at infinity. However, if $\hat{G}$ is analytic in an open cone $\mathcal{C}$ around $\mathbb{R} \setminus \{0\}$, then the evaluation of probability distribution functions and expectations can be reduced to the evaluation of integrals over $(0, +\infty)$, and an exponential change of variables $\xi = e^{i\omega+y}$, where $\omega \in [-\pi/2, \pi/2]$ is used to efficiently evaluate integrals. In some cases, other conformal deformations are more efficient (see [9]). The same families of conformal deformations can be used to evaluate the solutions of the Cauchy problem for the fractional-parabolic equations of the form:

$$(\partial_t - |D_x|^\alpha)V(t,x) = 0,\ t < T, x \in \mathbb{R},$$

where $\alpha > 0$, and more general ones.

3.4. Calculation of Probability Distributions and Expectations (Prices) in Multi-Factor Lévy Models and Solution of the Cauchy Problems in $\mathbb{R}^n$

When n is moderate, X is SINH-regular, and G can be approximated by a function whose Fourier transform $\hat{G}$ is analytic in the union of a tube domain and cone $\mathcal{U}_G \subset iD + \mathcal{C}$, and can be efficiently calculated; then one can construct appropriate conformal deformations of the form (15) for each variable. The complexity of the scheme is $O((E \ln E)^n)$, which is not large if, e.g., $n = 2, 3, 4$. For exchange or basket options with payoffs $(e^{x_1} - e^{x_2} - K)_+$, $(e^{x_1} + e^{x_2} - K)_+$, $\hat{G}$ can be explicitly calculated in terms of the Beta function, and the values of the latter can be efficiently calculated using the same sinh-acceleration technique (see [6,7]).

4. Barrier Options in Lévy Models, and Boundary Problems for PDO with Constant Symbols

We consider in detail the 1D case, and in the end, outline extensions to the multi-factor case. We start with the basic notation and facts of the Wiener–Hopf factorization, and then formulate the results for barrier options.

4.1. Main Notation

- X: a Lévy process on $\mathbb{R}$;
- $(\Omega, \mathcal{F}, \{\mathcal{F}\}_{t\geq 0})$: the filtered measure space generated by X;
- $\mathcal{M}$: the set of all stopping times with respect to the filtration $\{\mathcal{F}\}_{t\geq 0}$;
- $\mathbb{Q}$: an EMM chosen for pricing;
- $\psi(\xi) = -i\mu\xi + \psi^0(\xi)$: the characteristic exponent of X;
- $h, h_\pm \in \mathbb{R}, h_- < h_+$: barriers;
- τ_h^- and τ_h^+: first entrance time by X into $(-\infty, h]$ and $[h, +\infty)$, respectively;
- $q > 0$: the discount rate;
- The (expected) present value of the perpetual stream $g(X_t)$ which is lost at time τ_h^-:

$$V_-(g;q;h;x) = \mathbb{E}^x\left[\int_0^{\tau_h^-} e^{-qt}g(X_t)dt\right];$$

- The (expected) present value of the perpetual stream $g(X_t)$ which is lost at time τ_h^+:

$$V_+(g;q;h;x) = \mathbb{E}^x\left[\int_0^{\tau_h^+} e^{-qt}g(X_t)dt\right];$$

- the (expected present) value of the perpetual stream $g(X_t)$ which is lost at $\tau_{h_+}^+ \wedge \tau_{h_-}^-$:

$$V(g;q;h_-,h_+;x) = \mathbb{E}^x\left[\int_0^{\tau_h^+\wedge\tau_{h_-}^-} e^{-qt}g(X_t)dt\right];$$

- T_q: exponentially distributed random variable of mean $1/q$, independent of X;
- $\bar{X}_t = \sup_{0\le s\le t} X_s$ and $\underline{X}_t = \inf_{0\le s\le t} X_s$—the supremum and infimum processes (defined pathwise, a.s.);
- Normalized EPV operators (normalized resolvents) under X, $\bar{X}$, and $\underline{X}$ calculate the (normalized) expected present value of the streams under X, $\bar{X}$ and $\underline{X}$:

$$
\begin{aligned}
(\mathcal{E}_q g)(x) &:= \mathbb{E}^{\mathbb{Q}}[g(X_{T_q})] := q\mathbb{E}^{\mathbb{Q}}\left[\int_0^{+\infty} e^{-qt} g(X_t)dt \mid X_0 = x\right]\\
(\mathcal{E}_q^+ g)(x) &:= \mathbb{E}^{\mathbb{Q}}[g(\bar{X}_{T_q})] := q\mathbb{E}^{\mathbb{Q}}\left[\int_0^{+\infty} e^{-qt} g(\bar{X}_t)dt \mid X_0 = \bar{X}_0 = x\right]\\
(\mathcal{E}_q^- g)(x) &:= \mathbb{E}^{\mathbb{Q}}[g(\underline{X}_{T_q})] := q\mathbb{E}^{\mathbb{Q}}\left[\int_0^{+\infty} e^{-qt} g(\underline{X}_t)dt \mid X_0 = \underline{X}_0 = x\right];
\end{aligned}
$$

- Wiener–Hopf factors: $\phi_q^+(\xi) = \mathbb{E}^{\mathbb{Q}}[e^{i\bar{X}_{T_q}}]$, $\phi_q^-(\xi) = \mathbb{E}^{\mathbb{Q}}[e^{i\underline{X}_{T_q}}]$

  Basic facts:

(1) $\phi_q^+(\xi)$ (resp., $\phi_q^-(\xi)$) admits (uniformly bounded) analytic continuation to the upper (resp., lower) half-plane.

(2) The EPV operators act in $L_\infty(\mathbb{R})$. If $\operatorname{supp} g \subset (-\infty, h]$, then $\operatorname{supp} \mathcal{E}_q^+ g \subset (-\infty, h]$. If $\operatorname{supp} g \subset [h, +\infty)$, then $\operatorname{supp} \mathcal{E}_q^- g \subset [h, +\infty)$.

(3) $\mathcal{E}_q = q(q+\psi(D))^{-1}$, $\mathcal{E}_q^+ = \phi_q^+(D)$, $\mathcal{E}_q^- = \phi_q^-(D)$. If there exist $\mu_- \le 0 \le \mu_+$, $\mu_- < \mu_+$, such that:

$$\mathbb{E}^{\mathbb{Q}}[e^{\beta X_{T_q}}] < \infty,\ \beta \in [-\mu_-, -\mu_+], \tag{37}$$

then:

(a) ψ admits (uniformly bounded) analytic continuation to the strip $S_{[\mu_-,\mu_+]} := \{\xi \mid \operatorname{Im}\xi \in [\mu_-, \mu_+]\}$ (meaning: analytic in the interior and continuous up to the boundary);

(b) $\phi_q^+(\xi)$ admits analytic continuation to the half-plane $\{\operatorname{Im}\xi \ge \mu_-\}$;

(c) $\phi_q^-(\xi)$ admits analytic continuation to the half-plane $\{\operatorname{Im}\xi \le \mu_+\}$;

(d) the action of the EPV operators extends to L_∞ and Sobolev spaces with exponential weights.

The properties (a)–(d) are used to study the asymptotics of prices of barrier options (solutions of the boundary problems) at the boundary [21,62,63], and develop efficient numerical methods for pricing barrier options, credit default swaps (CDSs) and lookbacks [10,61,64,65].

4.2. Wiener–Hopf Factorization

We use three equivalent versions of the Wiener–Hopf factorization

$$\mathbb{E}[e^{i\xi X_{T_q}}] = \mathbb{E}[e^{i\xi \bar{X}_{T_q}}]\mathbb{E}[e^{i\xi \underline{X}_{T_q}}], \xi \in \mathbb{R}; \tag{38}$$

$$\frac{q}{q+\psi(\xi)} = \phi_q^+(\xi)\phi_q^-(\xi), \xi \in \mathbb{R}; \tag{39}$$

$$\mathcal{E}_q = \mathcal{E}_q^- \mathcal{E}_q^+ = \mathcal{E}_q^+ \mathcal{E}_q^-. \tag{40}$$

Equation (39) is a special case of the initial form of the Wiener–Hopf factorization used in complex analysis since [66]; (40), with the interpretation of the EPV operators as PDO, is used in analysis (see, e.g., [26]). In both cases, the derivation is possible under certain regularity conditions on ψ. The probabilistic versions (38) and (40), hence, (39), hold for any Lévy process. In probability, the straightforward and short proof of (38) is based on

Lemma 1 ([67], Lemma 2.1, and [68], p. 81). *Let X and T_q be as above. Then:*

(a) the random variables $\bar{X}_{T_q}$ and $X_{T_q} - \bar{X}_{T_q}$ are independent; and

(b) the random variables $\underline{X}_{T_q}$ and $X_{T_q} - \bar{X}_{T_q}$ are identical in law.

Equation (40) is a special case of the next lemma derived in [21,22,49] under unnecessary restrictions on the process, and in [46], for any Lévy process. The proof below (in a shortened form) is borrowed from [46].

Theorem 2. *Let X be a Lévy process. Then, for any $g \in L_\infty(\mathbb{R})$ and $h \in \mathbb{R}$,*

$$V_-(g;q;h;\cdot) \;=\; q^{-1}\mathcal{E}_q^-\mathbf{1}_{(h,+\infty)}\mathcal{E}_q^+g, \tag{41}$$

$$V_+(g;q;h;\cdot) \;=\; q^{-1}\mathcal{E}_q^+\mathbf{1}_{(-\infty,h)}\mathcal{E}_q^-g. \tag{42}$$

Proof. Let X, $\bar{X}$ and $\underline{X}$ start at 0. Then: $\mathcal{E}_q\, g(x) = \mathbb{E}[g(x+X_{T_q})]$,

$$\mathcal{E}^+g(x) = \mathbb{E}[g(x+\bar{X}_{T_q})], \quad \mathcal{E}^-g(x) = \mathbb{E}[g(x+\underline{X}_{T_q})],$$

and, by definition:

$$\begin{aligned} V_-(g;q;h;x) \;&:=\; \mathbb{E}\left[\int_0^{\tau_h^-} e^{-qt}g(x+X_t)dt\right] \\ &=\; \mathbb{E}\left[\int_0^{\infty} \mathbf{1}_{x+\underline{X}_t>h}e^{-qt}g(x+X_t)dt\right] \\ &=\; q^{-1}\mathbb{E}[g(x+X_T)\mathbf{1}_{x+\underline{X}_T>h}]. \end{aligned}$$

Applying Lemma 1, we continue:

$$\begin{aligned} V_-(q;h;x) \;&=\; q^{-1}\mathbb{E}[g(x+\underline{X}_T+X_T-\underline{X}_T)\mathbf{1}_{x+\underline{X}_T>h}] \\ &=\; q^{-1}\mathbb{E}[\mathbf{1}_{x+\underline{X}_T>h}\mathcal{E}_q^+g(x+\underline{X}_T)] \\ &=\; q^{-1}\mathcal{E}_q^-\mathbf{1}_{(h,+\infty)}\mathcal{E}_q^+g(x). \end{aligned}$$

This proves (41). Proof of (42) is by symmetry. □

Remark 2. *If (37) is satisfied, one can allow for exponentially increasing measurable $g(x)$ if:*

$$|g(x)| \le C(e^{-\mu'_-x}+e^{-\mu'_+x}), \tag{43}$$

where $\mu_- \le \mu'_- \le \mu'_+ \le \mu_+$, and

$$q+\psi(-i\mu'_\pm) > 0. \tag{44}$$

For the proof of Theorem 2 in this case, it suffices to consider a non-negative measurable function g. We approximate g with the sequence $g_n(x) = \min\{g(x), n\}$, apply Theorem 2 to g_n and justify passage to the limit using the dominated convergence theorem.

The boundary problem for $V_-(q;x;h)$, in the PDO notation, is

$$(q+\psi(D_x))V_-(g;q;h;x) = g(x),\ x > h, \tag{45}$$

subject to $V_-(g;q;h;x) = 0, x \le h$. Under certain regularity conditions on ψ and g, the standard analytical technique [26] can be applied, and the existence and uniqueness of solutions in a class of bounded sufficiently regular functions proved. In [21], we derived

$$V_-(g;q;h;\cdot) = q^{-1}\phi_q^-(D)\mathbf{1}_{(h,+\infty)}\phi_q^+(D)g, \tag{46}$$

which is identical to (41); the (ACP)-condition was used. Thus, under a weak regularity condition, the expectation of the stochastic integral is the unique solution of the corresponding boundary problem, and vice versa.

The following theorem is a trivial corollary of Theorem 2 and (40).

Theorem 3. *Let X be a Lévy process satisfying (37), and let G admit the representation $G = q^{-1}\mathcal{E}_q g$, where g is a measurable function satisfying (43) and (44).*
Then:

(a) $(\mathcal{E}_q^-)^{-1}G := \mathcal{E}_q^+ g$ *and* $(\mathcal{E}_q^+)^{-1}G := \mathcal{E}_q^- g$ *are measurable functions satisfying (43);*
(b) *For any* $h \in \mathbb{R}$:

$$\begin{aligned} V_{-,\text{inst}}(G;q;x;h) &:= \mathbb{E}^{\mathbb{Q}}\left[e^{-q\tau_h^-}G(X_{\tau_h^-}) \mid X_0 = x\right] = (\mathcal{E}_q^- \mathbf{1}_{(-\infty,h]}(\mathcal{E}_q^-)^{-1}G)(x), \\ V_{+,\text{inst}}(G;q;x;h) &:= \mathbb{E}^{\mathbb{Q}}\left[e^{-q\tau_h^+}G(X_{\tau_h^+}) \mid X_0 = x\right] = (\mathcal{E}_q^+ \mathbf{1}_{(-\infty,h]}(\mathcal{E}_q^+)^{-1}G)(x). \end{aligned}$$

4.3. Single Barrier Options

Let X be a Lévy process on $\mathbb{R}$ under an EMM $\mathbb{Q}$ chosen for pricing, $r \geq 0$ the riskless rate, and h the barrier. We consider the case of the down-and-out option, which expires if X enters $(-\infty, h]$ before the maturity date T. If the barrier is not breached, the payoff is $G(X_T)$. Applying Fubini's theorem and (41), we calculate the Laplace transform of the price $V_{n.t.d.}(G;h;T,x)$ with respect to T:

$$\begin{aligned} \tilde{V}_{\text{n.t.}}(G;h;q,x) &= \int_0^{+\infty} e^{-qT}V_{n.t.d.}(G;h;T,x)dT \\ &= e^{-rT}q^{-1}\mathbb{E}^{\mathbb{Q}}\left[\mathbf{1}_{\underline{X}_{T_q}>h}G(X_{T_q}) \mid X_0 = x\right] \\ &= e^{-rT}q^{-1}(\mathcal{E}_q^-\mathbf{1}_{(h,+\infty)}\mathcal{E}_q^+G)(x). \end{aligned}$$

Applying the inverse Laplace transform and using (41) and (46), we obtain, for any $\sigma > 0$:

$$V_{\text{n.t.}}(G;h;T;x) = \frac{e^{-rT}}{2\pi i}\int_{\text{Re}\, q=\sigma} e^{qT}q^{-1}(\mathcal{E}_q^-\mathbf{1}_{(h,+\infty)}\mathcal{E}_q^+G)(x)dq \tag{47}$$

$$= \frac{e^{-rT}}{2\pi i}\int_{\text{Re}\, q=\sigma} e^{qT}q^{-1}(\phi_q^-(D)\mathbf{1}_{(h,+\infty)}\phi_q^+(D)G)(x)dq. \tag{48}$$

The representation (47) is derived in [46] for arbitrary Lévy process. In [21,22], (48), is derived solving the corresponding boundary problem

$$(\partial_t + L - q)V(t,x) = 0, \quad x > h,\ t < T, \tag{49}$$

$$V(T,x) = G(x), \tag{50}$$

$$V(t,x) = G_b(x), \quad x \leq h,\ t < T, \tag{51}$$

under certain regularity condition on V; X satisfies the (ACT)-condition—the ambiguity of the specification of the payoff at (T,x), $x \leq h$, is irrelevant because Lévy processes are stochastically continuous, and, therefore, $\mathbb{Q}[\tau_h^- = T] = 0$. Thus, under weak regularity conditions (the condition on the process is stronger than in the time-independent case), the expectation of the stochastic expression defining the price is the unique solution of the corresponding boundary problem, and vice versa.

In [21,22,46,61,64], more general classes of single-barrier options, with payoff streams during the lifetime of the option, and non-zero payoffs $G_b(X_{\tau_h^-})$ at time $\tau_h^- < T$ are considered and pricing formulas derived. In [10], similar general formulas are derived when the payoff G_b depends on t and x.

4.4. Numerical Realizations

If G_b depends on t and x, the explicit formula is a quadruple integral, the explicit form of (48) is a triple integral, and in addition, one needs to evaluate the Wiener–Hopf factors for all dual variables arising in the pricing formula. Thus, efficient calculations are difficult. When one uses the Gaver–Stehfest algorithm (see, e.g., [10,61,64]), only a positive q appears, but the coefficients in the approximate Laplace inversion formula are very large. In the result, in many cases of interest, high precision arithmetic is necessary to ensure the stability of results not to mention their accuracy (see [61,64] for examples). One can avoid large coefficients using the method of lines. In the probabilistic interpretation, time to maturity is approximated by a sum of exponentially distributed random variables, which results in a sequence of time-independent problems (maturity randomization or Carr's randomization) (see [46,69] for efficient numerical realizations of this idea in applications to single and double barrier options, respectively). Appropriate conformal deformations of the lines of integration lead to faster and more accurate numerical algorithms [10,61]. We illustrate the choice of deformations with the simplest example of the no-touch digital option $V_{\text{n.t.}}(1;h;T;x)$ (the payoff at maturity is 1). Then, $\mathcal{E}_q^+ G = \phi_q^+(D)G = 1$, and (47)–(48) simplify. Assuming that (37) holds, we take a sufficiently small $\omega_0 < 0$, and write (48) as

$$V_{\text{n.t.}}(1;h;T;x) = \frac{e^{-rT}}{(2\pi)^2 i} \int_{\text{Re}\, q=\sigma} e^{qT} q^{-1} \int_{\text{Im}\, \xi=\omega_0} e^{i(x-h)\xi} \frac{\phi_q^-(\xi)}{-i\xi} d\xi\, dq. \quad (52)$$

The set of admissible $\omega_0 < 0$ is determined by the properties of the Wiener–Hopf factors; general formulas for the latter are well known. For efficient numerical realizations, the characteristic exponent should admit an analytic continuation to a strip $\{\text{Im}\, \xi \in (\lambda_-, \lambda_+)\}$, $\lambda_- < 0 < \lambda_+$, around the real axis [21]. Then, (see [21,64]), for any $q > 0$:

(I) There exist $\sigma_-(q) < 0 < \sigma_+(q)$ such that:

$$q + \psi(\eta) \notin (-\infty, 0], \quad \text{Im}\, \eta \in (\sigma_-(q), \sigma_+(q)); \quad (53)$$

(II) The Wiener–Hopf factor $\phi_q^+(\xi)$ admits analytic continuation to the half-plane $\{\text{Im}\, \xi > \sigma_-(q)\}$, and can be calculated as follows—for any $\omega_- \in (\sigma_-(q), \min\{\text{Im}\, \xi, \sigma_+(q)\})$:

$$\phi_q^+(\xi) = \exp\left[\frac{1}{2\pi i} \int_{\text{Im}\, \eta=\omega_-} \frac{\xi \ln(q+\psi(\eta))}{\eta(\xi-\eta)} d\eta\right]; \quad (54)$$

(III) The Wiener–Hopf factor $\phi_q^-(\xi)$ admits analytic continuation to the half-plane $\{\text{Im}\, \xi < \sigma_+(q)\}$, and can be calculated as follows—for any $\omega_+ \in (\max\{\text{Im}\, \xi, \sigma_-(q)\}, \sigma_+(q))$:

$$\phi_q^-(\xi) = \exp\left[-\frac{1}{2\pi i} \int_{\text{Im}\, \eta=\omega_+} \frac{\xi \ln(q+\psi(\eta))}{\eta(\xi-\eta)} d\eta\right]. \quad (55)$$

Analytic continuation with respect to q, ξ is possible, and conformal deformation of the contours of integration are possible as well.

Assume that X is SINH-regular. Then, a numerical realization of (52) is designed choosing deformations $\mathcal{L}^{(1)}$, $\mathcal{L}^{(2)}$ and $\mathcal{L}^{(3)}$ of the lines of integration $\{\text{Im}\, \xi = \omega_0\}$, $\{\text{Re}\, q = \sigma\}$ in (52) and the line of integration $\{\text{Im}\, \eta = \omega_+\}$ in (55). We use deformations of the form $\mathcal{L}^{(1)} = \chi_{\omega_1, b, \omega}(\mathbb{R})$, $\mathcal{L}^{(2)} = \chi_{\omega_1', b', \omega'}(\mathbb{R})$, where the function $\chi_{\omega_1, b, \omega}$ is defined by (15), and $\mathcal{L}^{(3)} = \chi_{L;\sigma, b_l, \omega_l}(R)$, where $\chi_{L;\sigma, b_l, \omega_l}$ is defined by

$$\chi_{L;\sigma, b_l, \omega_l}(y) = \sigma + i b_l \sinh(i\omega_l + y). \quad (56)$$

Since $x - h > 0$, the oscillating factor $e^{i(x-h)\xi}$ quickly decays if we choose $\omega > 0$; hence, the wings of the contour point upward. The contour $\mathcal{L}^{(2)}$ must be above $\mathcal{L}^{(1)}$ so that $\xi - \eta$ is separated from 0 for all $\xi \in \mathcal{L}^{(1)}$ and $\eta \in \mathcal{L}^{(2)}$. In particular, $\omega' \geq \omega$. The upper bound on ω' is implied by the requirement that $\mathcal{L}^{(2)}$ be in the domain of analyticity of ψ. Finally, $\mathcal{L}^{(3)}$ must be chosen so that $q + \psi(\eta) \notin (-\infty, 0]$ for all $\eta \in \mathcal{L}^{(2)}$ and $q \in \mathcal{L}^{(3)}$.

4.5. Double Barrier Options

The following procedure is borrowed from [69]. Consider knock-out double-barrier options with barriers $h_- < h_+$; the option expires worthless if X leaves (h_-, h_+) before maturity date T; if $X_t \in (h_-, h_+)$ until T, the option payoff at maturity is $G(X_T)$. Assuming the constant riskless rate, the option value at time 0 is:

$$V_{k.o.}(G; h_-, h_+; T, x) = e^{-rT}\mathbb{E}^{\mathbb{Q}}\left[\mathbf{1}_{\tau^-_{h_-}\wedge\tau^+_{h_+} > T} G(X_T)\right], \tag{57}$$

and the corresponding boundary problem is:

$$(\partial_\tau + \psi(D_x) + r)V(\tau, x) = 0, \quad \tau > 0,\ x \in (h_-, h_+), \tag{58}$$

$$V(0, x) = \mathbf{1}_{(h_-, h_+)}(x)G(x), \tag{59}$$

$$V(\tau, x) = 0, \quad \tau \geq 0,\ x \notin (h_-, h_+). \tag{60}$$

As in the case of single barrier options, the Laplace transform (maturity randomization) reduces the calculation of the expectation on the RHS of (57) to the evaluation of the perpetual stream

$$W_{k.o.}(G; h_-, h_+; q, x) = \mathbb{E}^{\mathbb{Q}}\left[\int_0^{\tau^-_{h_-}\wedge\tau^+_{h_+}} e^{-qt}G(X_t)dt\right], \tag{61}$$

which is abandoned at time $\tau^-_{h_-} \wedge \tau^+_{h_+}$. In [69], $W_{k.o.}(G; h_-, h_+; q, x)$ was evaluated as follows:

$$W_{k.o.}(G; h_-, h_+; q, x) = G^0(x) - G^1_+(x) - G^1_-(x) + G^2_+(x) + G^2_-(x) - G^3_+(x) - G^3_-(x) + \cdots \tag{62}$$

where:

$$G^0_+(x) = G^0(x)\big|_{[h_+, +\infty)}, \qquad G^0_-(x) = G^0(x)\big|_{(-\infty, h_-]},$$

$$G^n_+(x) = \mathcal{E}^-_q\left(\mathbf{1}_{(-\infty, h_-]}(x) \cdot ((\mathcal{E}^-_q)^{-1}G^{n-1}_-)(x)\right) \qquad \forall n \geq 1,$$

$$G^n_-(x) = \mathcal{E}^+_q\left(\mathbf{1}_{[h_+, +\infty)}(x) \cdot ((\mathcal{E}^+_q)^{-1}G^{n-1}_+)(x)\right) \qquad \forall n \geq 1.$$

In the case of HEJD, the series can be explicitly calculated [70,71]. For general Lévy processes, [69] uses a general numerical method based on the piece-wise linear interpolation of functions $G^n_\pm$, fast convolution and the refined version of the fast Fourier transform (FFT) technique developed earlier in [46]. It is demonstrated that the standard version of FFT and fractional FFT are either inaccurate or inefficient for many classes of Lévy processes.

In [64], the calculations are in the dual space and fractional-parabolic deformations of the contours of integration are used; the more efficient sinh-acceleration technique can be applied in the same vein.

Under additional regularity conditions on ψ, it is possible to prove that the boundary problem (58)–(60) has a unique solution in the class of bounded functions continuous on $[0, +\infty) \times (h_-, h_+)$. Then, if X satisfies the (ACT)-condition, the expectation given by (57) is this unique solution.

4.6. Regularity of Solutions of Boundary Problems

Assume that ψ^0 is an elliptic symbol of order $\nu \in (0, 2)$, hence, X is a pure jump process. Then, it is easy to prove (see [21]) that:

(a) If $\nu \in (1, 2)$ or $\nu \in (0, 1]$ and $\mu = 0$, then $\phi^\pm_q$ are elliptic symbols of order $\kappa^\pm = -\nu/2$;

(b) If $\nu = 1$, then $\phi^\pm_q$ are elliptic symbols of order $\kappa^\pm_q$, where $\kappa^\pm \in (-1, 0)$ depend on μ, and $\kappa^+_q + \kappa^-_q = -1$;

(c) If $\nu \in (0,1)$ and $\mu > 0$, then $\phi_q^+(\xi) = (q - i\mu\xi)^{-1}(1 + \alpha_q^+(\xi))$ and $\phi_q^-(\xi) = 1 + \alpha_q^-(\xi)$, where $\alpha_q^\pm$ are of order $-1 + \nu + \epsilon$, for any $\epsilon > 0$. Hence, $\kappa_q^+ := \mathrm{ord}\, \phi_q^+ = -1$, and $\kappa_q^- := \mathrm{ord}\, \phi_q^- = 0$;

(d) If $\nu \in (0,1)$ and $\mu < 0$, then $\phi_q^-(\xi) = (q - i\mu\xi)^{-1}(1 + \alpha_q^-(\xi))$ and $\phi_q^+(\xi) = 1 + \alpha_q^+(\xi)$, where $\alpha_q^\pm$ are of order $-1 + \nu + \epsilon$, for any $\epsilon > 0$. Hence, $\kappa_q^+ := \mathrm{ord}\, \phi_q^+ = 0$, and $\kappa_q^- := \mathrm{ord}\, \phi_q^- = -1$.

Straightforward calculations based on (a)–(d) show (see [21,62,63]) that the prices of down-and-out barrier options—hence, solutions of the corresponding boundary problems—are not smooth (and in the case of (c), discontinuous) at the barrier; the leading term of asymptotics as $x \downarrow h$ is calculated. In [63], the asymptotics of the derivatives of the price $V(t,x)$ (sensitivities) is calculated as well.

In cases (a), (b), (c), $V(\tau,x) \sim c(\tau)(x-h)^{\kappa_q^-}$, and $V_x(\tau,x) \to +\infty$ as $x \downarrow h$. In case (d), V_x is continuous up to the boundary but V_{xx} is unbounded.

The case of the BM with embedded jumps was not explicitly studied but the study can be performed in a similar vein. If the jump component is of order $\nu \in [1,2)$, then the solution is continuous up to the boundary but V_x is discontinuous; if $\nu \in (0,1)$, then V_x is continuous but V_{xx} is unbounded.

The case of the double-barrier options was not studied but since the study of the regularity at the boundary is easily localized (see [26]), one can easily prove that, for instance, in case (c), the solution is discontinuous at h_- but smooth at h_+; V_x is unbounded as $x \to h_-$, and V_{xx} is unbounded as $x \to h_+$.

4.7. The Case of Time-Dependent Boundaries

The regularity of solutions is an open problem. In this section, we formulate several natural hypotheses. Assuming that the boundary (or two boundaries) are piece-wise smooth, one can try to study the regularity of solutions localizing the problem. In the case of processes of order $\nu \in [1,2)$, the operator $A(D_t, D_x)$ is quasi-elliptic (the symbol $\hat{A}(\eta,\xi) = i\eta - i\mu\xi + \psi^0(\xi)$ satisfies $|\hat{A}(\eta,\xi)| \ge c(|\eta| + |\xi|^\nu) - C$, for $(\eta,\xi) \in \mathbb{R}^2$), and elliptic if $\nu = 1$; hence, localization can be performed as in the elliptic case [26]. If $\nu \in (1,2)$, one expects that, if t is fixed and (t,x) tends to a point (t,x_0) on a smooth part of the boundaries, the solution behaves as $c(t)|x - x_0|^{\nu/2}$. If $\nu = 1$, then the study becomes complicated. One expects that the asymptotics at the boundary can be naturally described in the coordinates $x = x' + f(t)$, which makes the boundary locally flat: $x' = h'$. For each point t in a small neighborhood of t_0, the asymptotics of the price $V(t,x')$ in the new coordinate system is of the form $c(t)|x'|^{\kappa^\pm(t)}$, where $\kappa^\pm(t) \in (0,1)$ ("-" for the lower boundary and "+" for the upper boundary) continuously depend on t and are defined by the "drift" $\mu'(t,h')$ in the new coordinate system and ψ^0 as in case (b) above.

If $\nu \in (0,1)$, then the localization itself becomes more difficult because the principal symbol $-i\eta - i\mu\xi$ is hyperbolic. Assuming that the localization is possible, one expects that the solution is smooth up to the boundary (or discontinuous at the boundary) when the vector field $-\partial_\tau + \mu\partial_x$ is transversal to the boundary and points to (or from, respectively) the boundary.

4.8. Multi-Factor Case

Consider first a boundary problem in $(0,+\infty) \times \mathbb{R}^{n-1}_{x'} \times (0,+\infty)$. In the analytical setting, one makes the Fourier transform with respect to x' and solves the family of problems on $\{\tau > 0, x_n > 0\}$ using the results above. However, then the dependence of the Wiener–Hopf factors on ξ' does not allow for the interpretation of the Wiener–Hopf factors as the symbols of the EPV operators, and one is forced to impose additional potentially unnecessary conditions in order to justify the results.

A natural alternative is to represent the payoff functions as sums of functions supported on direct products of the half-axis. Then, after an appropriate change of variables, each new problem is a problem on $(0,+\infty)^{n+1}$. Then, we make the Laplace transform with respect to τ and $x_j, j = 1, 2, \ldots, n-1$, and solve the resulting family of problem on $(0, +\infty)$ depending on $q \in (0, +\infty)^n$ exactly as in the case of Lévy processes on $\mathbb{R}$. The characteristic exponent of the one-dimensional process is $\psi(-iq, \xi_n)$, and the resulting analytical expression admits analytic continuation to $\{q \in \mathbb{C}^n \mid \operatorname{Re} q_j > 0\}$ (and wider regions, if the process is sufficiently regular).

In this way, one can derive explicit formulas for single barrier options, and the representation of the price of double-barrier options with flat parallel barriers. In the application to credit risk models (counterparty risk), important variations are problems with boundaries $x_j = h_j, j = n, n-1$, when $n = 2$, and even $j = n, n-1, n-2$, when $n = 3$. The case $n = 2$ can be solved modifying in the straightforward fashion the iterative procedure used in the case of two parallel flat boundaries. The proof and design of efficient numerical procedures become more involved but, at the theoretical level, the tools remain essentially the same. The case $n = 3$ becomes messier still but an iteration procedure can be designed, and convergence to the price proved as in the case of two boundaries.

The case of curved boundaries is similar to the case $n = 1$, and naturally, the results are at the level of hypotheses so far.

5. American Options and Free Boundary Problems

5.1. Basic Example

The European options can be exercised only at expiry: if $S_T = e^{X_T} \geq K$, then it is optimal to exercise the European call option and receive $e^{X_T} - K$; otherwise the option expires and is worthless, and the payoff is 0. An American option can be exercised at any moment until the expiry date, T. If the process X is Markovian, it is natural to expect that there is a subset B of the half-space $\{(t,x) \mid t \leq T\}$ such that the option is exercised the first time $(t, X_t) \in B$. B is called the exercise region and the part of the boundary ∂B that is in the open half-plane $\{(t,x) \mid t < T\}$ is called the early exercise boundary. In the simplest cases of the American option put and call options, the early exercise regions are connected, and of the form $\{(t,x) \mid x \leq h_*(t),\ t < T\}$ and $\{(t,x) \mid x \geq h^*(t),\ t < T\}$, respectively.

In the Brownian motion case, the early exercise boundary and option value can be found by solving the corresponding free boundary problem. In addition to the terminal condition at $t = T$, one adds the value matching and smooth pasting condition at the early exercise boundary. Let $r \geq 0$ be the riskless rate, $\delta \geq 0$ the dividend rate, and X_t be a Lévy process under the risk-neutral measure chosen for pricing, with the infinitesimal generator L. For the American put option in the one-factor model, the free boundary problem is of the form

$$\begin{aligned}
(\partial_t + L - r + \delta)V(t,x) &= 0, \quad x > h(t),\ t < T, && (63)\\
V(T,x) &= (K - e^x)_+, && (64)\\
V(t,x) &= K - e^x, \quad x = h_*(t),\ t < T, && (65)\\
V_x(t,x) &= -e^x, \quad x = h_*(t),\ t < T, && (66)
\end{aligned}$$

and V is sought in the class of continuous bounded functions, which are of the class $C^{1,2}$ above the early exercise boundary, with the first derivative V_x continuous up to the boundary (these regularity conditions for the American option price are well known and proven). On the RHS of (65), we write $K - e^x$ instead of $(K - e^x)_+$ because it is non-optimal to exercise the option unless the payoff is non-negative. The free boundary problem for the American call option is similar. The equivalence of the pricing problems for the American put and call options and the corresponding free boundary problems is proven in the Brownian motion case and for Brownian motion with embedded compound Poisson process. For general Lévy processes, the equivalence is an open problem. Note that for processes with jumps, the value matching condition (65) becomes non-local:

$$V(t,x) = K - e^x, \quad x \leq h_*(t),\ t < T, \tag{67}$$

and the smooth pasting condition may fail. We will not use the smooth pasting condition. Instead, we look for the boundary which maximizes the solution of the problem (63), (64), (67).

5.2. Behavior of the Early Exercise Boundary near Maturity

The next subtle point is the limit $h^*(T-) := \lim_{t\to T} h_*(t)$. Designing numerical methods, one is tempted to assume that, as $t \to T$, the region in the state space U_t where the American option with the payoff $g(X_t)$ remains alive, tends to $\{x \mid g(x) \geq 0\}$. In the case of the put option above, this assumption translates into the condition $h^*(T-0) = \ln K$: the limit of the early exercise boundary at maturity (in the S-coordinate) is equal to the strike. This statement is correct if $r > 0$ and $\delta = 0$. If the stock pays dividends, then it is possible that $h_*(T-) < \ln K$; in the case $r = 0$, simple general no-arbitrage considerations spelled out by Merton prove that it is non-optimal to exercise the American put before maturity. Essentially the same no-arbitrage argument proves that it is non-optimal to exercise the American call on non-dividend paying stock before maturity. Starting with [72,73], the behavior of the critical stock price near maturity for American options in diffusion models has been studied in a number of publications (see the bibliography in [74]). It is proved, in particular, that if the stock pays dividends, then, depending on the parameters σ^2, r, δ, the limit of the early exercise boundary for the American call in the Black–Scholes model is above the strike.

In [21,40], it is proven that in the presence of *positive* jumps, the early exercise boundary for the American put without dividends is separated from the strike by a margin. In [21], the result is obtained for KoBoL, NTS and NIG models using the Wiener–Hopf factorization technique. In [40], the proof is based on the calculation of $\theta(t,x) := V_t(t,x)$ of out-of-the-money options, at expiry (in the case of the put, in the region $x > \ln K$, in the case of the call, in the region $x < \ln K$). It was proved that, in the presence of jumps, $\theta_{\text{call}}(T-,x) < 0$ for $x < \ln K$ (in the diffusion case, $\theta_{\text{call}}(T-,x) = 0$ for $x < \ln K$). Using this result and the put-call parity, it is proven that it is not optimal to exercise the American put without dividends up to expiry in the region where $rK < -\theta_{\text{call}}(T-,x)$. It is demonstrated that for parameters' values documented in empirical studies of financial markets, the margin is several percent of the strike or even more than a dozen percent. Hence, a numerical method is based on the condition $h_*(T-) = \ln K$ is expected to produce rather inaccurate results. In [24], a similar formula for θ of out-of-the-money options in one-factor Lévy driven quadratic term structure models (QTSMs) is presented but without the proof.

In [74], the results in [24,40] are generalized for wide classes of multi-factor Markov models with jumps, and the proofs are simplified. First, consider a European option with the (effective) payoff $g_+(X_T) = \max\{0, g(X_T)\}$; for a European call on a stock with the price process e^{X_t}, and strike K, $g(x) = e^x - K$. Let $F(x,dy)$ be the density of jumps of the underlying Markov process X_t and denote by $U_-(g) := \{x \mid g(x) < 0\}$ the out-of-the-money region. Denote by $V(g_+;x,\tau)$ the option price at time $\tau > 0$ to expiry and $X_{T-\tau} = x$, and by $C(g_+;x)$ the limit of the $-\theta$ of the option, at expiry:

$$C(g_+;x) = \lim_{\tau\to+0} \frac{V(g_+;x,\tau)}{\tau}. \tag{68}$$

For wide classes of payoffs and jump-diffusion models, the limit exists for $x \in U_-(g)$, and it is given by

$$C(g_+;x) = \int g_+(x+y)F(x,dy), \tag{69}$$

for almost all $x \in U_-(g)$ (see [74]). If (69) is applied to digital options, one modifies the definition of the payoff: $g(x) = 1$ in the in-the-money region, and $g(x) < 0$ in the out-of-the-money region.

Let $U_+(g) = U_-(-g)$ be the in-the-money region of the option, and Ω_τ be the optimal non-exercise region for the American option, in the x–space, at time τ to expiry. Clearly, $\Omega_\tau \supset U_-(g)$, hence, $\cup_{\tau>0}\Omega_\tau \supset U_-(g)$. However, it may be the case that it is non-optimal to exercise the option up to expiry even if x is in the in-the-money region $U_+(g)$. Let L be the infinitesimal generator of X. Define: $\Omega^J = \{x \in U_+(g) \mid Lg(x) + \mathcal{C}((-g)_+, x) > 0\}$. The following two theorems from [74] give the "lower bound" for the limit of the no-exercise region at maturity and approximate formulas for the option price close to maturity, which can be used at the first step of backward induction procedures for pricing American options, in wide classes of Markov models.

Theorem 4 ([74], Thm. 2.2). $\Omega_{+0} \supset \Omega^J \cup U_-(g)$.

Theorem 5 ([74], Thm. 2.3). *Let $Lg(x) + \mathcal{C}((-g)_+, x) \neq 0$. Then, as $\tau \to 0$:*

1. *For x in the out-of-the-money region, $U_-(g)$:*

$$V_{\mathrm{am}}(g_+; x, \tau) \sim \tau \mathcal{C}(g_+; x); \tag{70}$$

2. *For x in the in-the-money region $U_+(g)$:*

$$V_{\mathrm{am}}(g_+; x, \tau) \sim g(x) + \tau(Lg(x) + \mathcal{C}((-g)_+; x))_+. \tag{71}$$

5.3. Perpetual American Options in One-Factor Models or Stationary Free Boundary Problems on $\mathbb{R}$

The solution of the optimal stopping problem in Lévy models with infinite time horizon (equivalently, stationary free boundary problems) is the main block for the solution of the problems with finite time horizon using maturity randomization (equivalently, method of lines), in regime-switching models, and the approximation of more general Markov models with regime-switching Lévy models.

The first simple but crucial trick which makes it possible to give simple proofs and design efficient procedures is the reduction in the option to *acquire an instantaneous payoff* $G(X_t)$ to the option to *abandon the stream* $-g(X_t)$. If q is the discount rate, then we assume that $G = q^{-1}\mathcal{E}_q g$, where g satisfies (43) and (44). This implies that G is sufficiently regular, does not increase too quickly at infinity, and the discount rate is not too small. In the case of options with finite time horizon, using sufficiently small time intervals in the method of lines, the latter condition can be satisfied in all cases. In the case of the American put with the infinite time horizon, on the non-dividend paying stock, the representation $G = q^{-1}\mathcal{E}_q g$ is impossible but a representation $G = q^{-1}(\mathcal{E}_q^-)^{-1} g$ is possible and suffices for the proof.

Assuming that $G = q^{-1}\mathcal{E}_q g$, for any stopping time τ:

$$\mathbb{E}^x\left[e^{-q\tau}G(X_\tau)\right] = q^{-1}\mathcal{E}_q g(x) + V_{\mathrm{ex}}(-g; \tau; x),$$

where:

$$V_{\mathrm{ex}}(f; \tau; x) = \mathbb{E}^x\left[\int_\tau^{+\infty} e^{-qt} f(X_t) dt\right].$$

Hence, an optimal time to exercise the American option with the payoff function G is an optimal time to abandon the stream $\{-g(X_t)\}_{t\geq 0}$, and vice versa.

Theorem 6. *Let the following conditions hold:*

(i) X is a Lévy process with the non-trivial infimum process;
(ii) g is a non-decreasing stream that changes sign;
(iiii) Bounds (37), (43) and (44) hold.

Then:

(a) There exists h such that:

$$\mathcal{E}_q^+ g(x) \leq 0, \quad x \leq h, \quad \text{and} \quad \mathcal{E}_q^+ g(x) \geq 0, \quad x \geq h. \tag{72}$$

(b) τ_h^-, the entry time into $(-\infty, h]$, is an optimal exit time in class $\mathcal{M}$.

(c) *The option value can be represented as the EPV of the stream $g(X_t) - U(X_t)$, where U is a non-decreasing function vanishing above h.*
(d) *If g is not monotone but (i) holds, then τ_h^- is an optimal exit time in the class of stopping times of the threshold type.*

Proof. (a,d) are immediate from the representation

$$V_{\text{ex}}(g;\tau_h^-;x) = q^{-1}\mathcal{E}_q^- \mathbf{1}_{(h,+\infty)}\mathcal{E}_q^+ g(x).$$

The expectation operator $\mathcal{E}_q^-$ being positive, $V_{\text{ex}}(g;\tau_h^-;x)$ is maximized when the function $\mathbf{1}_{(h,+\infty)}\mathcal{E}_q^+ g$ is maximized. Hence, all negative values of $\mathcal{E}_q^+ g$ must be set to 0.

(b,c) are directly derived (see [21,49,75]) verifying the conditions of the following general lemma; the proof is much shorter and simpler than purely probabilistic proofs. However, we need to assume that X satisfies the (ACP)-property. □

Lemma 2 ([21,49,75]). *Let τ_B be the first entrance time into a Borel set B. Let B be a Borel set such that:*

$$W_{\text{ex}}(g;\tau_B;\cdot) := (q-L)V_{\text{ex}}(g;\tau_B;\cdot) \text{ is universally measurable} \quad (73)$$
$$W_{\text{ex}}(g;\tau_B;f;x) = g(x), \quad x \in \mathbb{R}\setminus B, \quad \text{a.e.} \quad (74)$$
$$W_{\text{ex}}(g;\tau_B;x) \geq g(x), \quad x \in B, \quad \text{a.e.} \quad (75)$$
$$W_{\text{ex}}(g;\tau_B;x) \geq 0, \quad \forall\, x. \quad (76)$$

Then, τ_B maximizes $V_{\text{ex}}(g;\tau_B;\cdot)$ in the class $\mathcal{M}$.

Proof. Let τ be a stopping time. Then, using Dynkin's formula and (74)–(76), we obtain:

$$\begin{aligned} V_{\text{ex}}(g;\tau_B;x) &= \mathbb{E}^x\left[\int_0^\tau e^{-qt}(q-L)V_{\text{ex}}(g;\tau_B;X_t)dt\right] \\ &\quad +\mathbb{E}^x\left[e^{-q\tau}V_{\text{ex}}(g;\tau_B;X_\tau)\right] \\ &\geq \mathbb{E}^x\left[\int_0^\tau e^{-qt}g(X_t)dt\right]. \end{aligned}$$

With $\tau = \tau_B$, we obtain the equality, which means that τ_B is optimal. □

Remark 3. *In [51], we proved optimality in the class of all stopping times for a wide class of non-monotone payoffs, under the assumption that the jump density is completely monotone. In [52], the results are applied to solve a game-theoretical problem. In [76], the American options with lookback features are studied in cases when the exercise region is discontinuous.*

5.4. Good and Bad News Principles and the Failure of the Smooth Pasting Condition

In [77], for special cases, and in [47,49,78] in the general case, the following interpretation of the stopping rule (b) and its mirror reflection for the option to abandon a non-increasing stream were given:

GOOD NEWS PRINCIPLE. Abandon an increasing stream (exit) when the EPV of the stream under the supremum process becomes negative.

BAD NEWS PRINCIPLE. Acquire an increasing stream (entry) when the EPV of the stream under the infimum process becomes positive.

FAILURE OF THE SMOOTH PASTING CONDITION [21,75]. If $\phi_q^-(\xi) \sim c > 0$ as $\xi \to \pm\infty$, $\mathcal{E}_q^+ g$ is smooth at h, $\mathcal{E}_q^+ g(h) = 0$, and $(\mathcal{E}_q^+ g)'(h) > 0$, then h is the optimal exit boundary but the value function has a kink at h. For the option to abandon a non-increasing stream, replace $\phi_q^-, \mathcal{E}_q^-$ with $\phi_q^+, \mathcal{E}_q^+$.

In terms of atoms of the pdf of $\bar{X}_{T_q}$ and $\underline{X}_{T_q}$, the failure of the smooth pasting condition was reformulated and proven in [79].

5.5. American Options with Finite Time Horizon or Non-Stationary Boundary Problems on $\mathbb{R}$

We have an optimal stopping problem: for $t < T$, find:

$$V(t,x) = \sup_{\tau\in\mathcal{M},\tau\le T} \mathbb{E}^{\mathbb{Q}}_t\Big[e^{-r(\tau-t)}G(X_\tau) \mid X_t = x\Big]. \tag{77}$$

Carr's randomization [80] or method of lines approximates $V(t,x)$ with a sequence of perpetual American options. The maturity period is divided into N subintervals, using points $0 = t_0 < t_1 < \cdots < t_N = T$. Each sub-period $[t_s, t_{s+1}]$ is replaced with an exponentially distributed random maturity period T_s with mean $\Delta_s = t_{s+1} - t_s$, $s = 0,1,\dots,N-1$. The random variables $T_s, s = 0,\dots,N-1$ and the process X are assumed to be independent. Typically, one takes $\Delta_s = T/N$ for all s. If no optimization decision is involved, then we can say that the deterministic maturity date T is replaced with $T' = T_0 + T_1 + \cdots + T_{N-1}$. If all $T_s \sim \mathrm{Exp}\,N/T$, then T' is *Erlang*-distributed. The same idea can be applied to price barrier and lookback options, with *flat boundaries*. When optimizing decisions are involved, an accurate formulation of the approximate optimal stopping problem becomes more involved: a time-consistent exercise rule must take into account realizations of $T^s := T_0 + T_1 + \cdots + T_s$, $s = 0,1,\dots$, hence, the rule must be updated after the arrival of each T^s. The convergence of Carr's randomization procedure for American options in wide classes of Lévy models is proven in [81]. The paper [80] solves the sequence in the BM model using explicit formulas for the boundary problem for second-order differential operators; this technique cannot be applied to other Lévy models. Furthermore, ref. [80] formulates an equivalent form of maturity randomization as the approximation of the American option with a finite-time horizon by the American option with the Erlang-distributed maturity date, and stated that Richardson's extrapolation can be applied. Both statements are false for American options but hold for barrier options. The convergence of Carr's randomization approximation is proven in [71] for wide classes of Markov processes (additional conditions on the process are necessary), and Richardson's extrapolation of arbitrary order is justified in [63] for the value function and its derivatives.

The proof of optimality of the solution of the sequence and pricing procedure simplify if, at each step, the option is reduced to the option to abandon a stream. Then, the conditions of the basic theorems for perpetual options can be easily verified for each option in the sequence. This is the idea of the solution in [21,38,40,49]. For simplicity, consider equidistant dates, and set $\Delta = T/N$, $q = 1/\Delta + r$.

The backward induction procedure is as follows.

1. Set $V_0 = G_+$ (the payoff at maturity).
2. In the cycle $s = 1,2,\dots,N$, find V_s as the solution to the optimal stopping problem:

$$V_s(x) = \sup_{\tau\in\mathcal{M}} \mathbb{E}^{\mathbb{Q}}\left[e^{-q\tau}G(X_\tau) + \frac{1}{\Delta}\int_0^\tau dt\, e^{-qt}V_{s-1}(X_t) \mid X_0 = x\right]. \tag{78}$$

3. V_N is Carr's randomization approximation to time-0 option price.

The reduction to the sequence of exit problems is as follows.

Set $W_s = V_s - G$, and notice that the maximization of V_s is equivalent to the maximization of W_s. To reformulate the optimal stopping problem in terms of W_s, we find function g_q such that:

$$G(x) = q^{-1}\mathbb{E}^{\mathbb{Q}}\left[\int_0^{+\infty} e^{-qt}g_q(X_t)dt \mid X_0 = x\right].$$

For the put, $g_q(x) = qK - (q + \psi(-i))e^x$, where ψ is the characteristic exponent of the Lévy process X. We have:

$$-G(x) + \mathbb{E}^{\mathbb{Q}}[e^{-q\tau}G(X_\tau)] = -\mathbb{E}^{\mathbb{Q},x}\left[\int_0^\tau e^{-qt}g_q(X_t)dt \mid X_0 = x\right],$$

therefore, the optimal stopping problem in terms of W_s is:

$$W_s(x) = \sup_{\tau \in \mathcal{M}} \mathbb{E}^{\mathbb{Q}}\left[\int_0^\tau e^{-qt}(f(X_t) + \frac{1}{\Delta}W_{s-1}(X_t))dt \mid X_0 = x\right], \tag{79}$$

where $f(x) = -g_q(x) + (1/\Delta)G(x)$. Assuming that $r + \psi(-i) \geq 0$, the function:

$$f(x) \quad = \quad -(qK - (q + \psi(-i))e^x) + \frac{1}{\Delta}(K - e^x) = -rK + (r + \psi(-i))e^x$$

is non-decreasing (the case of the put option). The function $W_0 = (-G)_+$ is also non-decreasing. Using Theorem 6, we find Carr's randomization approximation h_1 to the early exercise boundary, and prove that W_1 is non-decreasing and vanishes below h_1.

In the cycle $s = 1, 2, \ldots, N-1$, using Theorem 6 and the induction assumptions: W_s is non-decreasing and vanishes below h_s, we find the approximation to the early exercise boundary h_{s+1} and W_{s+1} and prove that W_{s+1} is non-decreasing and vanishes below h_{s+1}. Finally, $V_s = W_s + G$, $s = 1, 2, \ldots, N$. See ([49], Chapt. 13) for details.

5.6. Shape of the Early Exercise Boundary and Smooth Pasting Condition

Assume that X is the process of finite variation, with non-zero drift μ. Then, one can conjecture that far from maturity, the early exercise boundary is almost flat; hence, the smooth pasting condition fails if the vector field $\partial_t + \langle \mu, \partial_x \rangle$ is transversal to the boundary and points from the boundary. One can also conjecture that the smooth pasting condition fails everywhere but this is far from evident in view of the fact that the behavior of the early exercise boundary at maturity can be very irregular, especially in the multi-factor case. One cannot exclude cases when, at some parts of the free boundary, the vector field $\partial_t + \langle \mu, \partial_x \rangle$ is transversal to the boundary and points toward the boundary. At these parts, the smooth pasting condition would hold.

6. Barrier Options and American Options in Regime-Switching Lévy Models and Systems of Pseudo-Differential Equations, Approximation of Stochastic Volatility Models and Models with Stochastic Interest Rate

Let $\mathcal{M}$ be a finite state Markov chain with transition rates λ_{jk}, $j, k = 1, 2, \ldots, M$. Set $\Lambda_j = \sum_{k \neq j} \lambda_{jk}$. For each j, let $X^{(j)}$ be a Lévy process on $\mathbb{R}$ with the characteristic exponent ψ_j and infinitesimal generator L_j under a measure $\mathbb{Q}_j$. The riskless rate q_j, instantaneous payoffs G_j, streams of payoffs g_j and early exercise boundary h_j depend on the state. In the case of barrier options, h_j are given, and in the case of American options, h_j are chosen, solving the optimal stopping problem. The infinitesimal generator of the model is a matrix PDO $\mathcal{L} = -\psi(D)$, where:

$$\psi(\xi) = \mathrm{diag}\,(\psi_j(\xi) + \Lambda_j)_{j=1}^M - [\lambda_{jk}]_{j \neq k}).$$

If there are no barriers, then, evidently, one can solve the Cauchy problem as in the scalar case; the only difference is that the matrix exponential $\exp[-(T-t)(\psi(\xi) + \mathrm{diag}\,[q_j]_{j=1}^M])$ appears, and the choice of appropriate contour deformations is more involved. Similarly, in the case of $h_1 = \cdots = h_M$, one can use the matrix form of the Wiener–Hopf factorization, and repeat the calculations (in the PDO form) which we used in the no-regime switching case. However, if M is large, then even these theoretically straightforward and simple methods are very difficult for efficient numerical realization. Furthermore, the straightforward methods outlined above are not applicable if the boundaries are different in different states. This is the case when stochastic volatility models and models with stochastic interest rates are approximated by Markov modulated Lévy models and/or American options are priced.

The idea of the approximation is as follows [82–86]. The action of the infinitesimal generator of the Markov process is discretized—replaced by the infinitesimal generator $\mathcal{L}_{disc}$ of the Markov chain with an infinite number of states, and then truncated. At the boundary of the truncated discretized state space, it is necessary to impose appropriate conditions so that the truncated operator is the infinitesimal generator of a Markov chain, without killing. Hence, Dirichlet boundary conditions may not be used. Instead, in order to avoid killing, a kind of reflection condition must be imposed: transition rates from each point at the boundary to points in the truncated discretized state space must change. If $\mathcal{L}_M$ is a diffusion, the discretization and boundary conditions are straightforward; in the case of processes with jumps, constructions are more involved.

In a number of publications [50,82–87], we used the following simple iteration procedure, which we outline below in the case of down-and-out options: the option is exercised in state j when $X^{(j)}$ breaches the barrier h_j from above. Numerical examples demonstrated the stability of the procedure even for $M > 2000$ [86].

A problem with finite time horizon is reduced to a sequence of boundary problems with infinite time horizon using the method of lines (Carr's randomization). The regularity conditions are as follows. For each j:

(i) $X^{(j)}_t$ satisfies the (ACP)-property (needed in the case of American options only);
(ii) g_j is measurable, non-negative, and does not grow too fast at infinity;
(iiii) $(q_j + \Lambda_j - L_j)G_j$ is continuous, monotone, and does not grow too fast at infinity.
(iv) In each state, the rate of growth is controlled by conditions (37), (43), with $q_j + \Lambda_j$ in place of q_j.

Note that the general results for American options are obtained for non-negative g_j. However, we have optimal stopping results in the non-regime switching case when the payoff function g may assume negative values. The same technique can be used in regime-switching models, hence, the procedure can be generalized to the case of non-monotone payoff functions, under certain conditions on the payoffs and processes. The case of barrier options with payoff functions that change sign can be reduced to the case of non-negative payoff functions using the linearity of the expectation operator.

At each step of the backward induction, we use the following block:

I. Reduce the pricing problem to the problem of evaluation of a perpetual stream (in different states, the payoff streams are different).
II. Assuming that the value functions in each state but state j are known, calculate the state-j option value.
III. In the case of American options, calculate the approximation to the early exercise boundary in state j.
IV. Using this conditional result as a guide, construct an iteration scheme for all states, and prove that the value functions converge to some limits.

The algorithm for perpetual American options (used as a block at each time step in the backward induction procedure) is as follows.

I. Choose the grid $\vec{x}$ (it might be necessary to use different grids in different states).
II. In the cycle $j = 1, \dots, M$, calculate the initial approximation $V_{j,0}$ to the option value:

$$V_{j,0}(\vec{x}) = E^{\mathbb{Q}_j}\left[\int_0^{\tau_j^-} e^{-(q_j+\Lambda_j)t} g_j(X_t)dt \mid X_0 = \vec{x}\right]$$

III. In the cycle $\ell = 0, 1, \dots,$ for each $j = 1, 2 \dots, M$, calculate:

$$V_{j,\ell+1}(\vec{x}) = E^{\mathbb{Q}_j}\left[\int_0^{\tau_j^-} e^{-(q_j+\Lambda_j)t}\left(g_j(X_t) + \sum_{k\neq j} \lambda_{jk} V_{k,\ell}(X_t)\right)dt \mid X_0 = \vec{x}\right]$$

Stop when $\|V_{\cdot,\ell+1} - V_{\cdot,\ell}\| \le \epsilon$, where ϵ is the error tolerance.

The limit as $\ell \to \infty$ exists because the sequence of option values in each state is increasing (the non-negativity of g_j is needed for the proof).

7. Affine and Quadratic Term Structure Models

7.1. Affine Processes

Affine processes are used in models with stochastic volatility and stochastic interest rates. The adjective affine is explained by two *almost equivalent* properties:

(1) The characteristic function of the transition density is of the form of an exponential function of an affine function of factors of the models with the coefficients depending on time to maturity $\tau = T - t$ and spectral parameter ξ:

$$\mathbb{E}[e^{i\langle\xi, X_T\rangle} \mid X_t = x] = \exp[\langle A(\tau,\xi), x\rangle + B(\tau,\xi)], \tag{80}$$

where $\langle a, b\rangle = \sum_j a_j b_j$ [88]. If the stochastic interest rate r_t is modeled as an affine function of the factors of the model, the state space must be enlarged as in the probabilistic version of the Feynman–Kac formula, and an additional factor $Y_t = \int_0^t r_s ds$ IS added. The extended model remains affine, and the representation (80) becomes possible.

(2) The coefficients of the stochastic differential equation (SDE) defining the process are affine functions of the state variable.

Property (1) allows one to calculate expectations $V(t,x) = \mathbb{E}[G(X_T) \mid X_t = x]$ as in the case of Lévy models. However, given SDE, it is necessary to (1) prove that the representation of the form (80) exists; and (2) calculate the matrix function $A(\tau, x)$ and vector-function $B(\tau, x)$.

The formal proof is straightforward [89]. *Assuming that the Feynman–Kac theorem holds*, write down the Cauchy problem

$$(\partial_t + \mathcal{L})V(t,x) = 0, t < T, x \in \mathcal{D}_0, \tag{81}$$

$$V(T,x) = e^{i\langle\xi,x\rangle}, \tag{82}$$

and where $\mathcal{D}_0$ and $\mathcal{L}$ are the interior of the state space $\mathcal{D}$ of the process and infinitesimal generator, respectively, substitute anzatz (80) into (81)–(82) and reduce the calculation of (A, B) to the solution of the generalized system of Riccati equations associated with the model. The proof in the general case is unknown (see [2] for the discussion and bibliography).

An incomplete list of outstanding problems for affine models, which can be regarded as mathematical problems for a general well-defined class of PDO, is as follows (each of the problems is solved for particular models or certain subclasses of affine models but the complete answers, in full generality, are unknown—for a short overview of the extant results, see [2]):

I. Prove the equivalence of (1) and (2) in the general case or for as wide a class of SDE with affine coefficients as possible. The difficulty stems, in particular, from the fact that the state space is of the form $(\mathbb{R}_+)^m \times \mathbb{R}^{n-m}$, the infinitesimal generator degenerates at the boundary and the term of order 0 ("electric potential") is an bounded affine function;

II. Prove the Feynman–Kac theorem for the (backward) Cauchy problem and more general boundary problems.

III. Derive general conditions for the explosion of the solution of the boundary problem (81) and (82) as $t \to T$.

IV. Study the domain of analyticity of the characteristic function (80) and its behavior at infinity, hence, the applicability of the conformal deformation technique. For partial results, see [2,15].

V. Derive conditions on the parameters of affine interest rate models which ensure that the solution of (81) and (82) with $\xi = 0$ (price of the discount bond) is bounded by 1. For partial results, see [23,25].

VI. Study the asymptotics of the solution of the (backward) Cauchy problem with the terminal condition $V(T,x) = G(x)$, as $T \to +\infty$. For partial results based on the eigenfunction expansion technique, see [90–93]. The main block is the generalized eigenfunction expansion of the essentially non-self-adjoint quadratic Hamiltonian. This is a special case of the same procedure in quadratic term structure models [91].

7.2. Wishart Models

Factors are naturally organized as a matrix rather than vector. One may regard Wishart models as square root models where the positive scalar stochastic factor is replaced by a positive-definite matrix. See [94,95] for a list of references. The outstanding problems are the same as for affine models.

7.3. Quadratic Term Structure Models (QTSM)

In the pure diffusion case, the infinitesimal generator of QTSM is of the form:

$$L = \langle \kappa(\theta - x), \partial_x \rangle + \frac{1}{2}\langle A\partial_x, \partial_x \rangle + \langle Bx, x \rangle + \langle d, x \rangle + d_0, \tag{83}$$

where κ is an anti-stable matrix with real entries, matrix A is positive definite, B is semi-definite (both with real entries), $\theta, d \in \mathbb{R}^n$, $d_0 \in \mathbb{R}$. The characteristic function is an exponential of a quadratic function of x, with the coefficients depending on time to maturity τ and the spectral parameter ξ and can be calculated solving the associated system of generalized Riccati equations. The generalized eigenfunction expansion can be calculated as well [90,91]. It is interesting that, for the parameters of QTSM documented in real financial markets, L is essentially non-self-adjoint, hence, not diagonalizable.

If the diffusion part of the infinitesimal generator is replaced with a PDO, then the exact calculation of the characteristic function and generalized eigenfunction expansion is not known. For asymptotic approximations, see [24,92].

7.4. Systems of Affine and Quadratic Term Structure Models

To the best of my knowledge, there are no general results in this direction, although in view of a huge body of publications in quantitative finance, one expects that some special cases have been considered.

8. Conclusions

In this paper, non-standard features of several basic problems arising in finance, insurance and economics are explained, and a group of related efficient methods (analytical and numerical) are outlined. The objective is to calculate the prices of contingent claims. In the general economic framework, the price V of a contingent claim is the expectation of a certain stochastic expression. In the case of Markov models, V is a function of time and the spot value $x = X_t$ of the underlying source of uncertainty. The formal application of Dynkin's formula leads to a boundary problem for an integro-differential (pseudo-differential) equation of the form $\partial_t + L_X$, where L_X is the infinitesimal generator of X (we allow for processes with killing/birth, hence, in the case of the Lévy model and the constant interest rate r, $L_X = -\psi(D_x) - r$, where $\psi(\xi)$ is the characteristic exponent of X). However, in the majority of cases of interest, the rigorous proof of the theorem of Feynman–Kac type is lacking. In the paper, several basic situations where proofs are available are considered in more detail; more general results are only outlined.

The following general features of the boundary problems discussed in this paper may be of general interest to specialists in PDE and PDO.

1. In models with jumps, boundary conditions are non-local, whereas the standard boundary and co-boundary problems for PDE, PDO and fractional differential equations are local. See, e.g., [3,4,96,97]. Therefore, (1) one of the standard approaches to boundary problems, namely reduction to the boundary, cannot be used to reduce the dimension of the problem; (2) numerical methods which do not take the non-locality

of the boundary conditions into account properly produce sizable errors; if the time horizon is large, the relative errors are, typically, very large.

2. In popular pure diffusion models such as the Heston model, the operator degenerates at the boundary. The degeneration is sufficiently regular so that the generalization of the Boutet de Monvel calculus for degenerate elliptic operators [3] is applicable in a number of situations (interestingly, the infinitesimal generator in the Heston model is one of the basic examples in [3]). Formally, one can apply this calculus to boundary problems in models with jumps provided that the characteristic exponent of the jump part is a rational function. However, such an application would require the approximation of non-local boundary conditions by local ones. For a reduction to the boundary in applications to the Heston model and other basic diffusion models, see [98] and the bibliography therein.
3. The degeneration and non-locality of the infinitesimal generators are the sources of fundamental difficulties for a rigorous proof of the Feynman–Kac theorem. One must establish certain regularity conditions of the solution for the proof. For applications of Dynkin's formula, conditions are weaker than for applications of Ito's formula, and, in some cases, general regularity results [3,5] can be used. However, the author is unaware of any general proof.
4. In the case of Lévy models (PDO with constant symbols) and problems with flat boundaries, the probabilistic version of the Wiener–Hopf factorization technique can be used to derive an explicit formula for the price in the form of oscillatory integrals, and the analytic form of the same technique used to derive the same formula for the unique solution of the corresponding boundary problem thereby proving the Feynman–Kac theorem. In the case of problems with a curved boundary, the general regularity results and the proof of the Feynman–Kac theorem are unknown.
5. The proof and study of regularity are especially non-trivial if the infinitesimal operator $L = -\psi(D_x)$ of a Lévy model is an elliptic PDO of order $\nu \in (0,1)$. Models of this kind are documented in the majority of empirical studies (if Lévy models are calibrated to the real data). We outlined approaches to study general boundary problems with operators of the form $\partial_t - \psi(D_x) - r$.
6. In popular Lévy models, the solutions of the boundary problems are irregular at the boundary, hence numerical methods that (implicitly or explicitly) assume that the solution is more smooth than it is inevitably produce large errors.
7. If the infinitesimal operator of a Lévy model is an elliptic PDO of order $\nu \in (0,1)$, then the smooth pasting principle for free boundary may fail.
8. In many cases, the free boundary is discontinuous at the terminal date, which implies that a numerical method that assumes the continuity is bound to be inaccurate.

In the paper, proofs of facts 4–8 and several related efficient numerical methods for the solution are outlined. The main blocks are as follows.

1. The interpretation of operators in the operator form from the Wiener–Hopf factorization as expectation operators under supremum and infimum processes, and explicit formulas for solutions of basic boundary problems under very mild restrictions on operators and boundary conditions, in the case of flat boundaries.
2. The interpretation allows one to prove the convergence of general algorithms for pricing options:
    (a) With finite time horizon (stationary boundary problems) using maturity randomization (method of lines);
    (b) In regime-switching models;
    (c) Approximations of models with a stochastic interest rate and stochastic volatility by regime-switching models (systems of boundary problems);
    (d) Options with non-monotone and discontinuous payoffs, with applications to game-theoretical problems.

3. In the interest of brevity, the refined version of the fast Fourier transform (FFT) and inverse Fourier transform (iFFT) introduced in [46,69] to price single and barrier options in Lévy models is not described in the paper. The refined version can be used to accurately evaluate options of various kinds. The advantage of refined FFT and iFFT as compared to existing versions, including the fractional FFT, stems from the flexibility of choices of grids of different length in the dual and state spaces, in each block of the numerical procedure. The final result is calculated using an almost optimal number of the standard FFT and iFFT blocks of the same (smaller) size.
4. Instead, in the paper, we describe very fast and accurate methods for the numerical evaluation of integrals, in dimensions 1–4, based on the conformal deformation technique. The main idea is close to the idea of the saddle point method but the families of the contour deformations that we use allow one to relatively easily construct deformations with respect to several variables. The joint deformation of several contours above can be regarded as a further step in the realization of a general program of study of the efficiency of combinations of one-dimensional inverse transforms for high-dimensional inversions outlined in [99,100] with additional twists: the calculation of the Wiener–Hopf factors, which is necessary to price lookback and barrier options. The authors of [99,100] consider three main different one-dimensional algorithms for the numerical realization of the Bromwich integral (i) Fourier series expansions with the Euler summation; (ii) combinations of Gaver functionals; and (iii) deformation of the contour in the Bromwich integral, and discuss various methods of multi-dimensional inversion based on combinations of these three basic blocks. Our results imply that, for the purposes of multi-dimensional inversion, the class of deformations must be enlarged. In particular, in some practically important situations, deformations close to the steepest descent such as Talbot's deformation $q = r\theta(\cot\theta + i), -\pi < \theta < \pi$ [101] are not applicable, and one must resort to seemingly less efficient deformations. Note that the general conformal deformation technique that we develop is especially efficient in the case of highly oscillatory integrals.

We are grateful to the anonymous referee for the suggestion to include a short review of other methods. Naturally, essentially all methods developed to solve boundary problems for PDE and PDO can be used to price contingent claims, and many of these methods (if not all—the literature is huge) are used. We list and explain sources and types of errors of several groups of methods.

1. "THE HILBERT TRANSFORM METHOD" is used in backward induction procedures to price options of several types. Calculations are in the dual space. At each time step, operators of the form $\mathcal{F}\mathbf{1}_{(-\infty,h)}\mathcal{F}^{-1}$ and $\mathcal{F}\mathbf{1}_{(h,+\infty)}\mathcal{F}^{-1}$ are expressed in terms of the Hilbert transform, and the latter is realized using the fast Hilbert transform. See [102–104] and the bibliographies therein. In these papers and other papers where the fast Hilbert transform is used, the grids of the same length in the state and dual spaces are used, which is presented as an advantage of the fast Hilbert transform approach. However, in many cases, the choice of grids of equal size leads to either very large errors or unnecessarily long grids and a large CPU time. See [46] for details and the explanation on how to use grids of various length in order to efficiently control discretization and truncation errors. Efficient methods of the numerical Fourier inversion described in the paper can be adjusted to the Hilbert transform. See [10] for an efficient numerical realization of operators $\mathcal{F}\mathbf{1}_{(-\infty,h)}\mathcal{F}^{-1}$ and $\mathcal{F}\mathbf{1}_{(h,+\infty)}\mathcal{F}^{-1}$, applicable when these operators are applied only once. For an efficient numerical realization of operators generalizing the Hilbert transform, which is applicable in backward induction procedures (double spiral method), see [105].
2. VARIATIONS OF THE STRAIGHTFORWARD APPLICATION OF THE FOURIER AND INVERSE FOURIER TRANSFORM to European options, equivalently, the solution of Cauchy problems for parabolic PDO on the real line, namely, COS method and Carr–Madan method mentioned in Section 3.2, introduce additional unnecessary errors. The so-called Lewis–Lipton formula is the standard Fourier inversion formula with the

prefixed line of integration. The choice of the line is non-optimal in the majority of cases; the conformal deformation method is much faster and more accurate. See [2] for numerical examples.

3. IN THE CONV METHOD [106,107], at each step of backward induction, an extremely inefficient interpolation procedure for the approximation of the value function at each time step is applied. In probabilistic terms, continuous distributions are approximated by discrete distributions supported on a uniform grid, a dual grid is chosen, and the calculations at each time step are reduced to the composition of FFT, multiplication by the array of values of the characteristic function, and iFFT. The procedure is simple but the errors are large. See [46] for the detailed analysis.
4. THE COS METHOD is applicable (and has been applied) to price options of various kind. The essence of the method is an approximation of the kernel of the transition operator by a linear combination of cosines (hence the name). I find it difficult to find a sound mathematical argument in favor of this approximation. On the contrary, it is possible to indicate additional sources of errors and produce examples which demonstrate the inefficiency of COS. In backward induction procedures, the errors of COS accumulate very quickly, and pricing barrier options with even a moderate time horizon (0.5Y) is, essentially, impossible. See numerical examples in [2,13,14,16–20,31].
5. IN THE PROJ METHOD, the transition density of a random variable (equivalently, the kernel of the transition operator) is projected on a B-spline basis. See [20] for the bibliography, the discussion about the relative efficiency of COS, PROJ, the method in [46] which does not use an approximation of the transition kernel, and for an efficient procedure for the calculation of the projection coefficients using the sinh-acceleration. Note that the error of the approximation of the transition kernel is in the H^2-norm; hence, if the transition density has large derivatives or is non-smooth at the origin, which is the case of the VG model and Lévy models of order ν close to 0, then the errors of PRPJ can be very large.
6. APPROXIMATIONS OF VALUE FUNCTIONS AT EACH TIME STEP AND FILTERING. In [16,20,46,69] (see also the bibliographies therein), the value function at each time step is approximated by piece-wise polynomials. In view of the irregularity of value functions near the boundary discussed in the paper, such an approximation introduces an error which can be controlled. See [16,60]. The approximation can be interpreted as a spectral filter, which is used in a number of publications to increase the speed of convergence. In [108], ad hoc spectral filters are used to increase the convergence of the integrals: "When Fourier techniques are employed to specific option pricing cases from computational finance with non-smooth functions, the so-called Gibbs phenomenon may become apparent. This seriously impacts the efficiency and accuracy of the pricing. For example, the Variance Gamma asset price process gives rise to algebraically decaying Fourier coefficients, resulting in a slowly converging Fourier series. We apply spectral filters to achieve faster convergence. Filtering is carried out in Fourier space; the series coefficients are pre-multiplied by a decreasing filter, which does not add significant computational cost. Tests with different filters show how the algebraic index of convergence is improved." The quoted statement is correct. However, spectral filters are designed to regularize the results. The regularization of value functions results in serious errors in regions of paramount importance for risk management: near barrier and strike, close to maturity and for long dated options. For instance, close to the barrier or default boundary, the value can be overvalued or undervalued manifold. This remark is applicable to the applications of spectral filtering in [109] as well. Note that the conformal deformation technique allows one to eliminate the Gibbs phenomenon without sacrificing accuracy, and at a small CPU cost.
7. APPROXIMATION OF SMALL JUMPS COMPONENT BY A DIFFUSION. Cont and Volchkova [110] approximated the small jump component by a diffusion. In the result, a PDO of order $\nu < 2$ is replaced with the sum $L^\epsilon = \epsilon\partial_x^2 + \mu\partial_x + L_J$, where

$\epsilon > 0$ is small, $\mu \in \mathbb{R}$ and L_J is an integral operator with the kernel of class L_1; for an accurate approximation, the peak of the kernel has to be very high. After that, a standard implicit–explicit finite difference scheme is used to price barrier options. It is evident that if the infinitesimal generator L is a PDO of order $\nu < 2$, hence, the derivative of the solution can be unbounded near the boundary, an approximation of L by L^ϵ must lead to sizable errors near the boundary because the solution of the boundary problem becomes smooth up to the boundary. Numerous numerical examples in [111] have demonstrated the inaccuracy of the Cont–Voltchkova method. Note that the methods in [111] resemble but are less efficient than the method in [46,69].

8. THE APPROXIMATION OF KOBOL AND OTHER PROCESSES OF INFINITE ACTIVITY BY HEJD MODEL. In [38,40], the author constructed an HEJD model (without a special label attached) whilst keeping in mind to try such an approximation. For pricing American and barrier options, the advantage of HEJD is a simple explicit formula for the Wiener–Hopf factors in the case of positive values of the spectral parameter, derived in [38,40]. Unfortunately, for wide regions in the parameter space and large values of the spectral parameter which arise if a small time step in a backward induction procedure is used, an accurate approximation requires the use of HEJD with very large parameter values and high precision arithmetic is necessary. The reason is the same as in the case of the Cont–Voltchkova method. Due to this inefficiency, the author did not mention approximation of KoBoL by HEJD. Later, such an approximation was used in a number of publications, e.g., [112,113]. For a typical set of parameter values of KoBoL and moderate maturities, such an approximation can be very inaccurate at the distance of up to several percent of barrier. See [46] for numerical examples that illustrate the inefficiency of HEJD approximation. The problem of a large spectral parameter can be partially resolved using Richardson's extrapolation. In applications to pricing barrier options, the convergence of Richardson's extrapolation of arbitrary order is proven in [63]. Note that the technique of conformal deformations [10] is more efficient than approximation by HEJD, even in cases when the approximation is reasonably accurate.
9. APPROXIMATION OF UNDERLYING JUMP-DIFFUSIONS WITH CONTINUOUS TIME MARKOV CHAINS. In [82–84,86], in the models with stochastic interest rates and/or stochastic volatility, the dynamics of additional factors is approximated by continuous time Markov chains. In the PDE language, a part of the infinitesimal generator is discretized; the result is a regime-switching Lévy model. At the first (discretization) step, a Markov chain with the infinite number of states (infinite grid) appears; at the second step, the infinite grid is truncated, and transition rates in a vicinity of the "boundary" of the truncated grid are adjusted so that the Markov chain remains the Markov chain without killing. Thus, the Dirichlet condition must be avoided. In the diffusion case, the adjustment can be interpreted as the discretization of the high contact condition $\partial_y^2 V = 0$; in the jump-diffusion case, the "discretized boundary condition" is non-local and more involved. Later, in a number of publications starting with [114], the approximation-by-continuous time Markov chain was used for more general Markov processes in 1D. In some publications, even the dynamics of Lévy factors was approximated by a continuous time Markov chain. Such an approximation is rather inefficient, especially if the tails decay slowly, and/or in the presence of barriers. Furthermore, in related publications, the discretized Dirichlet condition is used, which leads to significant errors of backward induction procedures with many steps.
10. EIGENFUNCTION EXPANSION APPROACH. In the case of diffusion models on the real line, there is a significant body of results obtained by V. Linetsky and their students (see, e.g., [115] and the bibliography therein). In [90,91,116], the generalized eigenfunction expansion is derived for solutions of the Cauchy problems in multi-factor models.
11. ASYMPTOTIC METHODS. Due to the irregularity of solutions near the boundary, the asymptotic formulas are reasonably accurate only in a rather small vicinity of the

boundary [62]: hence, they are rather useless for numerical purposes (although useful for the qualitative analysis). The same is true for asymptotics near maturity (in terms of time to maturity, short time asymptotics). The conformal deformation method can be used to calculate solutions close to maturity with high accuracy. For long time asymptotics, efficient methods can be derived using the eigenfunction expansion technique [92]. Note that there is a large body of the literature devoted to the study of the asymptotics of implied volatility close to maturity and far from maturity.

12. FAST GAUSS TRANSFORM [117,118] can be efficiently used in certain diffusion models, and models with jumps of a special structure.
13. For applications of finite elements to option pricing, see [119].
14. APPROXIMATIONS BASED ON PURELY PROBABILISTIC METHODS. The literature is huge. A typical feature is that the convergence of a method is proven without error bounds. A typical example is the Cont–Voltchkova method [110]. The proof of convergence is given; however, as the numerical examples in [111] demonstrate, in many cases, it is necessary to use extremely fine and long grids to satisfy the error tolerance of the order of one percent, at a very large CPU cost.
15. MONTE-CARLO SIMULATIONS. The version of the Monte Carlo simulations that is closest to the methods of the present paper is based on the evaluation of the cumulative probability distribution function (cpdf) on an appropriate grid and interpolation. In applications to finance, the idea was used for the first time in [120]. Note that in [120] FFT and in a number of papers since FFT is used. As numerical examples in [31] demonstrate, the evaluation of the cpdf of Lévy processes using FFT leads to inaccurate results. The conformal deformations technique allows one to design much more accurate Monte-Carlo simulation procedures [9,31].

Funding: This research received no external funding.

Institutional Review Board Statement: Not applicable.

Informed Consent Statement: Not applicable.

Data Availability Statement: Not applicable.

Conflicts of Interest: The authors declare no conflict of interest.

References

1. Heath, D.; Schweizer, M. Martingales vs. PDEs in Finance: An equivalence result with examples. *Ann. Appl. Prob.* **2000**, *37*, 347–357.
2. Levendorskiĭ, S. Pitfalls of the Fourier Transform Method in Affine Models, and Remedies. *Appl. Math. Financ.* **2016**, *23*, 81–134. [CrossRef]
3. Levendorskiĭ, S. Degenerate Elliptic Equations. In *Mathematics and Its Applications*; Kluwer Academic Publishers Group: Dordrecht, The Netherlands, 1993; Volume 258.
4. Levendorskiĭ, S.; Paneyakh, B. Degenerate elliptic equations and boundary value problems. In *Encyclopeadia of Mathematical Sciences*; Agranovich, M., Shubin, M., Eds.; Springer: Berlin, Germany, 1994; Volume 63, pp. 131–202.
5. Feehan, P.M.N.; Pop, C. Degenerate-elliptic operators in mathematical finance and higher-order regularity for solutions to variational equations. *Adv. Differ. Equ.* **2015**, *20*, 361–432.
6. Levendorskiĭ, S. Conformal Pseudo-Asymptotics and Special Functions. Working Paper. 2016. Available online: http://ssrn.com/abstract=2713494 (accessed on 20 January 2022).
7. Levendorskiĭ, S. Fractional-Parabolic Deformations with Sinh-Acceleration. Working Paper. 2016. Available online: http://ssrn.com/abstract=2758811 (accessed on 20 January 2022).
8. Boyarchenko, S.; Levendorskiĭ, S. New Families of Integral Representations and Efficient Evaluation of Stable Distributions. 2018. Available online: https://ssrn.com/abstract=3172884 (accessed on 20 January 2022).
9. Boyarchenko, S.; Levendorskiĭ, S. Conformal Accelerations Method and Efficient Evaluation of Stable Distributions. *Acta Appl. Math.* **2020**, *169*, 711–765. [CrossRef]
10. Boyarchenko, S.; Levendorskiĭ, S. Static and Semi-Static Hedging as Contrarian or Conformist Bets. *Math. Financ.* **2020**, *3*, 921–960. [CrossRef]
11. Carr, P.; Geman, H.; Madan, D.; Yor, M. The fine structure of asset returns: An empirical investigation. *J. Bus.* **2002**, *75*, 305–332. [CrossRef]

12. Sato, K. Lévy processes and infinitely divisible distributions. *Cambridge Studies in Advanced Mathematics*; Cambridge University Press: Cambridge, UK, 1999; Volume 68.
13. Boyarchenko, S.; Levendorskiĭ, S. New Efficient Versions of Fourier Transform Method in Applications to Option Pricing. Working Paper. 2011. Available online: http://ssrn.com/abstract=1846633 (accessed on 20 January 2022).
14. Boyarchenko, S.; Levendorskiĭ, S. Efficient variations of Fourier transform in applications to option pricing. *J. Comput. Financ.* **2014**, *18*, 57–90. [CrossRef]
15. Levendorskiĭ, S. Efficient Pricing and Reliable Calibration in the Heston Model. *Int. J. Theor. Appl. Financ.* **2012**, *15*, 125050. [CrossRef]
16. de Innocentis, M.; Levendorskiĭ, S. Pricing Discrete Barrier Options and Credit Default Swaps Under Lévy Processes. *Quant. Financ.* **2014**, *14*, 1337–1365. [CrossRef]
17. Boyarchenko, M.; Levendorskiĭ, S. Ghost Calibration and Pricing Barrier Options and Credit Default Swaps in Spectrally One-Sided Lévy Models: The Parabolic Laplace Inversion Method. *Quant. Financ.* **2015**, *15*, 421–441. [CrossRef]
18. Innocentis, M.; Levendorskiĭ, S. Calibration and Backtesting of the Heston Model for Counterparty Credit Risk. Working Paper. 2016. Available online: http://ssrn.com/abstract=2757008 (accessed on 20 January 2022).
19. Innocentis, M.; Levendorskiĭ, S. Calibration Heston Model for Credit Risk. *Risk* **2017**, 90–95. Available online: https://www.risk.net/risk-management/credit-risk/5330021/calibrating-heston-for-credit-risk (accessed on 20 January 2022).
20. Boyarchenko, S.; Levendorskiĭ, S.; Kirkby, J.; Cui, Z. SINH-acceleration for B-spline projection with option pricing applications. *arXiv* **2022**, arXiv:2109.08738.
21. Boyarchenko, S.; Levendorskiĭ, S. Non-Gaussian Merton-Black-Scholes Theory. *Advanced Series on Statistical Science & Applied Probability*; World Scientific Publishing Co.: River Edge, NJ, USA, 2002; Volume 9.
22. Boyarchenko, S.; Levendorskiĭ, S. Barrier options and touch-and-out options under regular Lévy processes of exponential type. *Ann. Appl. Probab.* **2002**, *12*, 1261–1298. [CrossRef]
23. Levendorskiĭ, S. Consistency conditions for affine term structure models. *Stoch. Process. Their Appl.* **2004**, *109*, 225–261. [CrossRef]
24. Levendorskiĭ, S. Pseudo-diffusions and quadratic term structure models. *Math. Financ.* **2005**, *15*, 393–424. [CrossRef]
25. Levendorskiĭ, S. Consistency conditions for affine term structure models. II. Option pricing under diffusions with embedded jumps. *Ann. Financ.* **2006**, *2*, 207–224. [CrossRef]
26. Eskin, G. Boundary Value Problems for Elliptic Pseudodifferential Equations. In *Translations of Mathematical Monographs*; American Mathematical Society: Providence, RI, USA, 1981; Volume 9.
27. Boyarchenko, S.; Levendorskiĭ, S. Option pricing for truncated Lévy processes. *Int. J. Theor. Appl. Financ.* **2000**, *3*, 549–552. [CrossRef]
28. Bertoin, J. Lévy Processes. In *Cambridge Tracts in Mathematics*; Cambridge University Press: Cambridge, UK, 1996; Volume 121.
29. Samorodnitsky, G.; Taqqu, M. *Stable Non-Gaussian Random Processes*; Chapman and Hall: New York, NY, USA, 1994.
30. Zolotarev, V. *One-Dimensional Stable Distributions, American Mathematical Society Translations of Mathematical Monographs*; Translations of the Original 1983 Russian; American Mathematical Society: Providence, NJ, USA, 1986; Volume 63.
31. Boyarchenko, S.; Levendorskiĭ, S. Sinh-Acceleration: Efficient Evaluation of Probability Distributions, Option Pricing, and Monte-Carlo Simulations. *Int. J. Theor. Appl. Financ.* **2019**, *22*, 1950011. [CrossRef]
32. Koponen, I. Analytic approach to the problem of convergence of truncated Lévy flights towards the Gaussian stochastic process. *Phys. Rev. E* **1995**, *52*, 1197–1199. [CrossRef]
33. Rosinski, I. Tempering stable processes. *Stoch. Proc. Appl.* **2007**, *117*, 677–707. [CrossRef]
34. Barndorff-Nielsen, O. Processes of Normal Inverse Gaussian Type. *Financ. Stochastics* **1998**, *2*, 41–68. [CrossRef]
35. Barndorff-Nielsen, O.; Levendorskiĭ, S. Feller Processes of Normal Inverse Gaussian type. *Quant. Financ.* **2001**, *1*, 318–331. [CrossRef]
36. Madan, D.; Milne, F. Option pricing with V.G. martingale components. *Math. Financ.* **1991**, *1*, 39–55. [CrossRef]
37. Merton, R. Option pricing when underlying stock returns are discontinuous. *J. Financ. Econ.* **1976**, *3*, 125–144. [CrossRef]
38. Levendorskiĭ, S. Pricing of the American Put under Lévy Processes. Research Report MaPhySto, Aarhus. 2002. Available online: http://www.maphysto.dk/publications/MPS-RR/2002/44.pdf (accessed on 20 January 2022).
39. Lipton, A. Assets with jumps. *Risk* **2002**, 149–153. Available online: https://www.risk.net/derivatives/1530269/assets-jumps (accessed on 20 January 2022).
40. Levendorskiĭ, S. Pricing of the American put under Lévy processes. *Int. J. Theor. Appl. Financ.* **2004**, *7*, 303–335. [CrossRef]
41. Kou, S. A jump-diffusion model for option pricing. *Manag. Sci.* **2002**, *48*, 1086–1101. [CrossRef]
42. Asmussen, S. *Ruin Probabilities*; Number 2 in Advanced Series on Statistical Science and Applied Probability; World Scientific: River Edge, NJ, USA, 2000.
43. Asmussen, S.; Avram, F.; Pistorius, M. Russian and American put options under exponential phase-type Lévy models. *Stoch. Process. Their Appl.* **2004**, *109*, 79–111. [CrossRef]
44. Kuznetsov, A. Wiener-Hopf factorization and distribution of extrema for a family of Lévy processes. *Ann. Appl. Prob.* **2010**, *20*, 1801–1830. [CrossRef]
45. Kuznetsov, A.; Kyprianou, A.; Pardo, J. Meromorphic Lévy processes and their fluctuation identities. *Ann. Appl. Probab.* **2012**, *22*, 1101–1135. [CrossRef]

46. Boyarchenko, M.; Levendorskiĭ, S. Prices and sensitivities of barrier and first-touch digital options in Lévy-driven models. *Int. J. Theor. Appl. Financ.* **2009**, *12*, 1125–1170. [CrossRef]
47. Boyarchenko, S.; Levendorskiĭ, S. General Option Exercise Rules, with Applications to Embedded Options and Monopolistic Expansion. *Contrib. Theor. Econ.* **2006**, *6*, 2. [CrossRef]
48. Boyarchenko, S.; Levendorskiĭ, S. Optimal stopping made easy. *J. Math. Econ.* **2007**, *43*, 201–217. [CrossRef]
49. Boyarchenko, S.; Levendorskiĭ, S. *Irreversible Decisions Under Uncertainty (Optimal Stopping Made Easy)*; Springer: Berlin, Germany, 2007.
50. Boyarchenko, S.; Levendorskiĭ, S. Exit Problems in Regime-Switching Models. *J. Math. Econ.* **2008**, *44*, 180–206. [CrossRef]
51. Boyarchenko, S.; Levendorskiĭ, S. Optimal stopping in Lévy models, with non-monotone discontinuous payoffs. *SIAM J. Control. Optim.* **2011**, *49*, 2062–2082. [CrossRef]
52. Boyarchenko, S.; Levendorskiĭ, S. Preemption games under Lévy uncertainty. *Games Econ. Behav.* **2014**, *88*, 354–380. [CrossRef]
53. Boyarchenko, S.; Levendorskiĭ, S. *On Rational Pricing of Derivative Securities For a Family of Non-Gaussian Processes*; Preprint 98/7; Institut für Mathematik, Universität Potsdam: Potsdam, Germany, 1998. Available online: http://opus.kobv.de/ubp/volltexte/2008/2519 (accessed on 20 January 2022).
54. Boyarchenko, S.; Levendorskiĭ, S. *Generalizations of the Black-Scholes Equation for Truncated Lévy Processes*; Working Paper; University of Pennsylvania: Philadelphia, PA, USA, 1999.
55. Stenger, F. *Numerical Methods Based on Sinc and Analytic Functions*; Springer: New York, NY, USA, 1993.
56. Carr, P.; Madan, D. Option valuation using the Fast Fourier Transform. *J. Comput. Financ.* **1999**, *2*, 61–73. [CrossRef]
57. Fang, F.; Oosterlee, C. A novel pricing method for European options based on Fourier-Cosine series expansions. *SIAM J. Sci. Comput.* **2008**, *31*, 826–848. [CrossRef]
58. Fang, F.; Oosterlee, C. Pricing early-exercise and discrete barrier options by Fourier-cosine series expansions. *Numer. Math.* **2009**, *114*, 27–62. [CrossRef]
59. Fang, F.; Jönsson, H.; Oosterlee, C.; Schoutens, W. Fast valuation and calibration of credit default swaps under Lévy dynamics. *J. Comput. Financ.* **2010**, *14*, 57–86. [CrossRef]
60. Levendorskiĭ, S.; Xie, J. Pricing of Discretely Sampled Asian Options Under Lévy Processes. Working Paper. 2012. Available online: http://papers.ssrn.com/abstract=2088214 (accessed on 20 January 2022).
61. Boyarchenko, S.; Levendorskiĭ, S. Efficient Laplace inversion, Wiener-Hopf factorization and pricing lookbacks. *Int. J. Theor. Appl. Financ.* **2013**, *16*, 1350011. [CrossRef]
62. Boyarchenko, M.; de Innocentis, M.; Levendorskiĭ, S. Prices of barrier and first-touch digital options in Lévy-driven models, near barrier. *Int. J. Theor. Appl. Financ.* **2011**, *14*, 1045–1090. [CrossRef]
63. Levendorskiĭ, S. Convergence of Carr's Randomization Approximation Near Barrier. *SIAM FM* **2011**, *2*, 79–111.
64. Levendorskiĭ, S. Method of Paired Contours and Pricing Barrier Options and CDS of Long Maturities. *Int. J. Theor. Appl. Financ.* **2014**, *17*, 1450033. [CrossRef]
65. Levendorskiĭ, S. Ultra-Fast Pricing Barrier Options and CDSs. *Int. J. Theor. Appl. Financ.* **2017**, *20*, 1750033. [CrossRef]
66. Wiener, N.; Hopf, E. Über eine Klasse singulärer Integralgleichungen. *Sitzungsberichte Der Preuss. Ischen Akad. Der Wiss.-Math.-Phys. Kl.* **1931**, *30*, 696–706.
67. Greenwood, P.; Pitman, J. Fluctuation identities for Lévy processes and splitting at the maximum. *Adv. Appl. Probab.* **1980**, *12*, 57–90. [CrossRef]
68. Rogers, L.; Williams, D. *Diffusions, Markov Processes, and Martingales. Volume 1. Foundations*, 2nd ed.; John Wiley & Sons, Ltd.: Chichester, UK, 1994.
69. Boyarchenko, M.; Levendorskiĭ, S. Valuation of continuously monitored double barrier options and related securities. *Math. Financ.* **2012**, *22*, 419–444. [CrossRef]
70. Boyarchenko, S. Two-Point Boundary Problems and Perpetual American Strangles in Jump-Diffusion Models. Working Paper. 2006. Available online: http://ssrn.com/abstract=896260 (accessed on 20 January 2022).
71. Boyarchenko, M.; Boyarchenko, S. Double Barrier Options in Regime-Switching Hyper-Exponential Jump-Diffusion Models. *Int. J. Theor. Appl. Financ.* **2011**, *14*, 1005–1044. [CrossRef]
72. Barles, G.; Bardeaux, J.; Romano, M.; Samsoen, N. Critical stock price near expiration. *Math. Financ.* **1995**, *5*, 77–95. [CrossRef]
73. Lamberton, D. Critical price for an American option near maturity. In *Seminar on Stochastic Analysis, Random Fields and Applications*; Dozzi, M., Russo, F., Eds.; Birkhauser: Basel, Switzerland, 1995.
74. Levendorskiĭ, S. American and European Options in Multi-Factor Jump-Diffusion Models, Near Expiry. *Financ. Stochastics* **2008**, *12*, 541–560. [CrossRef]
75. Boyarchenko, S.; Levendorskiĭ, S. Perpetual American options under Lévy processes. *SIAM J. Control. Optim.* **2002**, *40*, 1663–1696. [CrossRef]
76. Boyarchenko, S.; Levendorskiĭ, S. Perpetual American Options with Disconnected Exercise Regions in Lévy Models. Working Paper. 2014. Available online: http://ssrn.com/abstract=2472905 (accessed on 20 January 2022).
77. Boyarchenko, S. Irreversible Decisions and Record Setting News principles. *Am. Econ. Rev.* **2004**, *94*, 557–568. [CrossRef]
78. Boyarchenko, S.; Levendorskiĭ, S. American options: The EPV pricing model. *Ann. Financ.* **2005**, *1*, 267–292. [CrossRef]
79. Alili, L.; Kyprianou, A. Some remarks on first passage of Lévy process, the American put and pasting principle. *Ann. Appl. Probab.* **2005**, *15*, 2062–2080. [CrossRef]

80. Carr, P. Randomization and the American put. *Rev. Financ. Stud.* **1998**, *11*, 597–626. [CrossRef]
81. Bouchard, B.; Karoui, N.E.; Touzi, N. Maturity randomization for stochastic control problems. *Ann. Appl. Prob.* **2005**, *15*, 2575–2605. [CrossRef]
82. Boyarchenko, S.; Levendorskiĭ, S. American options in Lévy models with stochastic interest rates. *J. Comput. Financ.* **2009**, *12*, 1–30. [CrossRef]
83. Boyarchenko, S.; Levendorskiĭ, S. American Options in Regime-Switching Models with Non-Semibounded Stochastic Interest Rates. 2007. Available online: http://ssrn.com/abstract=1015410 (accessed on 20 January 2022).
84. Boyarchenko, S.; Levendorskiĭ, S. American Options in Lévy Models with Stochastic Volatility. 2007. Available online: http://ssrn.com/abstract=1031280 (accessed on 20 January 2022).
85. Boyarchenko, S.; Levendorskiĭ, S. American Options in the Heston Model with Stochastic Interest Rates. Working Paper. 2007. Available online: http://ssrn.com/abstract=1031282 (accessed on 20 January 2022).
86. Boyarchenko, S.; Levendorskiĭ, S. American Options in the Heston Model with Stochastic Interest Rate and its Generalizations. *Appl. Mathem. Finance* **2013**, *20*, 26–49. [CrossRef]
87. Boyarchenko, S.; Levendorskiĭ, S. American options in regime-switching models. *SIAM J. Control. Optim.* **2009**, *48*, 1353–1376. [CrossRef]
88. Duffie, D.; Filipović, D.; Singleton, K. Affine processes and applications in Finance. *Ann. Appl. Probab.* **2002**, *13*, 984–1053.
89. Duffie, D.; Pan, J.; Singleton, K. Transform Analysis and Asset Pricing for Affine Jump Diffusions. *Econometrica* **2000**, *68*, 1343–1376. [CrossRef]
90. Boyarchenko, N.; Levendorskiĭ, S. Eigenfunction Expansion Method in Multi-Factor Models. Working Paper. 2004. Available online: https://ssrn.com/abstract=627642 (accessed on 20 January 2022).
91. Boyarchenko, N.; Levendorskiĭ, S. The eigenfunction expansion method in multi-factor quadratic term structure models. *Math. Financ.* **2007**, *17*, 503–539. [CrossRef]
92. Boyarchenko, N.; Levendorskiĭ, S. Asymptotic Pricing in Term Structure Models Driven by Jump-Diffusions of Ornstein-Uhlenbeck Type. Working Paper. 2006. Available online: http://ssrn.com/abstract=890725 (accessed on 20 January 2022).
93. Boyarchenko, S.; Levendorskiĭ, S. Gauge Transformations in the Dual Space, and Pricing and Estimation in the Long Run in Affine Jump-Diffusion Models. Working Paper. 2019. Available online: https://ssrn.com/abstract=3504029 (accessed on 20 January 2022).
94. Gnoatto, A.; Grasselli, M. The explicit Laplace transform for the Wishart process. *J. Appl. Prob.* **2014**, *51*, 640–656. [CrossRef]
95. Keller-Ressel, M. Moment Explosions and Long-Term Behavior of Affine Stochastic Volatility Models. *Math. Financ.* **2010**, *21*, 73–78. [CrossRef]
96. Boutet de Monvel, L. Boundary problems for pseudo-differential operators. *Acta Math.* **1971**, *126*, 11–51. [CrossRef]
97. Samko, S.G.; Kilbas, A.A.; Marichev, O.I. *Fractional Integrals and Derivatives: Theory and Applications*; Gordon and Breach Science Publishers: New York, NY, USA; London, UK, 1993.
98. Itkin, A.; Muravey, D. Semi-analytic pricing of double barrier options with time- dependent barriers and rebates at hit. *Front. Math. Financ.* **2022**, *1*, 53–79. [CrossRef]
99. Abate, J.; Valko, P. Multi-precision Laplace inversion. *Int. J. Numer. Methods Eng.* **2004**, *60*, 979–993. [CrossRef]
100. Abate, J.; Whitt, W. A unified framework for numerically inverting Laplace transforms. *INFORMS J. Comput.* **2006**, *18*, 408–421. [CrossRef]
101. Talbot, A. The accurate inversion of Laplace transforms. *J. Inst. Math. Appl.* **1979**, *23*, 97–120. [CrossRef]
102. Feng, L.; Linetsky, V. Pricing discretely monitored barrier options and defaultable bonds in Lévy process models: A fast Hilbert transform approach. *Math. Financ.* **2008**, *18*, 337–384. [CrossRef]
103. Fusai, G.; Germano, G.; Marazzina, D. Spitzer identity, Wiener-Hopf factorization and pricing of discretely monitored exotic options. *Eur. J. Oper. Res.* **2016**, *251*, 124–134. [CrossRef]
104. Fusai, G.; Kyriakou, I. General Optimized Lower and Upper Bounds for Discrete and Continuous Arithmetic Asian Options. *Math. Oper. Res.* **2016**, *41*, 531–559. [CrossRef]
105. Levendorskiĭ, S. Pricing arithmetic Asian options under Lévy models by backward induction in the dual space. *SIAM FM* **2018**, *9*, 1–27. [CrossRef]
106. Lord, R.; Fang, F.; Bervoets, F.; Oosterlee, C.W. A fast and accurate FFT-based method for pricing early-exercise options under Levy processes. *SIAM J. Sci. Comput.* **2008**, *10*, 1678–1705. [CrossRef]
107. Leentvaar, C.C.W.; Oosterlee, C.W. Multi-asset option pricing using a parallel Fourier-based techniques. *J. Comput. Fin.* **2008**, *12*, 1–26. [CrossRef]
108. Ruijter, M.J.; Oosterlee, C.W. On the application of spectral filters in a Fourier option pricing technique. *J. Comput. Fin.* **2015**, *19*, 76–106. [CrossRef]
109. Phelan, C.E.; Marazzina, D.; Fusai, G.; Germano, G. Hilbert transform, spectral filters and option pricing. *Ann. Oper. Res.* **2018**, *268*, 1–26. [CrossRef]
110. Cont, R.; Voltchkova, E. A finite difference scheme for option pricing in jump diffusion and exponential Lévy models. *SIAM J. Numer. Anal.* **2005**, *43*, 1596–1626. [CrossRef]
111. Kudryavtsev, O.; Levendorskiĭ, S. Fast and accurate pricing of barrier options under Lévy processes. *Financ. Stochastics* **2009**, *13*, 531–562. [CrossRef]

112. Asmussen, S.; Madan, D.; Pistorius, M.R. Pricing Equity Default Swaps under an approximation to the CGMY Lévy Model. *J. Comput. Financ.* **2008**, *11*, 79–93. [CrossRef]
113. Crosby, J.; Le Saux, N.; Mijatovič, A. Approximating Lévy processes with a view to option pricing. *Int. J. Theor. Appl. Financ.* **2011**, *13*, 69–91.
114. Mijatovič, A.; Pistotius, M. Continuously monitored barrier options under Markov processes. *Math. Financ.* **2013**, *23*, 1–13. [CrossRef]
115. Li, L.; Linetsky, V. Optimal stopping and early exercise: An eigenfunction expansion approach. *Oper. Res.* **2013**, *61*, 625–646. [CrossRef]
116. Boyarchenko, S.; Levendorskiǐ, S. Efficient Pricing Barrier Options and CDS in Lévy Models with Stochastic Interest Rate. *Math. Financ.* **2017**, *27*, 1089–1123. [CrossRef]
117. Broadie, M.; Yamamoto, Y. Application of the Fast Gauss Transform to Option Pricing. *Manag. Sci.* **2003**, *49*, 1071–1088. [CrossRef]
118. Broadie, M.; Yamamoto, Y. A Double-Exponential Fast Gauss Transform Algorithm for Pricing Discrete Path-dependent Options. *Oper. Res.* **2005**, *53*, 764–779. [CrossRef]
119. Hilber, N.; Reichman, O.; Schwab, C.; Winter, C. *Computational Methods for Quantitative Finance. Finite Element Methods for Derivative Pricing*; Springer: Berlin, Germany, 2013.
120. Glasserman, P.; Liu, Z. Sensitivity estimates from characteristic functions. *Oper. Res.* **2010**, *58*, 1611–1623. [CrossRef]

MDPI AG
Grosspeteranlage 5
4052 Basel
Switzerland
Tel.: +41 61 683 77 34
www.mdpi.com

Mathematics Editorial Office
E-mail: mathematics@mdpi.com
www.mdpi.com/journal/mathematics

www.ingramcontent.com/pod-product-compliance
Lightning Source LLC
Chambersburg PA
CBHW041346210926
51850CB00013B/3006